Lecture Notes in Artificial Intelligence   11144

Subseries of Lecture Notes in Computer Science

LNAI Series Editors

Randy Goebel
University of Alberta, Edmonton, Canada
Yuzuru Tanaka
Hokkaido University, Sapporo, Japan
Wolfgang Wahlster
DFKI and Saarland University, Saarbrücken, Germany

LNAI Founding Series Editor

Joerg Siekmann
DFKI and Saarland University, Saarbrücken, Germany

More information about this series at http://www.springer.com/series/1244

Vicenç Torra · Yasuo Narukawa
Isabel Aguiló · Manuel González-Hidalgo (Eds.)

Modeling Decisions for Artificial Intelligence

15th International Conference, MDAI 2018
Mallorca, Spain, October 15–18, 2018
Proceedings

 Springer

Editors
Vicenç Torra (iD)
Maynooth University
Maynooth
Ireland

and

University of Skövde
Skövde
Sweden

Yasuo Narukawa
Department Management Science
Tamagawa University
Tokyo
Japan

Isabel Aguiló
University of the Balearic Islands
Palma de Mallorca
Spain

Manuel González-Hidalgo
University of the Balearic Islands
Palma de Mallorca, Baleares
Spain

ISSN 0302-9743   ISSN 1611-3349 (electronic)
Lecture Notes in Artificial Intelligence
ISBN 978-3-030-00201-5   ISBN 978-3-030-00202-2   (eBook)
https://doi.org/10.1007/978-3-030-00202-2

Library of Congress Control Number: 2018953174

LNCS Sublibrary: SL7 – Artificial Intelligence

This Springer imprint is published by the registered company Springer Nature Switzerland AG
The registered company address is: Gewerbestrasse 11, 6330 Cham, Switzerland

Preface

This volume contains papers presented at the 15th International Conference on Modeling Decisions for Artificial Intelligence, MDAI 2018, held on Mallorca, Spain, October 15–18, 2018. This conference followed MDAI 2004 (Barcelona, Spain), MDAI 2005 (Tsukuba, Japan), MDAI 2006 (Tarragona, Spain), MDAI 2007 (Kitakyushu, Japan), MDAI 2008 (Sabadell, Spain), MDAI 2009 (Awaji Island, Japan), MDAI 2010 (Perpignan, France), MDAI 2011 (Changsha, China), MDAI 2012 (Girona, Spain), MDAI 2013 (Barcelona, Spain), MDAI 2014 (Tokyo, Japan), MDAI 2015 (Skövde, Sweden), MDAI 2016 (Sant Julià de Lòria, Andorra), and MDAI 2017 (Kitakyushu, Japan) with proceedings also published in the LNAI series (Vols. 3131, 3558, 3885, 4617, 5285, 5861, 6408, 6820, 7647, 8234, 8825, 9321, 9880, and 10571).

The aim of this conference was to provide a forum for researchers to discuss different facets of decision processes in a broad sense. This includes model building and all kinds of mathematical tools for data aggregation, information fusion, and decision making; tools to help make decisions related to data science problems (including e.g., statistical and machine learning algorithms as well as data visualization tools); and algorithms for data privacy and transparency-aware methods so that data processing procedures and the decisions made as a result of them are fair, transparent, and avoid unnecessary disclosure of sensitive information.

The MDAI conference included tracks on the topics of (i) data science, (ii) data privacy, (iii) aggregation functions, (iv) human decision making, and (v) graphs and (social) networks.

The organizers received 43 papers from 15 different countries, 24 of which are published in this volume. Each submission received at least two reviews from the Program Committee and a few external reviewers. We would like to express our gratitude to them for their work. This volume also includes some of the plenary talks.

The conference was supported by the research group Scopia (Soft Computing, processament d'imatges i agregació), the University of Balearic Islands (UIB), the European Society for Fuzzy Logic and Technology (EUSFLAT), the Catalan Association for Artificial Intelligence (ACIA), the Japan Society for Fuzzy Theory and Intelligent Informatics (SOFT), and the UNESCO Chair in Data Privacy.

July 2018

Vicenç Torra
Yasuo Narukawa
Isabel Aguiló
Manuel González-Hidalgo

Organization

General Chairs

Isabel Aguiló   University of the Balearic Islands, Spain
Joan Torrens Sastre   University of the Balearic Islands, Spain

Program Chairs

Vicenç Torra   University of Skövde, Skövde, Sweden
Yasuo Narukawa   Tamagawa University, Tokyo, Japan

Advisory Board

Didier Dubois   Institut de Recherche en Informatique de Toulouse, CNRS, France
Lluis Godo   IIIA-CSIC, Spain
Kaoru Hirota   Beijing Institute of Technology; JSPS Beijing Office, China
Janusz Kacprzyk   Systems Research Institute, Polish Academy of Sciences, Poland
Sadaaki Miyamoto   University of Tsukuba, Japan
Michio Sugeno   Tokyo Institute of Technology, Japan
Ronald R. Yager   Machine Intelligence Institute, Iona Collegue, NY, USA

Program Committee

Isabel Aguiló   University of the Balearic Islands, Spain
Eva Armengol   IIIA-CSIC, Spain
Edurne Barrenechea   Universidad Pública de Navarra, Spain
Gloria Bordogna   Consiglio Nazionale delle Ricerche, Italy
Humberto Bustince   Universidad Pública de Navarra, Spain
Francisco Chiclana   De Montfort University, UK
Susana Díaz   Universidad de Oviedo, Spain
Josep Domingo-Ferrer   Universitat Rovira i Virgili, Spain
Jozo Dujmovic   San Francisco State University, CA, USA
Yasunori Endo   University of Tsukuba, Japan
Zoe Falomir   Universität Bremen, Germany
Katsushige Fujimoto   Fukushima University, Japan
Manuel González-Hidalgo   University of the Balearic Islands, Spain
Michel Grabisch   Université Paris I Panthéon-Sorbonne, France
Enrique Herrera-Viedma   Universidad de Granada, Spain

Aoi Honda	Kyushu Institute of Technology, Japan
Van-Nam Huynh	JAIST, Japan
Masahiro Inuiguchi	Osaka University, Japan
Simon James	Deakin University, Australia
Yuchi Kanzawa	Shibaura Institute of Technology, Japan
Petr Krajča	Palacky University Olomouc, Czech Republic
Marie-Jeanne Lesot	Université Pierre et Marie Curie (Paris VI), France
Xinwang Liu	Southeast University, China
Jun Long	National University of Defense Technology, China
Jean-Luc Marichal	University of Luxembourg, Luxembourg
Margalida Mas	University of the Balearic Islands, Spain
Sebastià Massanet	University of the Balearic Islands, Spain
Radko Mesiar	Slovak University of Technology, Slovakia
Andrea Mesiarová-Zemánková	Slovak Academy of Sciences, Slovakia
Arnau Mir	University of the Balearic Islands, Spain
Tetsuya Murai	Hokkaido University, Japan
Toshiaki Murofushi	Tokyo Institute of Technology, Japan
Guillermo Navarro-Arribas	Universitat Autònoma de Barcelona, Spain
Gabriella Pasi	Università di Milano Bicocca, Italy
Juan Vicente Riera	University of the Balearic Islands, Spain
Daniel Ruiz	University of the Balearic Islands, Spain
Sandra Sandri	Instituto Nacional de Pesquisas Espaciais, Brazil
Jaume Suñer	University of the Balearic Islands, Spain
László Szilágyi	Sapientia-Hungarian Science University of Transylvania, Hungary
Joan Torrens	University of the Balearic Islands, Spain
Aida Valls	Universitat Rovira i Virgili, Spain
Zeshui Xu	Southeast University, China
Yuji Yoshida	University of Kitakyushu, Japan

Local Organizing Committee Chairs

Isabel Aguiló	University of the Balearic Islands, Spain
Manuel González-Hidalgo	University of the Balearic Islands, Spain

Local Organizing Committee

Pedro Bibiloni	University of the Balearic Islands, Spain
Laura Fuentes	University of the Balearic Islands, Spain
Manuel González-Hidalgo	University of the Balearic Islands, Spain
Margalida Mas	University of the Balearic Islands, Spain
Sebastià Massanet	University of the Balearic Islands, Spain
Arnau Mir	University of the Balearic Islands, Spain
Juan Vicente Riera	University of the Balearic Islands, Spain
Daniel Ruiz	University of the Balearic Islands, Spain

| Jaume Suñer | University of the Balearic Islands, Spain |
| Joan Torrens | University of the Balearic Islands, Spain |

Additional Referees

Luis del Vasto
Jordi Casas
Vicent Costa
Sergio Martinez
Julian Salas
Javier Parra-Arnau

Supporting Institutions

The research group Scopia (Soft Computing, processament d'imatges i agregació)
The University of Balearic Islands (UIB)
The European Society for Fuzzy Logic and Technology (EUSFLAT)
The Catalan Association for Artificial Intelligence (ACIA)
The Japan Society for Fuzzy Theory and Intelligent Informatics (SOFT)
The UNESCO Chair in Data Privacy

Abstracts of Invited Talks

Consistency of Fuzzy Preference Relations

Gaspar Mayor

Universitat Illes Balears, Palma de Mallorca, Spain
gmayor@uib.es

In decision making the use of fuzzy preference relations to establish some degree of preference between any two alternatives is frequently adopted [3, 5, 12]. In order to design good decision making models some efforts on the characterization of consistency properties to avoid misleading solutions have been carried out [9, 14].

Transitivity has been a traditional requirement to characterize consistency in fuzzy contexts, i.e., when expert opinions are given by fuzzy preference relations. However, as it is pointed in [9] it is difficult to guarantee such consistency conditions in the process of decision making. In this work we present a type of consistency based on a t-norm and a t-conorm to guarantee consistency in the decision making process when fuzzy preference relations have been used. The proposed condition is quite general, which allows to include in it some of the definitions of consistency established by different authors.

Our objective is to obtain the degrees of preference $p_{i,j}$ with $i<j$ from the elemental preferences $p_{i,i+1}$ in such a way that the system is consistent, i.e.,

$$T(p_{ij}, p_{jk}) \leq p_{ik} \leq S(p_{ij}, p_{jk}) \qquad \forall i<j<k \qquad (1)$$

where T is a t-norm and S is a t-conorm. An appropriate type of multidimensional aggregation function is introduced to calculate $p_{i,j}$, $i<j$, from $p_{i,i+1}$.

References

1. Aguiló, I., Calvo, T., Fuster-Parra, P., Martín, J., Mayor, G., Suñer, J.: Preference structures: qualitative judgements based on smooth t-conorms. Inf. Sci. **366**, 165–176 (2016)
2. Calvo, T., Mayor, G., Suñer, J.: Associative globally monotone extended aggregation functions. In: Seising, R., Allende-Cid, H. (eds.) Claudio Moraga: A Passion for Multi-Valued Logic and Soft Computing. Studies in Fuzziness and Soft Computing, vol. 349, pp. 295–303. Springer, Cham (2017)
3. Chiclana, F., Herrera, F., Herrera-Viedma, E.: Integrating three representation models in fuzzt multipurpose decision making based on fuzzy preference relations. Fuzzy Sets Syst. **97**, 33–38 (1998)
4. Fishburn, P.: Preference structures and their numerical representations, Theoret. Comput. Sci. **217**(2), 359–383 (1999)
5. Fodor, J., Roubens, M.: Fuzzy Preference Modelling and Multi-criteria Decision Aid, Kluwer, Dordrecht (1994)

6. Fodor, J., Baets, B.: Fuzzy preference modelling: fundamentals and recent advances. In: Bustince, H., Herrera, F., Montero, J. (eds.) Fuzzy Sets and Their Extensions: Representation, Aggregation and Models. Studies in Fuzziness and Soft Computing, vol. 220, pp. 207–217. Springer, Heidelberg (2008)

7. Fodor, J.: Fuzzy preference relations based on differences. In: Greco, S., Marques Pereira, R.A., Squillante, M., Yager, R.R., Kacprzyk, J. (eds.) Preferences and Decisions. Studies in Fuzziness and Soft Computing, vol. 257, pp. 183–194. Springer, Heidelberg (2010)

8. Grabisch, M., Marichal, J.L., Mesiar, R., Pap, E.: Aggregation Functions. Cambridge University Press, Cambridge (2009)

9. Herrera-Viedma, E., Herrera, F., Chiclana, F., Luque, M.: Some issues on consistency of fuzzy preference relations. Eur. J. Oper. Res. **154**(1), 98–109 (2004)

10. Klement, E.P., Mesiar, R., Pap, E.: Triangular Norms. Kluwer, Dordrecht (2000)

11. Roubens, M., Vincke, Ph.: Preference Modelling. Lecture Notes in Economics and Mathematical Systems, vol. 250. Springer-Verlag, Berlin (1985)

12. Saaty, Th.L.: The Analytic Hierarchy Process. McGraw-Hill, New York (1980)

13. Tanino, T.: Fuzzy preference orderings in group decision making. Fuzzy Sets Syst. **12**, 117–131 (1984)

14. Torra, V., Narukawa, Y.: Modeling Decisions. Information Fusion and Aggregation Operators. Springer-Verlag, Heidelberg (2007)

15. Yager, R.R.: On ordered weighted averaging aggregation operators in multicriteria decision making. IEEE Trans. Syst. Man Cybern. **18**, 183–190 (1988)

Towards Distorted Statistics Based on Choquet Calculus

Michio Sugeno

Tokyo Institute of Technology, Japan

In this study we discuss statistics with distorted probabilities by applying Choquet calculus which we call 'distorted statistics'. To deal with distorted statistics, we consider distorted probability space on the non-negative real line. A (non-additive) distorted probability is derived from an ordinary additive probability by the monotone transformation with a generator. First, we explore some properties of Choquet integrals of non-negative, continuous and differentiable functions with respect to distorted probabilities. Next, we calculate elementary statistics such as the distorted mean and variance of a random variable for exponential and Gamma distributions. In addition, in the case of distorted exponential probability, we define its density function as the derivative of distorted exponential distribution function with respect to distorted Lebesgue measure.

Further, we deal with Choquet calculus of real-valued functions on the real line and explore their basic properties. Then, we consider distorted probability pace on the real line. We also calculate elementary distorted statistics for uniform and normal distributions. Finally, we compare distorted statistics with conventional skew statistics.

Assessing the Risk of Default Propagation in Interconnected Sectorial Financial Networks

Jordi Nin

BBVA Data & Analytics, Barcelona, Catalonia, Spain
jordi.nin@bbvadata.com

Systemic risk of financial institutions and sectorial companies relies on their inter-dependencies. The inter-connectivity of the financial networks has proven to be crucial to understand the propagation of default, as it plays a central role to assess the impact of single default events in the full system. Here, we take advantage of complex network theory to shed light on the mechanisms behind default propagation. Using real data from the financial company BBVA, we extract the network of client-supplier transactions between more than 140,000 companies, and their economic flows. In this talk, we introduce a basic computational model, inspired by the probabilities of default contagion, that allow us to obtain the main statistics of default diffusion given the network structure at individual and system levels. Achieved results show the exposure of different sectors to the default cascades, therefore allowing for a quantification and ranking of sectors accordingly. As we will show, this information is relevant to propose countermeasures to default propagation in specific scenarios.

Decision Making Tools with Semantic Data to Improve Tourists' Experiences

Aida Valls-Mateu

Department of Enginyeria Informàtica i Matemàtiques, Universitat Rovira i Virgili, Av Països Catalans, 26, 43007 Tarragona, Catalonia, Spain
aida.valls@urv.cat

The offices of management of touristic destinations are interested in providing a more user-centered experience that takes into account the personal interests of each visitor or group of visitors. Tourism is a key element of economic wealth of many places, therefore, improving the tourism experience may have a great impact not only on the visitor but also in the place.

In this kind of field, the objects of analysis are usually touristic activities (such as parks, museums, events, shopping malls, routes, sports, etc). The amount of options available at each possible destination is usually very large. Their description includes numerical data but also categorical one, sometimes provided as a list of keywords. Exploiting the semantics of those words is crucial to understand the tourist's interests and needs.

We will present two decision aiding methods that use domain ontologies to interpret the meaning of the keywords and help the managers and visitors to improve the touristic experience on a certain place.

References

1. Valls, A., Gibert, K., Orellana, A., Anton-Clavé, S.: Using Ontology-based Clustering to understand the Push and Pull factors for British tourists visiting a Mediterranean coastal destination. Info. Manag. (2017). Elsevier, Online April 2017
2. Moreno, A., Valls, A., Isern, D., Marin, L., Borràs, J.: SigTur/E-Destination: Ontology-based personalized recommendation of tourism and leisure activities. Eng. Appl. Artif. Intell. **26**(1), 633–651 (2013)

Improving Spatial Reasoning in Intelligent Systems: Challenges

Zoe Falomir

Bremen Spatial Cognition Center, University of Bremen, Germany
zfalomir@uni-bremen.de

Abstract. Here we tackle research on spatial thinking when facing two challenges: (i) describing scenes in natural language, and (ii) reasoning about perspectives for object recognition.

Regarding (i) addressing the following research questions is crucial: which kind of spatial features must intelligent systems use? Is location enough? And which kind of reference frames are suitable for communication? Deictic? Relative to the observer? Relative to the object? And what is the most salient object to describe? Intelligent systems must have common grounding with humans so that they can align representations and understand each other. Regarding (ii) addressing the following is decisive: how can we improve spatial perception in intelligent systems so that they can reason about object perspectives? Can tests done to people for measuring their spatial skills be used to model spatial logics?

On one side the challenge is to propose approaches to understand space and communicate about it as humans do. For that, intelligent systems (i.e. robots, tablets) can use computer vision and machine learning algorithms to analyse point clouds, recognise and describe scenes. On the other side the challenge is to propose approaches which solve spatial tests carried out to measure humans' intelligence and to apply these approaches in intelligent systems (i.e. computer games, robots) so that they can improve their spatial thinking, but also help improve humans' spatial thinking by providing them feedback.

Keywords: Qualitative spatial descriptors · Location · Spatial reasoning
Machine learning · Cognitive tests · Video games · Computer vision
Computational linguistics · Education · Spatial skills · Spatial cognition

Challenge I: Describing Scenes in Natural Language

Imagine the following scenario: It is 2056 and you have a robot at home to help you with your daily duties. One day you tell it: *Please, tidy the dining room.* To clarify, your robot asks back: *Should the new table be placed in front of the sofa or to the left*

The project *Cognitive Qualitative Descriptions and Applications* (CogQDA) funded by the University of Bremen is acknowledged. I also thank my collaborators in the described work.

of the armchair? And you answer: *Just leave it here on the left.* Or imagine another scenario in which you move to a new house and a decorator tutor application in your tablet helps you to arrange new furniture in a functional and fashionable way. Those situations would both involve spatial reasoning.

In the first scenario, your robot at home would need to understand the scene in order to talk about it, i.e. identifying the objects and their spatial locations in the living room. It also needs to identify that not all *left* locations are the same, but they depend on the reference frame used. In the second scenario, the decorator tutor would engage in human-computer interaction. It would need to produce natural language descriptions to provide the user with instructions, e.g. *if you locate the sofa on the corner, the room will look larger,* etc.

We envision a future where we humans communicate with automatic systems using language and these systems take decisions and interact with space accordingly. For that, these automatic systems need to describe the space using concepts that share a common reference frame with humans. But, how do we humans describe scenes i.e. in our home? What do we take as a reference to say where an object is located?

Qualitative spatial descriptors are based on reference systems which align with human perception and thus help establishing a more cognitive communication. Let us highlight that the spatial terms such as *in front of the sofa* and *to the left of the armchair* are qualitative and define a vague relation in space instead of a precise numerical location (e.g., [4]). The literature has studied the usefulness of using qualitative spatial descriptors in natural language: showing how people choose perspective and *relatum* to describe object arrangements in space [9] and showing how *salient* features are selected to describe objects depending on the context [5].

We outline here results of our experiments in cognitive scene description [2]: pieces of furniture in a 3D scene are detected and described according to its location using natural language based on qualitative spatial descriptors which are arranged according to reference frames and saliency determined after a cognitive study carried out to participants. Table 1 outlines some of our results.

Table 1. Scene narrative generated by *QSn3D* [2] where there are oriented and non-oriented pieces of furniture.

Photo	Language description	Logic description
	A.The biggest object (a wooden-chair) as the most salient object:	
	In the background, there is a wooden-chair on the right (oriented to the left). The wooden-chair has a white-table in the front.	is_categorized(object_0j, wooden-chair). location_wrt_observer(object_0j, right). distance_wrt_observer(object_0j, background). close_object(object_0j, object_1j). is_oriented(yes, object_0j, left). location_wrt_close_object(object_0j,object_1j, centre).
	B.The closest object to the observer (a white-table) as the most salient object: *In the foreground, there is a white-table in the centre. In the background, there is a wooden-chair on the right (oriented to the left).*	is_categorized(object_1j, white-table). location_wrt_observer(object_1j, centre). distance_wrt_observer(object_1j, foreground). close_object(object_1j, object_0j). is_oriented(no, object_1j, none).

A step further in this challenge is addressing human-machine interaction through dialogue, detecting changes in scenes and explaining them in a cognitive manner.

Challenge II: Spatial Reasoning About Object Perspectives

Pattern recognition and machine learning has demonstrated to be useful and effective in 3D object detection and recognition: a scene is discretised as a set of points floating in the air, called point clouds (see images in Table 1) and to recognize objects there, these points are put together again by learning different views of the object using machine learning methods [2]. Discretising space and then finding its continuity again is computationally very expensive and a challenge in AI and computer vision nowadays, as far as we are concerned.

So, in computer vision, pixels or cloud points do not automatically preserve space continuity. This contrast with situations, in real space, where if a change happens to an object side (dimension), it also affects the other dimensions automatically, preserving continuity. For example, when a cup handle breaks, we humans do not need to check from all perspectives to perceive the change in shape and depth, because we use continuity in space to infer that. The literature says that edge parallelism [6] is in our common sense from our childhood and that even young infants carry out physical reasoning taking into account continuity and solidity of objects [8].

Spatial reasoning is not an innate ability, since it has been shown that it can be trained [7] and showed a lasting performance [10]. Spatial reasoning skills correlate with success in Science, Technology, Engineering and Math (STEM) disciplines [11] and spatial ability has a unique role in the development of creativity or creative-thinking (measured by patents and publications) [3].

In cognitive science research, perceptual ability tests are carried out to people to measure how good are their spatial skills. And some of the problems intelligent systems must solve, are spatial problems which require spatial thinking such as inferring cross sections or canonical views of a 3D object, in order to recognize it. So, what can we learn from spatial cognition research that we can apply to computer vision and computer systems in general, so that the process of interacting with space is more 'intelligent or intuitive'?

Here we address that spatial thinking related to computer vision and to qualitative modelling leaded to the definition of a model for 3D object description which takes into account depth in the 3 canonical perspectives of the object at the same time [1] (see Fig. 1). Thus, it propagates changes in object volume, and it can also identify inconsistent descriptions. This model has been implemented in a video game which is being used at the moment in cognitive tests to see if the feedback provided is useful for people to improve their spatial reasoning skills. As future work, we intend to combine this cognitively-based knowledge approach with machine learning in order to improve object detection and reasoning about perspectives.

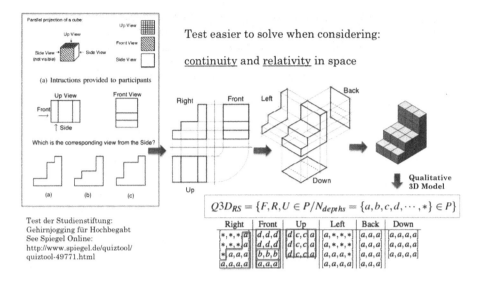

Fig. 1 Qualitative Descriptor for Reasoning about 3D perspectives.

References

1. Falomir, Z.: A qualitative model for reasoning about 3D objects using depth and different perspectives. In: Lechowski, T., Wałęga, P., Zawidzki, M. (eds.) LQMR 2015 Workshop. Annals of Computer Science and Information Systems, vol. 7, pp. 3–11. PTI (2015). https://doi.org/10.15439/2015F370
2. Falomir, Z., Kluth, T.: Qualitative spatial logic descriptors from 3D indoor scenes to generate explanations in natural language. Cogn. Process. **19**(2), 265–284 (2018). https://doi.org/10. 1007/s10339-017-0824-7
3. Kell, H.J., Lubinski, D., Benbow, C.P., Steiger, J.H.: Creativity and technical innovation: spatial ability's unique role. Psychol. Sci. **24**(9), 1831–1836 (2013). https://doi.org/10.1177/ 0956797613478615
4. Landau, B.: Update on "what" and "where" in spatial language: a new division of labor for spatial terms. Cogn. Sci. 1–30 (2016). https://doi.org/10.1111/cogs.12410
5. Mast, V., Falomir, Z., Wolter, D.: Probabilistic reference and grounding with PRAGR for dialogues with robots. J. Exp. Theor. Artif. Intell. **28**(5), 889–911 (2016). https://doi.org/10. 1080/0952813X.2016.1154611
6. Sinclair, N., de Freitas, E., Ferrara, F.: Virtual encounters: the murky and furtive world of mathematical inventiveness. ZDM Int. J. Math. Educ. **45**(2), 239–252 (2013). https://doi.org/ 10.1007/s11858-012-0465-3
7. Sorby, S.A.: Educational research in developing 3D spatial skills for engineering students. Int. J. Sci. Educ. **31**(3), 459–480 (2009). https://doi.org/10.1080/09500690802595839
8. Spelke, E.S., Breinlinger, K., Macomber, J., Jacobson, K.: Origins of knowledge **99**, 605–632 (1992)
9. Tenbrink, T., Coventry, K.R., Andonova, E.: Spatial strategies in the description of complex configurations. Discourse Process. **48**(4), 237–266 (2011)

10. Uttal, D., Meadow, N., Tipton, E., Hand, L., Alden, A., Warren, C., Newcombe, N.: The malleability of spatial skills: a meta-analysis of training studies. Psychol. Bull. **139**(2), 352–402 (2013). https://doi.org/10.1037/a0028446
11. Wai, J., Lubinksi, D., Benbow, C.P.: Spatial ability for STEM domains: aligning over 50 years of cumulative psychological knowledge solidifies its importance. J. Educ. Psychol. **101**(4), 817–835 (2009). https://doi.org/10.1037/a0016127

Contents

Clustering and Classification

Data Privacy and Security

Invited Paper

Graded Logic Aggregation

Jozo Dujmović[✉]

San Francisco State University, San Francisco, CA 94132, USA
jozo@sfsu.edu

Abstract. This paper summarizes basic properties of graded logic – a natural soft computing generalization of classical Boolean logic. Using graded logic aggregators we can build evaluation criteria and apply them in decision engineering. This paper is an extended summary that surveys key concepts of graded logic and graded logic aggregation.

Keywords: Graded logic · Soft computing · GCD · Evaluation
Logic aggregation · Boolean logic

1 A Soft Computing Generalization of Boolean Logic

The soft computing *graded logic* (GL) is a generalization of classical Boolean logic (BL) [1], that extends its domain from $\{0, 1\}^n$ to $[0, 1]^n$. GL is a form of fuzzy logic, and a system for development of graded logic aggregation structures. The goal of developing GL is to provide mathematical infrastructure for creating evaluation criteria that are used in decision engineering for modeling human reasoning in a way that is consistent with BL. Based on this goal, following are the necessary generalizations of BL, implemented in GL.

Anthropocentrism of Graded Logic Models. Logic is a key component of human reasoning. Therefore, GL should be based on observable properties of human reasoning, and it should serve for quantitative modeling of human reasoning. All GL functions must be provably present and recognizable in human reasoning. That also holds for all other humanoid properties of GL, like semantics, selective use of annihilators, compensativeness, partial truth, graded simultaneity/substitutability, and others.

Semantic Identity of All Variables. In the area of evaluation, GL functions are used as graded logic aggregators. These aggregators are not aggregating anonymous real numbers, but degrees of truth of precisely defined and semantically rich value statements (statements that assert a given degree of satisfaction of stakeholder's requirements). Consequently, all inputs, outputs, intermediate arguments, and all parameters must have clearly defined role, interpretation and meaning for stakeholders (decision makers). Semantic identity of all variables is a prerequisite for modeling human reasoning and decision processes in the area of evaluation.

The original version of this chapter has been revised: Minor errors in the text have been corrected. The correction to this chapter is available at https://doi.org/10.1007/978-3-030-00202-2_26

© Springer Nature Switzerland AG 2018
V. Torra et al. (Eds.): MDAI 2018, LNAI 11144, pp. 3–12, 2018.
https://doi.org/10.1007/978-3-030-00202-2_1

Expansion of Function Domain. Classical Boolean logic [1] is defined in vertices of the unit hypercube $\{0,1\}^n$. Soft computing generalizations of BL are defined in the whole unit hypercube $[0,1]^n$. Since $\{0,1\}^n \subset [0,1]^n$, the consequence of this obvious and natural generalization is that in vertices $\{0,1\}^n$ GL should behave same as BL. If that is satisfied, then GL can be considered a seamless generalization of BL. In BL truth is not graded: it is either completely present (1) or completely absent (0). In GL, truth x is partial, and exceptionally it can be fully present/absent: $x \in [0,1]$.

Expansion of Logic Domain. A GL logic aggregator $A(x_1, \ldots, x_n; \alpha)$ has (global) andness $\alpha = [n - (n+1)V]/(n-1)$, where $V = \int_{[0,1]^n} A(x_1, \ldots, x_n; \alpha) dx_1 \ldots dx_n$. This definition, introduced in [2], is based on the volume V of the fraction of hypercube under the surface of aggregator, and obviously it is applicable to all forms of conjunctive and disjunctive aggregators. The fundamental BL functions are conjunction (andness $\alpha = 1$, orness $\omega = 1 - \alpha = 0$), disjunction (andness $\alpha = 0$, orness $\omega = 1$), and negation. In all cases andness and orness are complementary: $\alpha + \omega = 1$. The primary GL functions are models of graded simultaneity (graded conjunction, $0 \leq \omega < \alpha \leq 1$), graded substitutability (graded disjunction, $0 \leq \alpha < \omega \leq 1$), neutrality (arithmetic mean, $\alpha = \omega = 1/2$), and negation (usually, standard negation $y = 1 - x$). The ranges of andness $0 \leq \alpha \leq 1$ and orness $0 \leq \omega \leq 1$ indicate the expansion of logic domain based on continuous transition from conjunction $x_1 \wedge \cdots \wedge x_n = \min(x_1, \ldots, x_n)$ to disjunction $x_1 \vee \cdots \vee x_n = \max(x_1, \ldots, x_n)$.

The volume V under the aggregator A inside the unit hypercube has the minimum value 0, yielding the maximum andness $\alpha_{max} = n/(n-1)$ that corresponds to the extreme model of simultaneity called *drastic conjunction*: $x_1 \hat{\wedge} \cdots \hat{\wedge} x_n = \lfloor x_1 \ldots x_n \rfloor$. The drastic conjunction is satisfied only in a single point where all arguments are completely satisfied $(x_1 = \cdots = x_n = 1)$. The maximum volume $V = 1$ yielding the maximum orness $\omega_{max} = n/(n-1)$ and the minimum andness $\alpha_{min} = -1/(n-1)$ corresponds to the extreme model of substitutability called *drastic disjunction*. The drastic disjunction is De Morgan dual of drastic conjunction: $x_1 \hat{\vee} \cdots \hat{\vee} x_n = 1 - \lfloor (1-x_1) \ldots (1-x_n) \rfloor$, and it is always completely satisfied, except in a single point where all arguments are 0. Therefore, GL provides the ultimate expansion of the logic domain in the range from drastic conjunction to drastic disjunction. The andness-parameterized continuous transition from drastic conjunction to drastic disjunction is a necessary property of all mathematical models of simultaneity and substitutability.

The Use of Semantic Domain. Traditional propositional logic ignores semantic aspects of human reasoning. In BL all truths are equivalent and equally important. However, in human reasoning, value statements and their degrees of truth have different degrees of importance. Ignoring importance of value statements is equivalent to excluding fundamental aspects of human reasoning. Consequently, GL must provide explicit modeling of importance. Like all other soft computing concepts, importance is a graded concept. It is also a compound concept because the percept of overall importance of an argument can be affected by multiple inputs (e.g. high andness, high orness, and a high relative importance with respect to other inputs of an aggregator)

Selective Inclusion/Exclusion of Annihilators. In BL, the annihilator of conjunctive operators is 0 and the annihilator of disjunctive operators is 1. In the case of conjunctive criterion, if an input argument supports the annihilator 0, it is called a mandatory requirement, and it must be satisfied. In the case of a disjunctive criterion, if an argument supports the annihilator 1, it is called the sufficient requirement because it is sufficient to satisfy a criterion. In human reasoning, however, there are conjunctive aggregators that do not support mandatory arguments, and disjunctive aggregators that do not support sufficient arguments. Consequently, GL must provide aggregators that selectively include annihilators. Aggregators that support annihilators are called *hard* and aggregators that do not support annihilators are called *soft*.

Selective Compensativeness of Logic Functions. BL functions are not compensative: both conjunction and disjunction are hard aggregators and bad conjunctive score caused by a 0 input cannot be compensated with good scores of other inputs. Similarly, good disjunctive score caused by an input equal to 1 is insensitive to values of other inputs. In human reasoning, however, we most frequently use compensative criteria where a bad score on one criterion can be compensated by a good score on another criterion, Thus, GL must provide the possibility to adjust the degree of compensativeness in the range from the full inclusion to the full exclusion.

Humanoid Properties of Logic Aggregators. In mathematical literature aggregators are usually defined with intention to create the most general family of functions. An extremely permissive definition of aggregator $A : [0, 1]^n \rightarrow [0, 1]$, $n > 1$ is based on nondecreasing monotonicity in all arguments $\mathbf{x} = (x_1, \ldots, x_n)$ and idempotency in boundary points: $A(0, \ldots, 0) = 0$, $A(1, \ldots, 1) = 1$; see [6, 7]. Obviously, these are the minimum possible restrictive conditions. Such aggregators can have discontinuities or oscillatory behavior of the aggregation function and/or its first derivatives. In addition, such aggregators can be false if no argument is false, and can be completely true if no argument is completely true. Such properties are not observable in human evaluation reasoning and are not desirable in GL models. Consequently, in GL, it is necessary to use the concept of *logic aggregator* $A(\mathbf{x}; \alpha)$ which is more restrictive than the general form of aggregator defined in mathematical literature. Since the basic aggregator $A(\mathbf{x}; \alpha)$ is a model of graded conjunction/disjunction, it is natural that it must be andness-directed (or orness-directed), i.e. parameterized with the global andness α (or global orness ω). The range of andness is $R_\alpha = [\alpha_{min}, \alpha_{max}]$, $R_\alpha^- = \,]\,\alpha_{min}, \alpha_{max}\,[$.

In GL, a definition of *basic logic aggregator* $A(\mathbf{x}; \alpha)$ is founded on the following restrictive conditions (which also hold for weighted logic aggregators $A(\mathbf{x}; \mathbf{W}, \alpha)$):

- Continuous function, nondecreasing in all arguments: $\mathbf{x} \leq \mathbf{y} \Rightarrow A(\mathbf{x}; \alpha) \leq A(\mathbf{y}; \alpha)$.
- Idempotency in extreme points: $\forall \alpha \in R_\alpha$, $A(0, \ldots, 0; \alpha) = 0$, $A(1, \ldots, 1; \alpha) = 1$.
- Sensitivity to positive truth: $\forall \alpha \in R_\alpha^-$, $\mathbf{x} > \mathbf{0} \Rightarrow A(\mathbf{x}; \alpha) > 0$.
- Sensitivity to incomplete truth: $\forall \alpha \in R_\alpha^-$, $\mathbf{x} < \mathbf{1} \Rightarrow A(\mathbf{x}; \alpha) < 1$.
- Parameterized, andness-directed (or orness-directed) continuous transition from the minimum andness (orness) α_{min} to the maximum andness (orness) α_{max}.
- Nonincreasing andness-monotonicity, and nondecreasing orness-monotonicity: $\partial A(\mathbf{x}; \alpha)/\partial \alpha \leq 0$, $\partial A(\mathbf{x}; 1 - \omega)/\partial \omega \geq 0$, $\alpha, \omega \in R_\alpha^-$.
- Discontinuities and oscillatory properties of $\partial A/\partial x_i$, $i \in \{1, \ldots, n\}$ are undesirable properties (exceptions are aggregators $\min(x_1, \ldots, x_n)$ and $\max(x_1, \ldots, x_n)$).
- Selective use of conjunctive annihilator 0 and disjunctive annihilator 1 (supported [or hard] or not supported [or soft], according to desired logic properties).

According to this definition, logic aggregators are located *between* the drastic conjunction and the drastic disjunction. The drastic conjunction and the drastic disjunction are GL functions but they do not satisfy the sensitivity to positive and incomplete truth and consequently do not have the status of logic aggregator. Generally, all GL functions and logic aggregators must have a proof of existence in human evaluation reasoning. Aggregators that cannot be found in human reasoning violate the fundamental requirement of anthropocentrism, and should not be used in decision models.

2   Graded Logic Conjecture and Graded Conjunction/Disjunction

The primary goal of GL is to answer the question "How do human beings aggregate subjective categories and which mathematical models describe this procedure adequately" [3]. In GL, the answer to this question is the following *graded logic conjecture*:

Human beings aggregate subjective categories by combining ten necessary and sufficient types of logic functions, which include nine aggregators derived from the graded conjunction/disjunction and the standard negation. In the order of increasing orness, these aggregators and negation are the following:

(1)   *Hyperconjunction (nonidempotent, annihilator = 0)*
(2)   *Full conjunction (idempotent, annihilator = 0)*
(3)   *Hard partial conjunction (idempotent, annihilator = 0)*
(4)   *Soft partial conjunction (idempotent, no annihilator)*
(5)   *Neutrality (idempotent, no annihilator)*
(6)   *Soft partial disjunction (idempotent, no annihilator)*
(7)   *Hard partial disjunction (idempotent, annihilator = 1)*
(8)   *Full disjunction (idempotent, annihilator = 1)*
(9)   *Hyperdisjunction (nonidempotent, annihilator = 1)*
(10)   *Standard negation ($x \mapsto 1 - x$)*

The full conjunction, full disjunction, and logic neutrality have fixed andness. All other conjunctive and disjunctive functions have adjustable andness, giving the possibility to fine-tune the intensity of simultaneity or substitutability while keeping the same type of idempotency and annihilator support. A desired level of andness is selected by decision maker, and consequently aggregators must be andness-directed (parameterized), and there must be a continuous transition from extreme andness to extreme orness.

These types of functions are necessary because each of them is provably used in human evaluation reasoning. They are sufficient because they completely cover all regions of the unit hypercube $[0, 1]^n$, including all possible combinations of conjunction/disjunction, hard/soft (annihilator support) properties, idempotency/nonidempotency, equal/different importance, and andness/orness intensity. Table 1 shows the range of andness for each type/category of aggregators and the subdivision

Table 1. Fifteen characteristic special cases of graded conjunction/disjunction.

Name	Andness	Property
Drastic conjunction	$\alpha = \frac{n}{n-1}$	Ultimate simultaneity: satisfied only if all arguments are completely satisfied
High hyperconjunction	$\frac{n2^n - n - 1}{(n-1)2^n} < \alpha < \frac{n}{n-1}$	The range of highest nonidempotent simultaneity, stronger than product t-norm and close to drastic conjunction
Medium hyperconjunction	$\alpha = \frac{n2^n - n - 1}{(n-1)2^n}$	Product t-norm. Nonidempotent hard simultaneity significantly stronger than the simultaneity of full conjunction
Low hyperconjunction	$1 < \alpha < \frac{n2^n - n - 1}{(n-1)2^n}$	Nonidempotent hard simultaneity stronger than the full conjunction and weaker than the product t-norm
Full conjunction	$\alpha = 1$	The minimum function (idempotent)
Hard partial conjunction	$\alpha_\theta \leq \alpha < 1$	The range of hard idempotent conjunctive aggregators with the minimum adjustable (threshold) andness α_θ, $1/2 < \alpha_\theta < 1$
Soft partial conjunction	$1/2 < \alpha < \alpha_\theta$	The range of soft idempotent conjunctive aggregators with adjustable andness
Neutrality	$\alpha = 1/2$	The arithmetic mean (idempotent)
Soft partial disjunction	$1 - \alpha_\theta < \alpha < 1/2$	The range of soft idempotent disjunctive aggregators with adjustable orness
Hard partial disjunction	$0 < \alpha \leq 1 - \alpha_\theta$	The range of hard idempotent disjunctive aggregators with the minimum adjustable threshold orness $\omega_\theta = \alpha_\theta$
Full disjunction	$\alpha = 0$	The maximum function (idempotent)
Low hyperdisjunction	$\frac{n + 1 - 2^n}{(n-1)2^n} < \alpha < 0$	Nonidempotent hard substitutability stronger than the full disjunction and weaker than the product t-conorm
Medium hyperdisjunction	$\alpha = \frac{n + 1 - 2^n}{(n-1)2^n}$	Product t-conorm. Nonidempotent hard substitutability significantly stronger than the substitutability of full disjunction
High hyperdisjunction	$\frac{-1}{n-1} < \alpha < \frac{n + 1 - 2^n}{(n-1)2^n}$	The range of highest nonidempotent substitutability, stronger than product t-conorm and close to drastic disjunction
Drastic disjunction	$\alpha = \frac{-1}{n-1}$	Ultimate substitutability: not satisfied only if all arguments are completely not satisfied; otherwise, completely satisfied

of hyperconjunctive and hyperdisjunctive aggregators [5, 8]. Let us also note that the continuous path from drastic conjunction to drastic disjunction is unifying idempotent means and nonidempotent t-norms/conorms in a single general logic aggregator called the *graded conjunction/disjunction* (GCD). The threshold andness is adjustable, $1/2 < \alpha_\theta < 1$. Most frequently, we use $\alpha_\theta = 3/4$, yielding the uniform GCD, where soft and hard aggregators have equal presence in the most frequent andness (or orness) range $\alpha \in [0, 1]$.

The range of hyperconjunction (and similarly, the range of hyperdisjunction) is divided in two areas and their border is the product t-norm (or the product t-conorm in the symmetric case of hyperdisjunction). According to Table 1, in the case of three variables $(n = 3)$ we have the following values of andness: for drastic conjunction $\alpha = 6/4$, for the product t-norm $\alpha = 5/4$, and for the pure conjunction $\alpha = 4/4$. This sequence indicates that the t-norm has central location between the full conjunction and the drastic conjunction. For other values of n, the situation is different, but the product t-norm is a reasonable aggregator for separating two characteristic regions of hyper-conjunction, at least for frequently used small values of n. The same conclusion holds for the dual product t-conorm in the area of hyperdisjunction.

3 Partitioning of Unit Hypercube

From the standpoint of idempotency, the unit hypercube $[0, 1]^n$ can be partitioned into three characteristic regions:

Region of nonidempotent hyperdisjunctive aggregators:

$$\max(x_1, \ldots, x_n) \leq A(x_1, \ldots, x_n; \omega) \leq 1, \quad 1 < \omega < \frac{n}{n-1}.$$

Region of idempotent conjunctive and disjunctive aggregators:

$$\min(x_1, \ldots, x_n) \leq A(x_1, \ldots, x_n; \omega) \leq \max(x_1, \ldots, x_n), \quad 0 \leq \omega \leq 1.$$

Region of nonidempotent hyperconjunctive aggregators:

$$0 \leq A(x_1, \ldots, x_n; \omega) \leq \min(x_1, \ldots, x_n), \quad \frac{-1}{n-1} < \omega < 0.$$

According to [4], the volumes under the full conjunction and the full disjunction are

$$V_{con}(n) = \int_{[0,1]^n} \min(x_1, \ldots, x_n) dx_1 \ldots dx_n = \frac{1}{n+1},$$

$$V_{dis}(n) = \int_{[0,1]^n} \max(x_1, \ldots, x_n) dx_1 \ldots dx_n = \frac{n}{n+1}.$$

These volumes are presented in Fig. 1, yielding the volume of hyperconjunctive and hyperdisjunctive regions of the unit hypercube $V_{hcon}(n) = V_{con}(n) = 1/(n+1)$, $V_{hdis}(n) = 1 - V_{dis}(n) = 1/(n+1) = V_{hcon}(n)$. Therefore, the volume of idempotent region is $V_{id}(n) = V_{dis}(n) - V_{con}(n) = 1 - V_{hcon}(n) - V_{hdis}(n) = (n-1)/(n+1)$ and the total volume of nonidempotent region is $V_{nid} = 1 - V_{id} = 2V_{hcon} = 2/(n+1)$.

If $n = 2$ then the volumes of idempotent, hyperdisjunctive, and hyperconjunctive regions are the same: $V_{id}(n) = V_{hcon}(n) = V_{hdis}(n) = 1/3$. However, this distribution quickly changes with increasing values of n. The volumes populated by idempotent and nonidempotent aggregators are presented in Figs. 1 and 2. These geometric properties

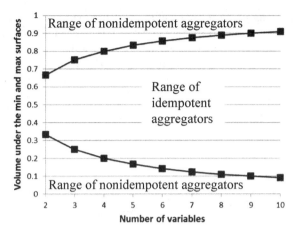

Fig. 1. The region of idempotent aggregators between the minimum and maximum functions

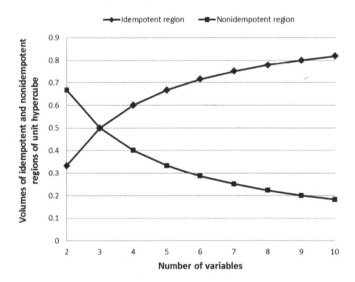

Fig. 2. The increasing volume of idempotent aggregators and the decreasing volume of nonidempotent aggregators

have significant consequences in logic. First, the regions of hyperconjunction and hyperdisjunction are shrinking as the number of arguments increases, while the complementary region of idempotent aggregators (means) increases. In the limit case $n \gg 1$, the idempotent aggregators fill the whole unit hypercube.

Taking into account the high visibility of t-norms/conorms in fuzzy logic and in mathematical literature, it is interesting to analyze the andness of all special cases of GCD, as shown in Fig. 3. In the area of idempotent aggregators the threshold andness and the threshold orness are adjustable [5]. In Fig. 3 we assume that $\alpha_\theta = \omega_\theta = 0.75$, yielding the uniform GCD (the default case where the regions of soft aggregators are the same as the regions of hard aggregators, and the symmetry of conjunctive and disjunctive aggregators supports De Morgan's laws). Between the pure conjunction and the pure disjunction we have four equal regions of soft and hard partial conjunction and partial disjunction. Inside each of these regions we can adjust andness/orness in the interval having the size of 0.25. In this way we can select the strength of aggregator within the same aggregator type.

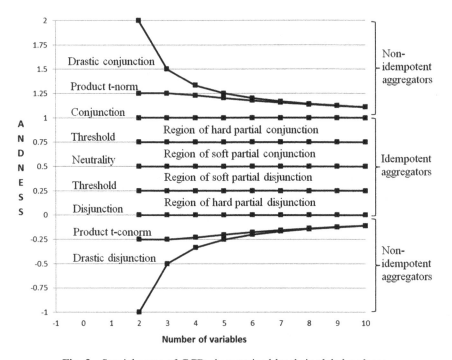

Fig. 3. Special cases of GCD characterized by their global andness.

Figure 3 illustrates fundamental properties of hyperconjunctive and hyperdisjunctive aggregators. As the number of arguments increases, the volume under the pure conjunction $V_{con}(n) = 1/(n+1)$ quickly decreases and the pure conjunction approaches t-norms and the drastic conjunction. Since the global andness is normalized so that its value is 1 for pure conjunction (and 0 for pure disjunction), it follows that the andness of drastic conjunction is a fast decreasing function of n and for more than 4 variables the difference between the product t-norm and the drastic conjunction becomes insignificant. The same properties are visible in the case of hyperdisjunction. In addition, as the number of variables increase, the product t-norm and its dual conorm approach the pure conjunction and the pure disjunction respectively. This opens the question of practical applicability of t-norms/conorms.

The geometric interpretations of the regions of hyperconjunction and hyperdisjunction show that hyperconjunctive and hyperdisjunctive aggregators lose their applicability and significance as they approach the drastic conjunction or drastic disjunction. Indeed, by definition, the drastic conjunction and drastic disjunction are not logic aggregators, because they are insensitive to positive and incomplete truth. These drastic functions are the borders of the world of logic aggregators and are not used in real life decision problems. Figure 3 shows that the t-norms/conorms quickly approach these borders. All hyperconjunctive aggregators with andness higher than the andness of product t-norm are very similar to each other and their applicability is extremely low. The space between t-norms and the pure conjunction is greater than the space between the space between the t-norm and the drastic conjunction.

4 Conclusions

In the area of aggregation, it is useful to differentiate two fundamentally different categories of aggregation problems:

1. *Mathematical problems of aggregation of anonymous real numbers* (numbers without semantic identity, explanation of role, and unit of measurement).
2. *Decision engineering problems of aggregation of arguments that have specific semantic identity* (a clearly defined meaning, role, linguistically labeled units of measurement, and quantifiable impact on attaining specific stakeholder's goals).

In the case of aggregation of anonymous real numbers, the goal is to study the widest possible class of aggregation functions [6, 7]. This is a legitimate goal in mathematics. Generally, applicability is not a noticeable goal of such studies, because each application area imposes restrictions and reduces the space for theoretical developments. Thus, it is both legitimate and desirable to develop general aggregation structures in semantic vacuum. However, from the standpoint of decision engineering, mathematical methods of aggregation of anonymous real numbers inevitably combine a specific degree of applicability and a specific degree of mathematical ballast that is not useful in decision engineering practice.

Unsurprisingly, aggregation methods are both important and frequent in decision engineering, where all arguments have clearly defined semantic identity. Typical examples of semantic identity are degrees of truth, degrees of importance, degrees of suitability for particular use, degrees of simultaneity, probability of specific desirable or undesirable events and consequences, etc. In all such cases, it is necessary to create aggregators that provide appropriate support for the semantic identity of arguments.

The point of departure in the decision engineering area is the fact that decisions are products of human mental activity. Consequently, applicable aggregators cannot be developed ignoring observable properties of human reasoning. Quite contrary, the study of applicable aggregators must begin with observing the patterns of human reasoning. Then, we must develop mathematical models that have expressive power to properly describe the observed reasoning patterns. This is the primary goal of graded logic: if the aggregated arguments are degrees of truth, then such reasoning patterns belong to logic, and their models naturally have the status of logic aggregators.

In the area of evaluation, all inputs and outputs have the semantic identity of degrees of truth of value statements (statements that assert the satisfaction of various requirements, specified by a decision engineer in order to contribute to the attainment of stakeholder's goals). Graded logic supports the area of logic aggregation, in a way that is necessary for solving evaluation problems in decision engineering.

Using the degree of truth between the complete truth and the complete falsity, the GL-based soft computing version of propositional logic operates inside the unit hypercube. In this region, it is necessary to have logic aggregators that are observable in human reasoning. Such aggregators must include the following: (1) models of simultaneity (conjunctive aggregators) and models of substitutability (disjunctive aggregators), (2) aggregators that support annihilators (hard aggregators) and aggregators that do not support annihilators (soft aggregators), (3) idempotent aggregators (supporting internality) modeled as means, and nonidempotent (hyperconjunctive and hyperdisjunctive aggregators) frequently modeled as t-norms/conorms, (4) aggregators that support adjustable importance of arguments, and aggregators that use fixed (equal) importance of arguments, and (5) general aggregators that have adjustable andness and special aggregators that have fixed andness. Transition between aggregators must be continuous and andness-directed, implemented using interpolation [5]. The fundamental aggregator that provides all requested features is the graded conjunction/ disjunction and the corresponding generalization of Boolean logic is the graded logic. A detailed presentation of GCD, GL, and a complete decision engineering framework, built using GL and verified in industrial applications, can be found in [8].

References

1. Boole, G.: An Investigation of the Laws of Thought, on Which are Founded the Mathematical Theories of Logic and Probabilities. Macmillan, London (1854)
2. Dujmović, J.: Weighted conjunctive and disjunctive means and their application in system evaluation. J. Univ. Belgrade EE Dept. Ser. Math. Phys. **483**, 147–158 (1974). http://www.jstor.org/stable/43667685
3. Zimmermann, H.-J.: Fuzzy Set Theory and Its Applications. Springer, New York (1996). https://doi.org/10.1007/978-94-010-0646-0
4. Dujmović, J.: Two integrals related to means. J. Univ. Belgrade EE Dept. Ser. Math. Phys. **412–460**, 231–232 (1973). http://www.jstor.org/stable/43668034
5. Dujmović, J.: Weighted compensative logic with adjustable threshold andness and orness. IEEE Trans. Fuzzy Syst. **23**(2), 270–290 (2015)
6. Beliakov, G., Pradera, A., Calvo, T.: Aggregation Functions: A Guide for Practitioners. Springer, Berlin (2007). https://doi.org/10.1007/978-3-540-73721-6
7. Grabisch, M., Marichal, J.-L., Mesiar, R., Pap, E.: Aggregation Functions. Cambridge University Press, Cambridge (2009)
8. Dujmović, J.: Soft Computing Evaluation Logic. Wiley, Hoboken (2018)

Aggregation Operators, Fuzzy Measures and Integrals

Coherent Risk Measures Derived from Utility Functions

Yuji Yoshida$^{(\boxtimes)}$

Faculty of Economics and Business Administration, University of Kitakyushu,
4-2-1 Kitagata, Kokuraminami, Kitakyushu 802-8577, Japan
yoshida@kitakyu-u.ac.jp

Abstract. Coherent risk measures in financial management are discussed from the view point of average value-at-risks with risk spectra. A minimization problem of the distance between risk estimations through decision maker's utility and coherent risk measures with risk spectra is introduced. The risk spectrum of the optimal coherent risk measures in this problem is obtained and it inherits the risk averse property of utility functions. Various properties of coherent risk measures and risk spectrum are demonstrated. Several numerical examples are given to illustrate the results.

1 Introduction

Risk measure is one of the most important concepts in economic theory, financial analysis, asset management and so on. In classical mean-variance portfolio models, the variance was used as a risk measure in asset management [9]. Recently drastic declines of asset prices are studied, and *value-at-risk* is used widely to estimate the risk of asset price decline in practical financial management [7]. Value-at-risk is defined by percentiles at a specified probability, however it does not have coherency. *Coherent risk measures* have been studied to improve the criterion of risks with worst scenarios [3], and several improved risk measures based on value-at-risks are proposed: For example, conditional value-at-risk, expected shortfall, entropic value-at-risk [6,11,12]. Kusuoka [8] gave a spectral representation for coherent risk measures, and Acerbi [1] and Adam et al. [2] demonstrated its applications to portfolio selection and so on. Cotter and Dowd [4] examined exponential type spectral measures, and Yaari [13] studied a relation between spectral measures and distortion risk measures. Emmer et al. [5] compared risk measures by their properties to find best risk measures. We discuss *what is the optimal coherent risk measure*, and then from Kusuoka [8] we give *an optimal risk spectrum* for coherent risk measures. In this paper we focus on the downside ranges of utility functions related to decision maker's risk sensitivity, and we adopt a *risk estimation through utility functions* as an optimization object for coherent risk measures. We obtain an optimal risk spectrum of coherent risk measures minimizing the distance between risk estimations through utility functions and coherent risk measures with risk spectra, and then the risk spectrum inherits the risk averse property of utility functions.

© Springer Nature Switzerland AG 2018
V. Torra et al. (Eds.): MDAI 2018, LNAI 11144, pp. 15–26, 2018.
https://doi.org/10.1007/978-3-030-00202-2_2

In Sect. 2 we introduce coherent risk measures and weighted average value-at-risks, and in Sect. 3 we deal with coherent risk measures with risk spectra and their properties. In Sect. 4 we discuss a coherent risk measure with risk spectrum which is nearest to risk estimations through utility functions. Then, as a weight of weighted average value-at-risks, the risk spectrum drives the risk aversion from decision maker's utility functions to coherent risk measures. In Sect. 5 we give several examples of utility functions and we observe the risk spectra and the weighted average value-at-risks.

2 Value-at-Risks and Coherent Risk Measures

Let Ω be a sample space and let P be a non-atomic probability on Ω. Let $\mathcal{X}$ be a family of integrable real-valued random variables X on Ω which have a differentiable cumulative distribution function $F_X(\cdot) = P(X < \cdot)$ and a density function $w(x) = \frac{d}{dx}F_X(x)$. For a random variable $X \in \mathcal{X}$ there exist an open interval I and a strictly increasing and continuous inverse function $F_X^{-1} : (0,1) \to I$. Then we have $\lim_{x \downarrow \inf I} F_X(x) = 0$ and $\lim_{x \uparrow \sup I} F_X(x) = 1$. Let $\mathbb{R} = (-\infty, \infty)$, and we put $w = 0$ on $\mathbb{R} \setminus I$ for simple representation. The *value-at-risk (VaR)* at a risk probability p is given by the percentile of the distribution function F_X:

$$\mathrm{VaR}_p(X) = \begin{cases} \inf I & \text{if } p = 0 \\ \sup\{x \in I \mid F_X(x) \le p\} & \text{if } p \in (0,1) \\ \sup I & \text{if } p = 1. \end{cases} \tag{1}$$

Then we have $F_X(\mathrm{VaR}_p(X)) = p$ and $\mathrm{VaR}_p(X) = F_X^{-1}(p)$ for $p \in (0,1)$. The *average value-at-risk (AVaR)* for a probability p is also given by

$$\mathrm{AVaR}_p(X) = \frac{1}{p}\int_0^p \mathrm{VaR}_q(X)\,dq \tag{2}$$

if $p \in (0,1]$ and $\mathrm{AVaR}_p(X) = \inf I$ if $p = 0$. Further we denote by $E(X) = \int X\,dP$ and $\sigma(X) = \sqrt{E((X - E(X))^2)}$ the expectation and the standard deviation respectively for random variables $X \in \mathcal{X}$. For the family $\mathcal{X}$, we assume the following (i) and (ii):

(i) There exists a strictly increasing function $\kappa : (0,1) \mapsto \mathbb{R}$ such that

$$\mathrm{VaR}_p(X) = E(X) + \kappa(p)\sigma(X) \tag{3}$$

for random variables $X \in \mathcal{X}$ and $p \in (0,1)$.
(ii) There exists a probability density function ψ on $\mathbb{R} \times [0, \infty)$ of means $E(X)$ and standard deviations $\sigma(X)$ of random variables $X \in \mathcal{X}$.

The following definitions are introduced to characterize risk measures.

Definition 1 ([3,8]). Let a map $\rho : \mathcal{X} \mapsto \mathbb{R}$.

(i) Two random variables $X(\in \mathcal{X})$ and $Y(\in \mathcal{X})$ are called *comonotonic* if $(X(\omega) - X(\omega'))(Y(\omega) - Y(\omega')) \geq 0$ holds for almost all $\omega, \omega' \in \Omega$.

(ii) ρ is called *comonotonically additive* if $\rho(X + Y) = \rho(X) + \rho(Y)$ holds for all comonotonic $X, Y \in \mathcal{X}$.

(iii) ρ is called *law invariant* if $\rho(X) = \rho(Y)$ holds for all $X, Y \in \mathcal{X}$ satisfying $P(X < \cdot) = P(Y < \cdot)$.

(iv) ρ is called *continuous* if $\lim_{n \to \infty} \rho(X_n) = \rho(X)$ holds for $\{X_n\} \subset \mathcal{X}$ and $X \in \mathcal{X}$ such that $\lim_{n \to \infty} X_n = X$ almost surely.

Definition 2 ([3]). A map $\rho : \mathcal{X} \mapsto \mathbb{R}$ is called a *coherent risk measure* if it satisfies the following (i)–(iv):

(i) $\rho(X) \geq \rho(Y)$ holds for $X, Y \in \mathcal{X}$ satisfying $X \leq Y$. (*monotonicity*)
(ii) $\rho(cX) = c\rho(X)$ holds for $X \in \mathcal{X}$ and $c \in \mathbb{R}_+$. (*positive homogeneity*)
(iii) $\rho(X + c) = \rho(X) - c$ holds for $X \in \mathcal{X}$ and $c \in \mathbb{R}$. (*translation invariance*)
(iv) $\rho(X + Y) \leq \rho(X) + \rho(Y)$ holds for $X, Y \in \mathcal{X}$. (*sub-additivity*)

For coherent risk measures, we can easily obtain the following *spectral representation* from [8].

Lemma 1. *Let $\rho : \mathcal{X} \mapsto \mathbb{R}$ be a law invariant, comonotonically additive, continuous coherent risk measure. Then there exists a probability measure λ on $[0, 1]$ such that*

$$\rho(X) = -\int_0^1 \mathrm{AVaR}_p(X)\, d\lambda(p) \tag{4}$$

for $X \in \mathcal{X}$.

3 Weighted Average Value-at-risks with Risk Spectra

Let $\mathcal{N}$ be a family of functions $\nu : (0, 1] \mapsto [0, \infty)$ such that $\int_0^1 \nu(p)\, dp = 1$ and $\nu(1) = \liminf_{p \uparrow 1} \nu(p)$, and let $\underline{\mathcal{N}}$ be a family of functions $\nu \in \mathcal{N}$ such that ν is non-increasing and right-continuous on $(0, 1)$. In the following lemma, we have another representation of (4).

Lemma 2. *For a probability measure λ on $[0, 1]$, we let*

$$\nu(q) = \int_q^1 \frac{1}{p}\, d\lambda(p)$$

for $q \in (0, 1)$ and $\nu(1) = \liminf_{q \uparrow 1} \nu(q)$. Then it holds that $\nu \in \underline{\mathcal{N}}$ and

$$\int_0^1 \mathrm{AVaR}_p(X)\, d\lambda(p) = \int_0^1 \mathrm{VaR}_q(X)\, \nu(q)\, dq$$

for $X \in \mathcal{X}$.

From Lemma 2, for a random variable $X \in \mathcal{X}$ and a function $\nu \in \mathcal{N}$, we also introduce weighted average value-at-risks with weight ν by

$$\mathrm{AVaR}_1^\nu(X) = \int_0^1 \mathrm{VaR}_q(X)\,\nu(q)\,dq. \tag{5}$$

Then ν is called a *risk spectrum* if $\nu \in \underline{\mathcal{N}}$. Further for a probability $p \in (0,1)$ we define *a weighted average value-at-risk with risk spectrum* ν on $(0,p)$ by

$$\mathrm{AVaR}_p^\nu(X) = \int_0^p \mathrm{VaR}_q(X)\,\nu(q)\,dq \Big/ \int_0^p \nu(q)\,dq. \tag{6}$$

From Lemmas 1 and 2, we obtain the following theorem.

Theorem 1

(i) Let $\rho : \mathcal{X} \mapsto \mathbb{R}$ be a law invariant, comonotonically additive, continuous coherent risk measure. Then there exists a risk spectrum $\nu \in \underline{\mathcal{N}}$ such that

$$\rho(X) = -\int_0^1 \mathrm{VaR}_q(X)\,\nu(q)\,dq = -\mathrm{AVaR}_1^\nu(X) \tag{7}$$

for $X \in \mathcal{X}$.

(ii) Let a function $\nu \in \underline{\mathcal{N}}$. Then $-\mathrm{AVaR}_p^\nu$ is a coherent risk measure on $\mathcal{X}$ for $p \in (0,1]$.

From Theorem 1 we focus on coherent risk measures in the form $-\mathrm{AVaR}_p^\nu$ with risk spectra $\nu \in \underline{\mathcal{N}}$, which is given in (6).

Proposition 1. Let a probability $p \in (0,1]$. The following (i) and (ii) hold.

(i) Let $\nu_1 \in \mathcal{N}$ and $\nu_2 \in \mathcal{N}$ satisfy

$$\nu_1(r)\nu_2(q) \le \nu_1(q)\nu_2(r) \qquad \text{for } q,r \in \mathbb{R} \text{ satisfying } 0 \le q < r \le p. \tag{8}$$

Then it holds that $\mathrm{AVaR}_p^{\nu_1}(X) \le \mathrm{AVaR}_p^{\nu_2}(X)$ for $X \in \mathcal{X}$.

(ii) It holds that

$$\sup_{\nu \in \underline{\mathcal{N}}} \mathrm{AVaR}_p^\nu(X) = \mathrm{AVaR}_p(X) \tag{9}$$

for $X \in \mathcal{X}$. Namely the maximum of weighted average value-at-risks is the average value-at-risk.

Remark. In Proposition 1(i), we have the following equivalence: Eq. (8) $\Longleftrightarrow$ ν_2/ν_1 is non-decreasing on $\{q \in (0,p) \mid \nu_1(q) > 0\}$. If ν_1 and ν_2 are piecewise differentiable, we have: Eq. (8) $\Longleftrightarrow$ $\nu_1'/\nu_1 \le \nu_2'/\nu_2$ on $\{q \in (0,p) \mid \nu_1(q) > 0\}$.

4 An Optimal Risk Spectrum Derived from Risk Averse Utility Functions

In the rest of this paper, we deal with a decision maker's risk averse utility functions $f : I \mapsto \mathbb{R}$ which are C^2-class and satisfy $f' > 0$ and $f'' \leq 0$ on I. Let a probability $p \in (0, 1]$. Under decision maker's utility function f, the average value-at-risks of random variables $X(\in \mathcal{X})$ for probabilities q over $(0, p)$ are estimated as the following non-linear form:

$$f^{-1}\left(\frac{1}{p}\int_0^p f(\mathrm{VaR}_q(X))\,dq\right) = f^{-1}(E(f(X) \mid X \leq \mathrm{VaR}_p(X)). \qquad (10)$$

Eq. (10) implies the estimated value of random variable X through utility function f on the downside range $(-\infty, \mathrm{VaR}_p(X))$. Now we discuss an optimization problem to find a weighted average value-at-risk (6) which is the nearest to the risk estimation (10). The risk estimation (10) on the downside range $(-\infty, \mathrm{VaR}_p(X))$ is related to the most risk sensitive parts of utility function f and it acquires decision maker's risky sense regarding random variable X. On the other hand, from Theorem 1 coherent risk measures are represented by (7). When we find coherent risk measures corresponding to (10), they should be related to the downside parts of X, i.e. $\{x \in \mathbb{R} \mid P(X \leq x) \leq p\} = \{\mathrm{VaR}_q(X) \mid 0 < q \leq p\}$. Therefore by Theorem 1(ii) we use coherent risk measures $-\mathrm{AVaR}_p^\nu$ defined by (6).

Optimization Problem 1. Find a risk spectrum $\nu \in \underline{\mathcal{N}}$ which minimizes the distance

$$\sum_{X \in \mathcal{X}} \left(f^{-1}\left(\frac{1}{p}\int_0^p f(\mathrm{VaR}_q(X))\,dq\right) - \mathrm{AVaR}_p^\nu(X)\right)^2 \qquad (11)$$

for $p \in (0, 1]$.

Solving Optimization Problem 1, the optimal risk spectrum ν, with which the coherent risk measure given in (6) has a kind of semi-linear properties such as Definition 2(ii) and (iii), can inherit decision maker's risk averse sense of the non-linear utility function f as a weighting function on $(0, p)$.

Theorem 2. *Let $\nu \in \mathcal{N}$ be a function given by*

$$\nu(p) = e^{-\int_p^1 C(q)\,dq} C(p) \qquad (12)$$

for $p \in (0, 1]$ with its component function

$$C(p) = \frac{\displaystyle\sum_{X \in \mathcal{X}} \sigma(X)\frac{f(\mathrm{VaR}_p(X)) - \frac{1}{p}\int_0^p f(\mathrm{VaR}_q(X))\,dq}{pf'\left(f^{-1}\left(\frac{1}{p}\int_0^p f(\mathrm{VaR}_q(X))\,dq\right)\right)}}{\displaystyle\sum_{X \in \mathcal{X}} \sigma(X)\left(\mathrm{VaR}_p(X) - f^{-1}\left(\frac{1}{p}\int_0^p f(\mathrm{VaR}_q(X))\,dq\right)\right)}. \qquad (13)$$

If ν is non-increasing, then ν is an optimal risk premium for Optimization Problem 1.

Sketch Proof of Theorem 2. Let $p \in (0,1)$. From (3), for $X \in \mathcal{X}$, we put $\mathrm{VaR}_p(X) = \mu + \kappa(p) \cdot \sigma$ with a mean $\mu = E(X)$ and a standard deviation $\sigma = \sigma(X)$. To discuss the minimization (11), by (6) we define

$$G(\underline{\nu}) = \sum_{X \in \mathcal{X}} \left(f^{-1}\left(\frac{1}{p} \int_0^p f(\mu + \kappa(q)\sigma)\, dq \right) - \frac{\int_0^p (\mu + \kappa(q)\sigma)\, \underline{\nu}(q)\, dq}{\int_0^p \underline{\nu}(q)\, dq} \right)^2$$

for risk spectra $\underline{\nu}$. Let ν be a risk spectrum attaining the minimum (11). Then $(1-t)\nu + t\varepsilon$ is also a risk spectrum for $t \in (0,1)$ and risk spectra ε. Hence we have

$$\lim_{t \downarrow 0} \frac{G((1-t)\nu + t\varepsilon) - G(\nu)}{t} = 0$$

for any risk spectrum ε. This follows

$$\sum_{X \in \mathcal{X}} \sigma \left(f^{-1}\left(\frac{1}{p} \int_0^p f(\mu + \kappa(q)\sigma)\, dq \right) - \frac{\int_0^p (\mu + \kappa(q)\sigma)\, \nu(q)\, dq}{\int_0^p \nu(q)\, dq} \right) = 0.$$

Therefore we obtain

$$\sum_{X \in \mathcal{X}} \sigma \left(f^{-1}\left(\frac{1}{p} \int_0^p f(\mu + \kappa(q)\sigma)\, dq \right) \int_0^p \nu(q)\, dq - \int_0^p (\mu + \kappa(q)\sigma)\, \nu(q)\, dq \right) = 0$$

for all $p \in (0,1)$. Differentiating this equation with respect to p, we get

$$\frac{\nu(p)}{\int_0^p \nu(q)\, dq} = C(p)$$

for all $p \in (0,1)$, where C is defined by (13). Thus we obtain (12) from this equation.

$\square$

With a probability density function ψ on $\mathbb{R} \times [0, \infty)$ of means μ and standard deviations σ of random variables X, (13) follows

$$C(p) = \frac{\displaystyle\iint_{\mathbb{R} \times (0,\infty)} \sigma \, \frac{f(\mu + \kappa(p)\sigma) - \frac{1}{p}\int_0^p f(\mu + \kappa(q)\sigma)\, dq}{p f'\left(f^{-1}\left(\frac{1}{p}\int_0^p f(\mu + \kappa(q)\sigma)\, dq \right) \right)} \, \psi(\mu, \sigma)\, d\mu d\sigma}{\displaystyle\iint_{\mathbb{R} \times (0,\infty)} \sigma \left((\mu + \kappa(p)\sigma) - f^{-1}\left(\frac{1}{p}\int_0^p f(\mu + \kappa(q)\sigma)\, dq \right) \right) \psi(\mu, \sigma)\, d\mu d\sigma}.$$

$$(14)$$

We can easily check the following results.

Proposition 2. *The optimal risk spectrum ν and its component function C in Theorem 2 have the following properties (i)–(v):*

(i) $C'(p) < 0$ *for* $p \in (0,1)$.
(ii) $0 < \nu(p) < C(p) \le \frac{1}{p}$ *for* $p \in (0,1)$.
(iii) $0 < \nu(1) = C(1) \le 1$.
(iv) $\lim_{p \to 0} C(p) = \infty$.
(v) *If* $f'' < 0$ *on* I, *then* $C(p) < \frac{1}{p}$ *for* $p \in (0,1)$.

5 Examples

In this section we give several examples for the results in the previous sections.

Example 1. Let a domain $I = \mathbb{R}$ and let f be a *risk neutral function* $f(x) = ax + b$ $(x \in \mathbb{R})$ with constants $a(> 0)$ and $b(\in \mathbb{R})$. Then it is trivial that f satisfies the conditions, and its optimal risk spectrum is given by $\nu(p) = 1$ with the component function $C(p) = \frac{1}{p}$. The corresponding weighted average value-at-risk is reduced to the *average value-at-risk*:

$$\text{AVaR}_p^\nu = \text{AVaR}_p \tag{15}$$

for $p \in (0,1]$.

Example 2. Let a domain $I = \mathbb{R}$ and let a *risk averse exponential utility function*

$$f(x) = \frac{1 - e^{-\tau x}}{\tau} \tag{16}$$

for $x \in \mathbb{R}$ with a positive constant τ. Then we can easily check (16) satisfies the conditions. Let $\mathcal{X}$ be a family of random variables X which have a normal distribution function with a density function

$$w(x) = \frac{1}{\sqrt{2\pi}\sigma} e^{-\frac{(x-\mu)^2}{2\sigma^2}} \tag{17}$$

for $x \in \mathbb{R}$, where the mean $\mu = E(X)$ and the standard deviation $\sigma = \sigma(X)$ of random variables X. Define the cumulative distribution function $\Phi : (-\infty, \infty) \to (0,1)$ of the standard normal distribution by

$$\Phi(x) = \frac{1}{\sqrt{2\pi}} \int_{-\infty}^{x} e^{-\frac{z^2}{2}} \, dz \tag{18}$$

for $x \in \mathbb{R}$, and define an increasing function $\kappa : (0,1) \mapsto (-\infty, \infty)$ by its inverse function

$$\kappa(p) = \Phi^{-1}(p) \tag{19}$$

for probabilities $p \in (0,1)$ (Fig. 1). Then we have value-at-risk $\text{VaR}_p(X) = \mu + \kappa(p)\sigma$ for $X \in \mathcal{X}$. Suppose $\mathcal{X}$ is a family of random variables X with a distribution function $\psi : \mathbb{R} \times (0, \infty) \mapsto [0, \infty)$ such that $\psi(\mu, \sigma) = \phi(\mu) \cdot \sqrt{\frac{2}{\pi}} e^{-\frac{\sigma^2}{2}}$ for $(\mu, \sigma) \in \mathbb{R} \times [0, \infty)$, where $\phi(\mu)$ is some probability distribution and $\sqrt{\frac{2}{\pi}} e^{-\frac{\sigma^2}{2}}$ is a chi distribution. From Theorem 2, the optimal risk spectrum for Optimization Problem 1 is given by

$$\nu(p) = e^{-\int_p^1 C(q)\,dq} C(p) \tag{20}$$

for $p \in (0,1]$, where the component function is given by

$$C(p) = \frac{1}{p} \cdot \frac{\iint_{\mathbb{R}\times(0,\infty)} \sigma \left(1 - \frac{1}{\frac{1}{p}\int_0^p e^{\tau\sigma(\kappa(p)-\kappa(q))}\,dq}\right) \psi(\mu,\sigma)\,d\mu d\sigma}{\iint_{\mathbb{R}\times(0,\infty)} \sigma \log\left(\frac{1}{p}\int_0^p e^{\tau\sigma(\kappa(p)-\kappa(q))}\,dq\right) \psi(\mu,\sigma)\,d\mu d\sigma}, \tag{21}$$

and this is also reduced to

$$C(p) = \frac{1}{p} \cdot \frac{\int_0^\infty \left(1 - \frac{1}{\frac{1}{p}\int_0^P e^{\tau\sigma(\kappa(p)-\kappa(q))}\,dq}\right)\sigma e^{-\frac{\sigma^2}{2}}\,d\sigma}{\int_0^\infty \log\left(\frac{1}{p}\int_0^P e^{\tau\sigma(\kappa(p)-\kappa(q))}\,dq\right)\sigma e^{-\frac{\sigma^2}{2}}\,d\sigma}. \tag{22}$$

Then for $\tau = 1$ Fig. 1 shows the concave utility function $f(x)$ given in (16) and the function $\kappa(p)$ given in (19). Using these functions, Fig. 2 illustrates the optimal risk spectrum $\nu(p)$ and its component function $C(p)$ given in (20) and (22).

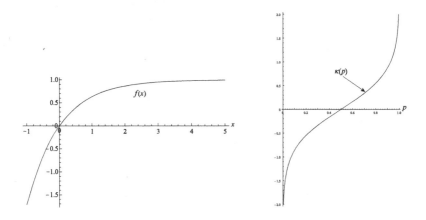

Fig. 1. Utility function $f(x)$ and function $\kappa(p) = \Phi^{-1}(p)$ in Example 2 ($\tau = 1$).

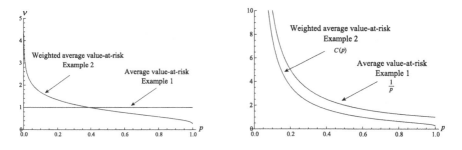

Fig. 2. The optimal risk spectrum $\nu(p)$ and its component function $C(p)$.

Example 3. Let a domain $I = (\alpha, \beta)$, where $-\infty < \alpha < \beta < \infty$, and let a *risk averse quadratic utility function*

$$f(x) = -a(x - c)^2 + b \tag{23}$$

for $x \in I$ with constants $a(> 0)$, $b(\in \mathbb{R})$ and $c(\in [\beta, \infty))$. Then we can easily check (23) satisfies the conditions. Let $\mathcal{X}$ be a family of random variables X which have a density function

$$w(x) = \max \left\{ \frac{1}{\sqrt{6}\sigma} - \left| \frac{x - \mu}{6\sigma^2} \right|, 0 \right\} \tag{24}$$

for $x \in \mathbb{R}$, where the mean $\mu = E(X)$ and the standard deviation $\sigma = \sigma(X)$. Then its distribution function is

$$F(x) = \int_{-\infty}^{x} w(z)\, dz = \frac{(x - \mu)^2}{12\sigma^2} - \frac{|x - \mu|}{\sqrt{6}\sigma} + \frac{1}{2}$$

for $x \in \mathbb{R}$, and then the value-at-risk $\mathrm{VaR}_p(X) = \mu + \kappa(p)\sigma$ is given by an increasing function (Fig. 3)

$$\kappa(p) = \begin{cases} 2\sqrt{3p} - \sqrt{6} & \text{if } 0 \le p \le \frac{1}{2} \\ -2\sqrt{3(1 - p)} + \sqrt{6} & \text{if } \frac{1}{2} < p \le 1. \end{cases} \tag{25}$$

Suppose $\mathcal{X}$ is a family of random variables X with a distribution function $\psi :$ $\mathbb{R} \times (0, \infty) \mapsto [0, \infty)$ such that $\psi(\mu, \sigma) = \max\{1 - |\mu|, 0\} \cdot \frac{1}{2\pi} e^{-\frac{\sigma^2}{16\pi}}$. From Theorem 2 we have the optimal risk spectrum $\nu(p) = e^{-\int_p^1 C(q)\, dq} C(p)$ for $p \in (0, 1]$, where the component function $C(p)$ is

$$\frac{1}{2p} \cdot \frac{\iint_{\mathbb{R} \times (0,\infty)} \sigma \frac{-(\mu + \kappa(p)\sigma - c)^2 + \frac{1}{p}\int_0^p (\mu + \kappa(q)\sigma - c)^2\, dq}{\sqrt{\frac{1}{p}\int_0^p (\mu + \kappa(q)\sigma - c)^2\, dq}} \psi(\mu, \sigma)\, d\mu d\sigma}{\iint_{\mathbb{R} \times (0,\infty)} \sigma \left((\mu + \kappa(p)\sigma - c) + \sqrt{\frac{1}{p}\int_0^p (\mu + \kappa(q)\sigma - c)^2\, dq} \right) \psi(\mu, \sigma)\, d\mu d\sigma}. \tag{26}$$

Let $a = 1, b = 100, c = 10, \alpha = -10$ and $\beta = 10$. Then the utility function is $f(x) = x(20 - x)$ for $x \in I = (-10, 10)$. Then Fig. 3 shows the utility function f and a function (25), Fig. 4 illustrates the optimal risk spectrum ν and its component function C in (26).

Example 4. Let a domain $I = (\alpha, \beta)$, where $-\infty < \alpha < \beta < \infty$, and let a *risk averse logarithmic utility function*

$$f(x) = a \log(x + c) + b \tag{27}$$

for $x \in (-c, \infty)$ with constants $a(> 0)$, $b(\in \mathbb{R})$ and $c = -\frac{b}{a}$ satisfying $\beta + c \le e\alpha + ec$. Then we can easily check f satisfies the conditions. Let $\mathcal{X}$ be the same family of random variables X in Sect. 3. From Theorem 2 we have the optimal risk spectrum $\nu(p) = e^{-\int_p^1 C(q)\, dq} C(p)$ for $p \in (0, 1]$, where the component function is

$$C(p) = \frac{1}{p} \cdot \frac{\iint_{\mathbb{R} \times (0,\infty)} \sigma \cdot \frac{\log(\mu + \kappa(p)\sigma + c) - \frac{1}{p}\int_0^p \log(\mu + \kappa(q)\sigma + c)\, dq}{e^{-\frac{1}{p}\int_0^p \log(\mu + \kappa(q)\sigma + c)\, dq}} \psi(\mu, \sigma)\, d\mu d\sigma}{\iint_{\mathbb{R} \times (0,\infty)} \sigma \left((\mu + \kappa(p)\sigma + c) - e^{\frac{1}{p}\int_0^p \log(\mu + \kappa(q)\sigma + c)\, dq} \right) \psi(\mu, \sigma)\, d\mu d\sigma}. \tag{28}$$

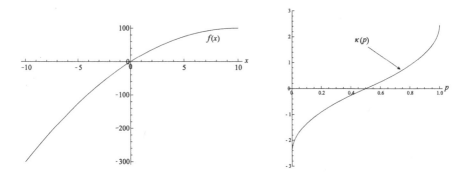

Fig. 3. Utility function $f(x)$ and function $\kappa(p)$ in Example 3.

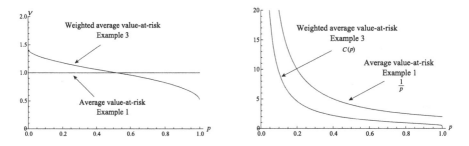

Fig. 4. The optimal risk spectrum ν and its component function C.

Let $a = 1, b = -\log \frac{10e + 100}{e - 1}, c = \frac{10e + 100}{e - 1}, \alpha = -10$ and $\beta = 100$. Then the utility function is

$$f(x) = \log \left(\frac{e - 1}{10e + 100} x + 1 \right) \tag{29}$$

for $x \in I = (-10, 100)$. Then Fig. 4 shows the utility function (29) and Fig. 5 illustrates the optimal risk spectrum ν and its component function C in (28).

Concluding Remarks

(i) Theorem 2 gives a method to combine *the theory of coherent risk measures* with *subjective risk averse decision making* via *weighted average value-at-risks*. In (11) we adopt risk estimations with risk averse utility as an optimization target for coherent risk measures. Another optimization target may be taken to find best coherent risk measures.

(ii) Average value-at-risks give coherent risk measures derived from risk neutral utilities (Example 1). From Proposition 1(ii) we have $\inf_{\nu \in \underline{\mathcal{N}}}(-\mathrm{AVaR}_p^\nu(X)) = -\mathrm{AVaR}_p(X)$, and therefore we find the average value-at-risk gives the lower bound of coherent risk measures derived from risk averse utilities.

(iii) Using Theorem 2, we can incorporate the decision maker's risk averse attitude into coherent risk measures as risk spectra of weighted average

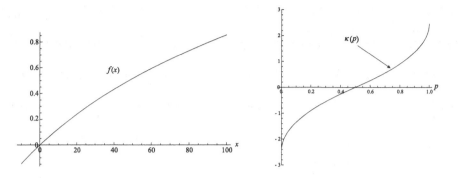

Fig. 5. Utility function $f(x)$ and function $\kappa(p)$ in Example 4.

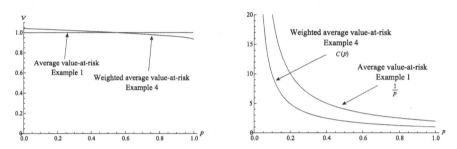

Fig. 6. The optimal risk spectrum ν and its component function C.

value-at-risks. This result will be applicable to subjective portfolio optimization in financial management and so on.

Acknowledgments. This research is supported from JSPS KAKENHI Grant Number JP 16K05282.

References

1. Acerbi, C.: Spectral measures of risk: a coherent representation of subjective risk aversion. J. Bank. Financ. **26**, 1505–1518 (2002)
2. Adam, A., Houkari, M., Laurent, J.-P.: Spectral risk measures and portfolio selection. J. Bank. Financ. **32**, 1870–1882 (2008)
3. Artzner, P., Delbaen, F., Eber, J.-M., Heath, D.: Coherent measures of risk. Math. Financ. **9**, 203–228 (1999)
4. Cotter, J., Dowd, K.: Extreme spectral risk measures: an application to futures clearinghouse margin requirements. J. Bank. Financ. **30**, 3469–3485 (2006)
5. Emmer, S., Kratz, M., Tasche, D.: What is the best risk measure in practice? A comparison of standard measures. J. Risk **18**, 31–60 (2015)
6. Javidi, A.A.: Entropic value-at-risk: a new coherent risk measure. J. Optim. Theory Appl. **155**, 1105–1123 (2012)
7. Jorion, P.: Value at Risk: The New Benchmark for Managing Financial Risk. McGraw-Hill, New York (2006)

8. Kusuoka, S.: On law-invariant coherent risk measures. Adv. Math. Econ. **3**, 83–95 (2001)
9. Markowitz, H.: Mean-Variance Analysis in Portfolio Choice and Capital Markets. Blackwell, Oxford (1990)
10. Pratt, J.W.: Risk aversion in the small and the large. Econometrica **32**, 122–136 (1964)
11. Rockafellar, R.T., Uryasev, S.: Optimization of conditional value-at-risk. J. Risk **2**, 21–41 (2000)
12. Tasche, D.: Expected shortfall and beyond. J. Bank. Financ. **26**, 1519–1533 (2002)
13. Yaari, M.E.: The dual theory of choice under risk. Econometrica **55**, 95–115 (1987)
14. Yoshida, Y.: A dynamic value-at-risk portfolio model. In: Torra, V., Narakawa, Y., Yin, J., Long, J. (eds.) MDAI 2011. LNCS (LNAI), vol. 6820, pp. 43–54. Springer, Heidelberg (2011). https://doi.org/10.1007/978-3-642-22589-5_6
15. Yoshida, Y.: An Ordered Weighted Average with a Truncation Weight on Intervals. In: Torra, V., Narukawa, Y., López, B., Villaret, M. (eds.) MDAI 2012. LNCS (LNAI), vol. 7647, pp. 45–55. Springer, Heidelberg (2012). https://doi.org/10.1007/978-3-642-34620-0_6

On k-$\oplus$-additive Aggregation Functions

Fateme Kouchakinejad$^{1(\boxtimes)}$, Anna Kolesárová^2, and Radko Mesiar^3

^{1} Department of Statistics, Faculty of Mathematics and Computer Science,
Shahid Bahonar University of Kerman, Kerman, Iran
kouchakinezhad@gmail.com
^{2} Institute of Information Engineering, Automation and Mathematics,
Faculty of Chemical and Food Technology, Slovak University of Technology,
Radlinského 9, 812 37 Bratislava, Slovakia
anna.kolesarova@stuba.sk
^{3} Department of Mathematics and Descriptive Geometry,
Faculty of Civil Engineering, Slovak University of Technology,
Radlinského 11, 810 05 Bratislava, Slovakia
radko.mesiar@stuba.sk

Abstract. To generalize the concept of k-maxitivity and k-additivity, we introduce k-$\oplus$-additive aggregation functions. We also characterize this kind of aggregation functions under some special conditions. Several examples are given to illustrate the new definitions.

Keywords: k-additivity · k-maxitivity
k-$\oplus$-additive aggregation function · Pseudo-addition

1 Introduction

Let $n \in \mathbb{N}$. Throughout the paper, the points in $[0,1]^n$ will be denoted by bold letters, $\mathbf{x} = (x_1, \ldots, x_n)$. Given $\mathbf{x}, \mathbf{y} \in [0,1]^n$, we write $\mathbf{x} \leqslant \mathbf{y}$ if $x_i \leqslant y_i$ for each $i \in \{1, \ldots, n\}$.

Consider a fixed finite space $X = \{1, \ldots, n\}$. Recall that a set function $m : 2^X \to [0,1]$ is called a capacity if it is monotone and $m(\emptyset) = 0$, $m(X) = 1$. The additivity of a capacity m makes it a probability measure, determined by n values of singletons, $w_i = m(\{i\})$, $i = 1, 2, \ldots, n$, constrained by the condition $\sum_{i=1}^{n} w_i = 1$. The additivity of a probability measure excludes the possibility of interactions between single subsets of X. On the other hand, a general capacity m allows to model interaction of any group of subsets of X, but it requires the knowledge of $2^n - 2$ values $m(A)$, $A \neq X, \emptyset$. Grabisch [2] introduced the notion of k-additive capacities, $k \in \{1, \ldots, n\}$, to reduce the complexity of a general capacity m, but to allow to model interaction of some groups of subsets of X. Inspired by this idea, Kolesárová et al. [5] introduced k-additive aggregation functions. They also clarified the relation between k-additive capacities, their Owen (multilinear) extension [8] and k-additive aggregation functions.

© Springer Nature Switzerland AG 2018
V. Torra et al. (Eds.): MDAI 2018, LNAI 11144, pp. 27–34, 2018.
https://doi.org/10.1007/978-3-030-00202-2_3

Another property of aggregation functions is k-maxitivity. In [7], the authors introduced and discussed k-maxitive aggregation functions which can be seen as monotone extensions of k-maxitive capacities.

Both considered operations of addition and maximum can be seen as particular instances of pseudo-additions acting on $[0, \infty]$, see [4,10]. Therefore, as a generalization of concepts of k-additivity and k-maxitivity, it is natural to consider the concept of k-pseudo-additivity. Note that in the framework of capacities, some first attempts in this direction were already done in [6]. On the other hand, the idea of k-pseudo-additive aggregation functions is a new proposal developed in this paper.

The paper is organized as follows. In the next section, some preliminaries are given, recalling the concepts of k-maxitive and k-additive aggregation functions, among others. Section 3 is devoted to pseudo-additions which will be considered later. In Sect. 4, the main part of the this contribution, we introduce k-$\oplus$-additive aggregation functions and discuss some particular cases. Finally, some concluding remarks are added.

2   k-maxitive and k-additive Aggregation Functions

We start by recalling the definition of an aggregation function which has a key role in this paper.

Definition 1. [3] *Let n be a fixed natural number. A mapping $A : [0,1]^n \to [0,1]$ is called an (n-ary) aggregation function if it is monotone and satisfies the boundary conditions $A(\mathbf{0}) = A(0,\dots,0) = 0$ and $A(\mathbf{1}) = A(1,\dots,1) = 1$.*

Definition 2. [7] *Let $k \in \{1,\dots,n\}$. An aggregation function $A : [0,1]^n \to [0,1]$ is called k-maxitive if for any $\mathbf{x}_1,\dots,\mathbf{x}_{k+1} \in [0,1]^n$ there is a proper subset I of $\{1,\dots,k+1\}$ such that*

$$A\left(\bigvee_{j=1}^{k+1} \mathbf{x}_j\right) = A\left(\bigvee_{j \in I} \mathbf{x}_j\right).$$

Observe that the k-maxitivity of an aggregation function A can be defined equivalently by requiring the equality

$$\bigvee_{\substack{I \subseteq \{1,\dots,k+1\} \\ |I| \text{ is odd}}} A\left(\bigvee_{j \in I} \mathbf{x}_j\right) = \bigvee_{\substack{I \subseteq \{1,\dots,k+1\} \\ |I| \text{ is even}}} A\left(\bigvee_{j \in I} \mathbf{x}_j\right)$$

for any $\mathbf{x}_1,\dots,\mathbf{x}_{k+1} \in [0,1]^n$.

Remark 1. Observe that due to the fact that for n-ary vectors $\mathbf{x}_1, \mathbf{x}_2, \dots, \mathbf{x}_{n+1}$ there is always an index set $I \subseteq \{1,2,\dots,n+1\}$ with cardinality n such that $\bigvee_{i=1}^{n+1} x_i = \bigvee_{i \in I} x_i$ each n-ary aggregation function is n-maxitive (obviously, it can be maxitive with $k < n$). As an example of an n-maxitive aggregation function which is not k-maxitive for any $k \in \{1, \ldots, n-1\}$, one can recall the arithmetic mean.

Symmetric k-maxitive aggregation functions have the following simple representation, see [7].

Theorem 1. *A symmetric aggregation function* $A : [0,1]^n \to [0,1]$ *is* k-*maxitive if and only if there is an aggregation function* $B : [0,1]^k \to [0,1]$ *such that*

$$A(\mathbf{x}) = B(x_{\sigma(1)}, \ldots, x_{\sigma(k)}),$$

where $\sigma : X \to X$ *is a permutation such that* $x_{\sigma(1)} \geq \cdots \geq x_{\sigma(k)} \geq \cdots \geq x_{\sigma(n)}$.

Note that the same symmetric k-maxitive n-ary aggregation function A can be obtained from different k-ary aggregation functions B.

For example, the standard median $A = Med : [0,1]^3 \to [0,1]$ is a 2-maxitive aggregation function [7], and $Med(x_1, x_2, x_3) = Min(x_{\sigma(1)}, x_{\sigma(2)}) = P_2(x_{\sigma(1)}, x_{\sigma(2)})$ (here P_2 is the second projection).

Finally, we add a few words about k-maxitive capacities and k-maxitive aggregation functions. First, recall that a capacity m is called k-maxitive if for any subset $U \subseteq \{1, \ldots, n\}$ with $|U| > k$ there exists a proper subset V of U such that $m(V) = m(U)$. For $k > 1$, m is called proper k-maxitive if it is k-maxitive but not $(k-1)$-maxitive.

It is clear that for any k-maxitive aggregation function A the equation $m(U) = A(1_U)$ defines a k-maxitive capacity, i.e., A is a monotone extension of the k-maxitive capacity m. Also observe that several fuzzy integrals, such as the Sugeno integral [11] or the Shilkret integral [9] result into k-maxitive aggregation functions on $[0,1]$ once a k-maxitive capacity m is considered. For more information we refer to [7].

In the rest of this section, we review the property of k-additivity.

Definition 3. [5] *Let* $k, n \in \mathbb{N}$ *and let* $A : [0,1]^n \to [0,1]$ *be an aggregation function. Then* A *is called* k-*additive if for all collections* $\mathbf{x}_1, \ldots, \mathbf{x}_{k+1} \in [0,1]^n$ *such that also* $\sum_{i=1}^{k+1} \mathbf{x}_i \in [0,1]^n$ *we have*

$$\sum_{i=1}^{k+1}(-1)^{k+1-i} \left(\sum_{\substack{I \subseteq \{1,\ldots,k+1\} \\ |I|=i}} A \left(\sum_{j \in I} \mathbf{x}_j \right) \right) = 0. \tag{1}$$

Equivalently, if

$$\sum_{\substack{I \subseteq \{1,\ldots,k+1\} \\ |I| \text{ is odd}}} A \left(\sum_{j \in I} \mathbf{x}_j \right) = \sum_{\substack{I \subseteq \{1,\ldots,k+1\} \\ |I| \text{ is even}}} A \left(\sum_{j \in I} \mathbf{x}_j \right). \tag{2}$$

Observe that 1-additivity is just the standard additivity of aggregation functions. Moreover, it is not difficult to check that each k-additive aggregation function A is also p-additive for any integer $p > k$. The standard product $P : [0,1]^n \rightarrow [0,1]$, $P(x) = \prod_{i=1}^{n} x_i$ is a proper n-additive aggregation function [5] (i.e., it is not $(n-1)$-additive). Also, in this case any k-additive aggregation function $A : [0,1]^n \rightarrow [0,1]$ is related to a k-additive capacity m (see [2]) given by $m(U) = A(1_U)$. Note that, considering $k > 1$ and a proper k-additive capacity m, the related Choquet integral [1] with respect to m results into an aggregation function which, however, is not k-additive. In [5], the next important result was shown.

Theorem 2. *Let m be a k-additive capacity on X. Then the related Owen extension [8] $O_m : [0,1]^n \rightarrow [0,1]$ given by*

$$O_m(\mathbf{x}) = \sum_{I \subseteq \{1,\ldots,n\}} M_m(I) \prod_{i \in I} x_i.$$

is a k-additive aggregation function.

Recall that M_m is the Möbius transform of a capacity m and it is given, for any $U \subseteq X$, by

$$M_m(U) = \sum_{V \subseteq U} (-1)^{|U \setminus V|} m(V).$$

3   Pseudo-additions

We start this section with the definition of a pseudo-addition.

Definition 4. *[10] A binary operation $\oplus : [0,\infty]^2 \rightarrow [0,\infty]$ satisfying the following conditions is called pseudo-addition:*

1. *$x \oplus 0 = 0 \oplus x = x$;*
2. *$(x \oplus y) \oplus z = x \oplus (y \oplus z)$;*
3. *$x \leqslant x'$ and $y \leqslant y' \Rightarrow x \oplus y \leqslant x' \oplus y'$;*
4. *$x_n \rightarrow x$ and $y_n \rightarrow y \Rightarrow x_n \oplus y_n \rightarrow x \oplus y$.*

A pseudo-addition can be represented by a family of functions as follows:

Definition 5. *[10] Let $\{]a_k, b_k[: k \in \mathcal{K}\}$ be a family of disjoint open intervals in $[0,\infty]$ indexed by a countable set $\mathcal{K}$. For each $k \in \mathcal{K}$, associate a continuous and strictly increasing function $g_k : [a_k, b_k] \rightarrow [0,\infty]$ satisfying $g_k(a_k) = 0$. We say that a binary operation $\oplus$ on $[0,\infty]$ has a representation*

$$\{\langle]a_k, b_k[, g_k \rangle : k \in \mathcal{K}\}$$

if for all $x, y \in [0,\infty]$,

$$x \oplus y = \begin{cases} g_k^*(g_k(x) + g_k(y)) & \text{if } (x,y) \in [a_k, b_k]^2 \\ \max\{x, y\} & \text{otherwise,} \end{cases}$$

where g_k^* is the pseudo-inverse of g_k, defined by

$$g_k^*(x) = g_k^{-1}(\min\{x, g_k(b_k)\}).$$

Theorem 3. [10] *A binary operation on $[0, \infty]$ is a pseudo-addition if and only if it has a representation $\{\langle]a_k, b_k[, g_k \rangle : k \in \mathcal{K}\}$.*

Example 1. Basic types of pseudo-additions are as follows:
1. If $\mathcal{K} = \emptyset$ then $\oplus = \vee$ (max).
2. If $|\mathcal{K}| = 1,]a_1, b_1[=]0, \infty[$, and $g(\infty) = \infty$ then

$$x \oplus y = g^{-1}(g(x) + g(y)).$$

For example,
   i. for $g(x) = x$ we have $x \oplus y = x + y$,
   ii. for $g(x) = x^2$ we have $x \oplus y = \sqrt{x^2 + y^2}$.
3. If $|\mathcal{K}| = 1,]a_1, b_1[=]0, \infty[$, and $g(\infty) < \infty$ then

$$x \oplus y = g^{-1}(\min\{g(\infty), g(x) + g(y)\}).$$

For example, if $g(x) = \dfrac{2x}{x+1}$ we have

$$x \oplus y = \begin{cases} \dfrac{2xy + x + y}{1 - xy} & \text{if } xy < 1, \\ \infty & \text{otherwise} \end{cases}$$

(under the convention $0.\infty = 0$).
4. If $|\mathcal{K}| \geq 1$ and there is a $k \in \mathcal{K}$ such that $]a_k, b_k[=]0, 1[$, then $\oplus_{|_{[0,1]^2}} = S$, where S is a continuous Archimedean t-conorm generated by an additive generator $g : [0, 1] \to [0, \infty]$.
   i. S is strict if $g(1) = \infty$ and then

$$S(x, y) = g^{-1}(g(x) + g(y)).$$

For example, $S(x, y) = x + y - xy$ if $g(x) = -\log(1 - x)$.
   ii. S is nilpotent if $g(1) = 1$ and then

$$S(x, y) = g^{-1}(\min\{1, g(x) + g(y)\}).$$

For example, $S(x, y) = \min\{1, x + y\}$ if $g(x) = x$.

4 K-$\oplus$-additive Aggregation Functions

Now, we are ready to introduce k-$\oplus$-additive aggregation functions.

Definition 6. *An aggregation function $A : [0, 1]^n \to [0, 1]$ is k-$\oplus$-additive if for all $\mathbf{x}_1, \ldots, \mathbf{x}_{k+1} \in [0, 1]^n$ such that $\oplus_{j=1}^{k+1} \mathbf{x}_j \in [0, 1]^n$,*

$$\bigoplus_{\substack{I \subseteq \{1,\ldots,k+1\} \\ |I| \text{ is odd}}} A\left(\bigoplus_{j \in I} \mathbf{x}_j\right) = \bigoplus_{\substack{I \subseteq \{1,\ldots,k+1\} \\ |I| \text{ is even}}} A\left(\bigoplus_{j \in I} \mathbf{x}_j\right). \tag{3}$$

Evidently, k-$+$-additive aggregation functions are just k-additive aggregation functions. Similarly, k-$\vee$-aggregation functions are k-maxitive aggregation functions.

4.1 Archimedean Pseudo-additions

Here, we consider an $\oplus$ related to an additive generator $g : [0,\infty] \to [0,\infty]$ with $g(1) = 1$. Observe that if a pseudo-addition $\oplus$ is generated by an additive generator g, then, for any positive real constant c, cg is also an additive generator of $\oplus$. To ensure the uniqueness of additive generators, without loss of generality, we can always consider $g(1) = 1$.

Theorem 4. *An aggregation function $A : [0,1]^n \to [0,1]$ is k-$\oplus$-additive if and only if $B : [0,1]^n \to [0,1]$,*

$$B(\mathbf{x}) = g(A(g^{-1}(x_1),\ldots,g^{-1}(x_n))),$$

is a k-additive aggregation function. Then

$$A(\mathbf{x}) = g^{-1}(B(g(x_1),\ldots,g(x_n))).$$

Example 2. Let $g(x) = x^2$. Then an aggregation function $A : [0,1]^n \to [0,1]$ is 1-$\oplus$-additive if and only if

$$A(\mathbf{x}) = \sqrt{W_{\mathbf{w}}(x_1^2,\ldots,x_n^2)} = \sqrt{\sum_{i=1}^{n} w_i x_i^2},$$

i.e., if A is a weighted quadratic mean.

Note that $C(\mathbf{x}) = \sum_{i=1}^{n} w_i x_i^2$ is 2-$\oplus$-additive.

Definition 7. *Let m be a capacity on X and let g be a generator of a pseudo-addition $\oplus$. The g-Owen extension of m is defined as follows*

$$O_m^g(\mathbf{x}) = g^{-1}\left(\sum_{I \subseteq \{1,\ldots,n\}} M_m(I) \prod_{i \in I} g(x_i) \right).$$

Theorem 5. *Let m be a k-additive capacity. Then, the g-Owen extension $O_m^g : [0,1]^n \to [0,1]$ is a k-$\oplus$-additive aggregation function.*

Note that the related k-$\oplus$-additive capacity μ to $A = O_m^g$ is given by $\mu(E) = A(1_E) = g^{-1}(m(E))$.

Example 3. Consider $g(x) = x^2$ and $n = k = 2$. Let $m(\{1\}) = a$, $m(\{2\}) = b$, $a, b \in [0,1]$. Then

$$O_m^g(x_1, x_2) = \sqrt{ax_1^2 + bx_2^2 + (1 - a - b)x_1^2 x_2^2}.$$

4.2   Archimedean t-conorm-based Pseudo-additions

Now, let $\oplus_{|[0,1]^2} = S$ be a continuous Archimedean t-conorm generated by an additive generator $g : [0,1] \to [0,\infty]$. We have the following result.

Theorem 6. *Let S be a strict t-conorm and $B : [0,\infty]^n \to [0,\infty]$ an aggregation function on $[0,\infty]$. Then*

$$A = g^{-1} \circ B \circ g$$

is a k-$\oplus$-additive aggregation function on $[0,1]$ if and only if B is k-additive on $[0,\infty]$.

Note that B is a k-additive aggregation function on $[0,\infty]$ whenever it is a polynomial with degree at most k which is increasing on $[0,\infty]$ and $B(\mathbf{0}) = 0$. However, there are also some non-polynomial solutions, in contrast to the k-additive aggregation functions on $[0,1]$. For example, consider a function $B : [0,\infty]^n \to [0,\infty]$ given by

$$B(\mathbf{x}) = \begin{cases} 0 & \text{if } \mathbf{x} = \mathbf{0}, \\ \infty & \text{otherwise.} \end{cases}$$

Then B is additive (and thus k-additive for any $k \in \mathbb{N}$) aggregation function on $[0,\infty]$. The related k-$\oplus$-additive aggregation function A on $[0,1]$ is given by

$$A(\mathbf{x}) = \begin{cases} 0 & \text{if } \mathbf{x} = \mathbf{0}, \\ 1 & \text{otherwise,} \end{cases}$$

independently of the additive generator g (recall that this A is the strongest aggregation function on $[0,1]$, while B is the strongest aggregation function on $[0,\infty]$). Observe that A is 1-$\oplus$-additive if and only if 1 is an idempotent element of the pseudo-addition $\oplus$, i.e., if $1 \oplus 1 = 1$.

Theorem 7. *Let S be a nilpotent t-conorm and B a k-additive aggregation function on $[0,1]$. Then*

$$A = g^{-1} \circ B \circ g$$

is a k-$\oplus$-additive aggregation function on $[0,1]$.

Observe that, in contrast to Theorem 6, Theorem 7 gives a sufficient condition only. Consider, for example, $g(x) = x$, $x \in [0,1]$, i.e., the related nilpotent t-conorm S is the Łukasiewicz t-conorm given by $S_L(x,y) = \min\{1, x+y\}$. Let $A : [0,1]^n \to [0,1]$ be given by $A(\mathbf{x}) = \min\{1, \sum_{i=1}^{n} w_i x_i\}$, where the weights $w_1, \ldots, w_n$ are from the interval $[1,\infty]$ (with convention $0.\infty = 0$). Then A is 1-$\oplus$-additive (and thus k-$\oplus$-additive for each $k \in \mathbb{N}$) aggregation function on $[0,1]$, but the related B is not k-additive for each $k \in \mathbb{N}$.

5 Concluding Remarks

We have introduced new interesting classes of aggregation functions, namely pseudo-additive aggregation functions of order k (k-$\oplus$-additive aggregation functions). Our approach generalizes the k-additivity and the k-maxitivity of aggregation functions proposed and discussed in [5] and [7], respectively. Note that the k-$\oplus$-additive aggregation functions are closely related to k-$\oplus$-additive capacities discussed in [6], and in the case when $\oplus_{|[0,1]^2}$ is an Archimedean t-conorm, then to decomposable measures proposed by S. Weber [12], see also λ-measures of Sugeno [10]. Our approach opens several new problems for the further investigations, in particular a complete characterization of k-$\oplus$-additive aggregation functions for some special pseudo-additions $\oplus$, and their link to particular capacities.

Acknowledgment. The work on this contribution was supported by the grants APVV-14-0013 and APVV-17-0066. Fateme Kouchakinejad kindly acknowledges the support from Iran National Science Foundation: INSF.

References

1. Choquet, G.: Theory of capacities. Ann. Inst. Fourier. **5**, 131–295 (1953)
2. Grabisch, M.: k-order additive discrete fuzzy measures and their representation. Fuzzy Sets Syst. **92**, 167–189 (1997)
3. Grabisch, M., Marichal, J.L., Mesiar, R., Pap, E.: Aggregation Functions. Cambridge University Press, Cambridge (2009)
4. Klement, E.P., Mesiar, R., Pap, E.: Triangular Norms. Kluwer, Dordrecht (2000)
5. Kolesárová, A., Li, J., Mesiar, R.: k-additive aggregation functions and their characterization. Eur. J. Operat. Res. **265**, 985–992 (2018)
6. Mesiar, R.: k-order pan-discrete fuzzy measures. In: 7th IFSA World Congress, Prague, pp. 488–490 (1997)
7. Mesiar, R., Kolesárová, A.: k-maxitive aggregation functions. Fuzzy Sets Syst. **346**, 127–137 (2018). https://doi.org/10.1016/j.fss.2017.12.016
8. Owen, G.: Multilinear extensions of games. In: Roth, A.E. (ed.) The Shapley value. Essays in Honour of Lloyd S. Shapley, pp. 139–151. Cambridge University Press (1988)
9. Shilkret, N.: Maxitive measure and integration. Indag. Math. **33**, 109–116 (1971)
10. Sugeno, M., Murofushi, T.: Pseudo-additive measures and integrals. J. Math. Anal. Appl. **122**, 197–222 (1987)
11. Sugeno, M.: Theory of fuzzy integrals and its applications. Ph.D. thesis, Tokyo Institute of Technology (1974)
12. Weber, S.: ⊥-decomposable measures integrals for Archimedean t-conorms ⊥. J. Math. Anal. Appl. **101**, 114–138 (1984)

Constructing an Outranking Relation with Weighted OWA for Multi-criteria Decision Analysis

Jonathan Ayebakuro Orama and Aida Valls[✉]

Departament de Enginyeria Informàtica i Matemàtiques,
Universitat Rovira i Virgili, Av Paisos Catalans, 26,
43007 Tarragona, Catalonia, Spain
aida.valls@urv.cat

Abstract. Some decision aiding methods are based on constructing and exploiting outranking relations. An alternative *a* outranks another *b* if *a* is at least as good as *b (aSb)*. One well known method in this field is ELECTRE. The outranking relation is usually built by means of a weighted average (WA) of the votes given by a set of criterion with respect to the fulfilment of aSb. The value obtained represent the strength of the majority opinion. The WA operator can be observed to have sometimes an undesired compensative effect. In this paper we propose the use of other aggregation operators with different mathematical properties. In particular, we substitute the WA by three operators from the Ordered Weighted Average (OWA) family of operators because it permits to decide the degree of andness/orness that is used during the aggregation. The OWAWA (Ordered Weighted Average Weighted Average), WOWA (Weighted Ordered Weighted Average) and IOWA (Induced Ordered Weighted Average) operators are studied. They are capable to combine the importance given to each criterion with the conjunctive/disjunctive requirement applied in the definition of the outranking relation.

Keywords: Decision support systems · Outranking relations
Ordered Weighted Average

1 Introduction

Multiple Criteria Decision Aiding discipline studies systematic methods for complex decision problems concerning diverse and often contradictory criteria, by analyzing a set of possible alternatives in order to find the best one [1]. One of the most successful approaches nowadays is known as *outranking methods*. It is based on social choice models that copy the human reasoning procedure [4].

MCDA methods take a set of alternatives (i.e. potential solutions) and generate a ranking of the alternatives according to a set of criteria. Criteria are tools constructed for the evaluation of alternatives compared in terms of suitability based on the decision maker's needs. Each criterion corresponds to a point of view considered in the decision process. Outranking methods are characterized by being based on constructing preference relations between the alternatives by means of pairwise comparisons, instead of

© Springer Nature Switzerland AG 2018
V. Torra et al. (Eds.): MDAI 2018, LNAI 11144, pp. 35–47, 2018.
https://doi.org/10.1007/978-3-030-00202-2_4

aggregating directly the values given by the criteria. The aim is to build a binary outranking relation aSb, which means "a is at least as good as b" [1]. Each criterion is asked about its contribution to this outranking assertion and it provides a vote in favor or against to aSb. Votes must be aggregated in order to associate a value to aSb for all possible pairs of alternatives. There are two main methods known as PROMETHEE and ELECTRE. In this study, we focus on ELECTRE method as it strictly applies the concept of veto. Moreover, ELECTRE method has been widely acknowledged as an efficient decision aiding tool with successful applications in many domains [4].

ELECTRE uses a weighted average to merge all the votes supporting aSb and then it includes the opposite votes by using a veto procedure. Once the valued outranking relation is constructed, different exploitation procedures exist in order to derive a ranking from it [1]. The contribution of this paper is the use of other aggregation operators for merging the votes in favor of the outranking relation. In particular, we propose the use of OWA-like operators because they enable the definition of conjunctive/disjunctive policies of aggregation that may be more appropriate in some decision problems. The compensation problem of classic weighted average may be solved with the possibility of establishing a more appropriate and-like aggregation (to model simultaneity) or or-like operator (for replaceability). As we do not want to suppress the weights representing the voting power for each criterion, we propose the use of Weighted OWA operators like OWAWA, WOWA and IOWA.

The paper is structured as follows. Section 2 presents the different aggregation operators based on OWA that will be used in the study. Section 3 briefly outlies the ELECTRE method. Section 4 defines the new procedure for calculating the overall concordance. Section 5, makes an empirical analysis and comparison. Finally, Sect. 6 discusses the main conclusions of this study.

2   Weighted OWA Operators

Aggregation operators are mathematical formulations that map a set of n values R^n to a single value R and must satisfy certain properties (idempotency, monotonicity, etc.) [9]. The most popular aggregation operators are averaging operators. The simplest aggregation operator with weights is the weighted average (WA), where the source of the values (i.e. the evaluation criteria) are assigned weights to indicate its trade-off importance. Given a set of arguments $A = (a_1, \ldots, a_n)$ and a weighting vector V with weights $v_j \in [0, 1]$ associated with each argument source (i.e. criterion), such that $\sum_{j=1}^{n} v_j = 1$. The weighted average is defined as:

$$WA(A) = \sum_{j=1}^{n} v_j a_j \tag{1}$$

Differently, OWA [10] uses weights to provide a parameterized family of mean type aggregation operators. The main distinguishing feature of this operator is the reordering of arguments according to their values before weights are assigned. Given a set of arguments $A = (a_1, \ldots, a_n)$ and a weighting vector W with weights $w_j \in [0, 1]$, such that $\sum_{j=1}^{n} w_j = 1$. The ordered weighted average is defined as:

$$OWA(A) = \sum_{j=1}^{n} w_j b_j, \tag{2}$$

where b_j is the jth largest of the a_i.

An interesting fact about OWA is that weights are not given to the criteria but to the values. Thus, we can perform different aggregation policies (disjunctive or conjunctive) according to the decision maker (DM) needs. For example, the DM could assign weights in such a way that extreme arguments are regarded less than central arguments. In summary, the weights of OWA shows the importance of arguments in relation to the ordering of the arguments.

In some problems the DM is interested in carefully considering the weighting policies due to is significant impact on the results [3]. The use of OWA weights enables to model the andness/orness, which can be combined with usual WA weights for the different criteria. Next subsections introduce three different ways of combining them in OWA-like operators that exploit the advantages of both OWA and WA approaches.

2.1 OWAWA

In [6] the OWAWA operator is introduced as a generalization of the WA and the OWA operator.

An OWAWA operator is a mapping $A = (a_1, \ldots, a_n) \to R$, having an associated weighting vector V (WA), with $\sum_{i=1}^{n} v_i = 1$ and $v_i \in [0, 1]$ and a weighting vector W (OWA), with $\sum_{j=1}^{n} w_j = 1$ and $w_j \in [0, 1]$, such that:

$$OWAWA_\beta(A) = \beta \sum_{j=1}^{n} w_j b_j + (1 - \beta) \sum_{i=1}^{n} v_i a_i, \tag{3}$$

where b_j is the jth largest of the a_i and $\beta \in [0, 1]$.

The novel feature of the OWAWA operator is the ability to take into account the degree of importance of WA and OWA in specific situations. This is managed with the parameter β. As $\beta \to 1$, the importance of OWA increases while as $\beta \to 0$, the importance of WA increases. The OWAWA operator is monotonic, idempotent, commutative and bounded.

2.2 WOWA

The WOWA operator was introduced in [8] as a combination of the WA operator and the OWA operator by means of constructing a different weight that integrates the associated weighting system seen in WA, V, with the weighting according to ordering of OWA, W.

A WOWA operator is a mapping $A = (a_1, \ldots, a_n) \rightarrow R$ of dimension n where,

$$WOWA(A) = \sum_{j=1}^{n} \omega_j b_j, \qquad (4)$$

where b_j is the jth largest of the a_i and the weight ω_i is defined taking into account the importance of the sources of the arguments and their position after the reordering step, defined as $\omega_i = w^* \left(\sum_{j \leq i} v_{\sigma(j)} \right) - w^* \left(\sum_{j < i} v_{\sigma(j)} \right)$ with $\sum_{i=1}^{n} \omega_i = 1$.

w^* is a non-decreasing function that interpolates the points $\{(0,0)\} \cup \left\{ (i/n, \sum_{j \leq i} w_j) \right\} \forall i = 1, \ldots, n$. w^* is required to be a straight line when the points can be interpolated in this way. Moreover, w^* may be a regular monotonic non-decreasing quantifier $Q(x)$, with $Q(0) = 0, Q(1) = 1$ and if $x > y$ then $Q(x) \geq Q(y)$.

The WOWA operator is defined in such a way that it reduces to the OWA operator when $v_i = 1/n$ and reduces to the WA operator when $w_i = 1/n$. This shows that OWA and WA are special cases of the generalized WOWA operator.

2.3   IOWA

The last method to combine the two different sets of weights is by means of an induced ordered weighted averaging operator (IOWA). IOWA was introduced in [11] to introduce an additional variable that influences the ordering stage of OWA. The IOWA operator rather ordering arguments by their numeric values an ordered inducing variable is used to order the arguments. Then, IOWA operator is defined in terms of arguments in form of a two-tuple, called an OWA pair $\langle u_i, a_i \rangle$, where u_i is the order inducing variable of the i th argument and a_i is the argument variable of the i th argument. In the reordering step a_i is not used but u_i.

Given n arguments to be aggregated denoted as $A = (a_1, \ldots, a_n)$, the ordered arguments are obtained in a way such that b_j^u is the a value of the OWA pair having the j th largest u value. The IOWA operator can then be defined as:

$$IOWA(\langle u_1, a_1 \rangle, \ldots, \langle u_1, a_1 \rangle) = \sum_{j=1}^{n} \omega_j b_j^u \qquad (5)$$

In IOWA a tie occurs when two OWA pairs $\langle u_j, a_j \rangle, \langle u_k, a_k \rangle$ have equal order inducing variables, i.e. $u_j = u_k$. In this case, each OWA pair is replaced with an OWA pair having the same order-inducing variable u but an argument variable that is an average of the previous argument variables. This means that having $\langle u_j, a_j \rangle$ and $\langle u_k, a_k \rangle$ where $u_j = u_k$ they are replaced by $\langle (u_j = u_k), (a_j + a_k/2) \rangle$ in the aggregation process. The IOWA operator is idempotent, communicative, monotonic and bounded.

Using as inducing variable the vector V of importance of the criteria, we have another way of combining V and W.

3   Outranking Relations in the ELECTRE Methodology

The so-called outranking methods in the MCDA literature are based on conducting a pairwise comparison of alternatives with regards to each criterion. The goal of the comparison is to find out if it satisfies an outranking relation aSb meaning that alternative a is "at least as good as" alternative b.

The outranking relation S may be binary or valued. In this paper we study the case of valued outranking relations, having then $S = A \rightarrow [0, 1]$. The value assigned to S is usually denoted as credibility (on the outranking relation). In the ELECTRE method, to calculate the credibility of aSb, two conditions must hold:

1. Concordance condition: After pairwise comparison of a and b for each criterion, a majority of the criteria must support aSb. It accounts for the majority opinion.
2. Non-discordance condition: Ensures that among the minority no criteria strongly refutes aSb. It permits the right to veto (i.e. "respect to minorities").

The outranking concept explained above is inspired in voting models used in different theories of social election. It is similar to voting procedures applied United Nations Security Council, where some countries have the right to veto the majority opinion. Following this idea, in ELECTRE methodology, to calculate the credibility value of the outranking relation $\rho(a, b) \in [0, 1]$, the following steps are applied [1]:

1. Calculation of a partial concordance index for each criterion $c_j(a, b) \in [0, 1]$. In each criterion, two discrimination thresholds may be used to model the uncertainty of the decision maker: the indifference and the preference threshold.
2. Calculation of the overall concordance $c(a, b) \in [0, 1]$. It is calculated as a weighted average of $c_j(a, b)$ using as weights the voting power of each criterion. The resulting value represent the strength of the coalition of criteria being in favor of the outranking relation aSb.
3. Calculation of a partial discordance index for each criterion $d_j(a, b) \in [0, 1]$. The DM can give to some criteria the right to veto the majority opinion if there are essential reasons to refute it. In this case, the criteria has an associated veto threshold, such that larger differences of this threshold in favor of b will eliminate the possibility that option a outranks option b.
4. Calculation of the final credibility as:

$$
\rho(a, b) = \begin{cases} c(a, b) & \text{if } \forall_j d_j(a, b) \leq c(a, b) \\ c(a, b). \prod_{j \in J(a,b)} \frac{1 - d_j(a,b)}{1 - c(a,b)} & \text{otherwise} \end{cases} , \qquad (6)
$$

where $J(a, b)$ is the set of criteria for which the discordance is larger than the overall concordance.

Once the credibility matrix is obtained, an exploitation procedure is applied in order to establish a preference-based order among the alternatives. A simple ranking technique is known as Net Flow Score (NFS) procedure. NFS is based on the two evidences: strength and weakness. They are measured in the graph corresponding to the valued credibility matrix calculated in step 4. The strength of alternative a is defined as the sum of the credibility values of the output edges to the node a. The weakness of alternative a is defined as the sum of the credibility values of the input edges to the node a. In terms of outranking relations, the net flow score of an alternative a is defined in Eq. 7. A total ranking can be derived from the NFS, being the higher the score, the better.

$$NFS(a) = |b \in A : aSb| - |b \in A : bSa| \tag{7}$$

4   Using Weighted OWA in the Overall Concordance Calculation

Some previous works have considered a modification of the way that overall concordance is calculated in ELECTRE in different situations. The paper [7] looks at a situation where the extent to which a criterion surpasses the preference threshold can be reflected in a change in the importance of that criterion in the concordance calculation. In [2] the concordance index is modified to take into consideration three possible interactions between the criteria that modify each joint importance: mutual strengthening, mutual weakening and antagonistic. In both cases, the weights are modified but the overall *concordance index* for each pair a, b is calculated as the weighted average of the partial concordances indices.

Having $C = \{c_j(a,b)\}, j = 1 \ldots n$:

$$c(a,b) = WA(C) \tag{8}$$

In this paper we propose the substitution of the WA operator by a weighted OWA operator, presented in Sect. 2. The first proposal is using OWAWA operator that linearly combines both the result of WA and the result of OWA. In this case, the parameter beta must be defined by the user. This parameter allows to base the result most on the criteria importance weights or on the and/or weights.

$$c(a,b) = OWAWA_\beta(C) \tag{9}$$

The second proposal consists in using IOWA operator with the criteria importance V used as order-inducing variable. In this case, the values provided by the most important criteria will the ones assigned to the first weights of the OWA vector W.

$$c(a,b) = IOWA(\langle V, C \rangle) \tag{10}$$

The third approach uses the WOWA operator which generates a new weighting vector from the V and W.

$$c(a, b) = WOWA(C) \tag{11}$$

5 Experiments

The OWA-based outranking construction proposed has been tested with two different datasets. To evaluate the differences produced by the different operators, we compare the ranking obtained using the Net Flow Score. A minimum credibility of 0.8 in the outranking relation is used in this procedure. The correlation between different rankings is calculated to see how new proposals are able to integrate both sets of weights.

5.1 Finding a Hotel

The first case study poses the problem of making lodging arrangements to attend a congress in Jyväskylä (Finland). The DM wishes to make a choice from six hotel alternatives all in proximity to the congress site. The choice will be made based on the criteria and weights listed below. The data used in this case study is given in Tables 1 and 2. Six hotels have been evaluated using 6 criteria: **C01**- Distance to the congress site, **C02**- Distance to the city center, **C03**- Sports facilities, **C04**- Restaurants available, **C05**- Category and **C06**- Services provided (wifi, laundry, etc.). Two first criteria are minimized (−) and the rest are maximized (+).

Table 1. Hotels performance table

	C01−	C02−	C03+	C04+	C05+	C06+
Alexandra	1600.0	300.0	2.0	3.0	4.0	5.0
Sokos	1700.0	400.0	2.0	2.0	4.0	5.0
Cumulus	1700.0	550.0	4.0	0.0	3.0	3.0
Scandic	600.0	350.0	3.0	2.0	4.0	2.0
Kampus	1550.0	610.0	4.0	0.0	3.0	2.0
Alba	110.0	1300.0	1.0	1.0	3.0	4.0

Table 2. Criteria parameters

	C01	C02	C03	C04	C05	C06
Indifference	200.0	100.0	0.0	0.0	0.0	1.0
Preference	700.0	300.0	1.0	1.0	0.0	1.0
Weight	0.1	0.3	0.3	0.05	0.15	0.1

Two sets of OWA weights have been considered for this study: a disjunctive policy with $w_c = (0.408, 0.169, 0.130, 0.109, 0.096, 0.088)$ and a conjunctive policy with weights $w_d = (0.028, 0.083, 0.139, 0.194, 0.25, 0.306)$. These weights were obtained from the use of a regular monotonic non-decreasing quantifier, as proposed in [7].

To establish the disjunctive policy, the quantifier $Q(x) = \sqrt{x}$ is used, while for the conjunctive policy $Q(x) = x^2$.

Next tables show the overall concordance values obtained with the three combined operators proposed in this paper to merge the partial concordance indices (Table 3).

Table 3. Outranking values with weighted average (WA)

	Alexa	Sokos	Cumulus	Scandic	Kampus	Alba
Alexa	1.0	1.0	0.7	0.6	0.7	0.9
Sokos	0.95	1.0	0.7	0.6	0.7	0.9
Cumulus	0.475	0.625	1.0	0.55	1.0	0.85
Scandic	0.85	0.9	0.7	1.0	0.7	0.842
Kampus	0.4	0.535	1.0	0.46	1.0	0.75
Alba	0.2	0.2	0.4	0.2	0.4	1.0

For OWAWA, three values of β have been tested. In orange, concordances higher than 0.8 are highlighted, as they are the ones used in the NFS ranking procedure. We show only some results for the disjunctive version of the operators (and WA as reference) for space limitations (Tables 4, 5, 6 and 7). For conjunctive policies, we have observed that the values of the outranking matrix are much lower than with the rest, finding very few values above the threshold of 0.8 and leading to rankings with many ties.

Table 4. Outranking values with OWA disjunctive (OWAd)

	Alexa	Sokos	Cumulus	Scandic	Kampus	Alba
Alexa	1.0	1.0	0.912	0.816	0.912	0.912
Sokos	0.912	1.0	0.912	0.816	0.912	0.912
Cumulus	0.6095	0.6745	1.0	0.642	1.0	0.816
Scandic	0.816	0.912	0.912	1.0	0.912	0.85632
Kampus	0.577	0.6355	1.0	0.603	1.0	0.707
Alba	0.577	0.577	0.816	0.577	0.816	1.0

Table 5. Outranking values with OWAWA beta = 0.5 disjunctive (OWAWA.5d)

	Alexa	Sokos	Cumulus	Scandic	Kampus	Alba
Alexa	1.0	1.0	0.806	0.708	0.806	0.906
Sokos	0.936	1.0	0.806	0.708	0.806	0.906
Cumulus	0.54225	0.64975	1.0	0.596	1.0	0.833
Scandic	0.833	0.906	0.806	1.0	0.806	0.84916
Kampus	0.4885	0.58525	1.0	0.5315	1.0	0.7285
Alba	0.3885	0.3885	0.608	0.3885	0.608	1.0

Table 6. Outranking values with IOWA disjunctive (IOWAd)

	Alexa	Sokos	Cumulus	Scandic	Kampus	Alba
Alexa	1.0	1.0	0.7115	0.609	0.7115	0.8975
Sokos	0.912	1.0	0.7115	0.609	0.7115	0.8975
Cumulus	0.463125	0.607375	1.0	0.53525	1.0	0.8095
Scandic	0.8095	0.8975	0.7115	1.0	0.7115	0.83805
Kampus	0.391	0.520825	1.0	0.4487	1.0	0.707
Alba	0.205	0.205	0.423	0.205	0.423	1.0

Table 7. Outranking values with WOWA disjunctive (WOWAd)

	Alexa	Sokos	Cumulus	Scandic	Kampus	Alba
Alexa	1.0	1.0	0.83666	0.7746	0.83666	0.94868
Sokos	0.97467	1.0	0.83666	0.7746	0.83666	0.94868
Cumulus	0.68351	0.78561	1.0	0.73455	1.0	0.92195
Scandic	0.92195	0.94869	0.83666	1.0	0.83666	0.9172092
Kampus	0.63246	0.72435	1.0	0.67329	1.0	0.86602
Alba	0.44722	0.44722	0.63245	0.44722	0.63245	1.0

Table 8. Net Flow Score for each hotel and each method

	Alexa	Sokos	Cumulus	Scandic	Kampus	Alba
WA	0	0	1	3	0	−4
OWAd	3	3	−3	3	−4	−2
OWAc	1	−1	0	0	0	0
OWAWA.3d	0	0	1	3	0	−4
OWAWA.3c	1	0	0	1	0	−2
OWAWA.5d	2	2	−2	5	−3	−4
OWAWA.5c	1	1	0	0	0	−2
OWAWA.7d	2	2	−2	5	−3	−4
OWAWA.7c	1	−1	0	0	0	0
IOWAd	0	0	1	3	0	−4
IOWAc	3	0	−4	3	−4	2
WOWAd	2	2	−2	5	−2	−5
WOWAc	1	0	0	1	0	−2

The NFS values (Eq. 7) given to each hotel are shown in Table 8. Their ranking positions (1...6) are given in Table 9, with the hotels in the best positions highlighted.

We can see that in most of the cases Alba is in the worst position, although with a conjunctive policy the worst is Sokos. The best position is given to Scandic or Alexa

Table 9. Rank position of each hotel according to its NFS

	Alexa	Sokos	Cumulus	Scandic	Kampus	Alba
WA	4	4	2	1	4	6
OWAd	2	2	5	2	6	4
OWAc	1	6	3,5	3,5	3,5	3,5
OWAWA.3d	4	4	2	1	4	6
OWAWA.3c	1,5	4	4	1,5	4	6
OWAWA.5d	2,5	2,5	4	1	5	6
OWAWA.5c	1,5	1,5	4	4	4	6
OWAWA.7d	2,5	2,5	4	1	5	6
OWAWA.7c	1	6	3,5	3,5	3,5	3,5
IOWAd	4	4	2	1	4	6
IOWAc	1,5	4	5,5	1,5	5,5	3
WOWAd	2,5	2,5	4,5	1	4,5	6
WOWAc	1,5	4	4	1,5	4	6

(ndra), sometimes in a tie. Sokos is also in the best position when a disjunctive policy is used. The case of Sokos hotel is quite interesting because its position is the least stable of all hotels.

A look at the criteria weights show criteria C02 and C03 to have a combined 60% of the total importance assigned to criteria, as such hotels like Scandic with good performance values on C02 and C03 have better positions in WA. It can also be observed that Alexandra hotel has no low score in any criterion and some high values, thus it is the winner in case of conjunctive policies. It is worth to notice that Sokos is the worst when using OWA disjunctive, but when including the criteria weights it improve its position, as it is good in C02 and C03, as said before. Thus, the operators balance both weighting vectors. Kampus is in an intermediate position with WA because it has bad scores in non-relevant criteria, but when including the and/or weights, it goes to worst positions because of its low score in C04 and C06.

In order to measure the similarity between the rankings, Table 10 gives the Spearman rho correlation between the 3 results obtained using a single set of weights (WA, OWAd and OWAc) with respect to the use of the two sets of weights together. Table 10 also indicates the operator that gives a highest correlation (most similar ranking, with correlation higher than 0.95) for each of the proposed methods.

We can see that the disjunctive policies with OWAWA with low beta and IOWA give similar results to the WA. Rankings similar to OWA with disjunctive weights are obtained with OWAWA also disjunctive and high beta (OWA-like), as expected. Also OWAWA with high beta reproduces the ranking of OWA for the conjunctive case. An interesting observation is that WOWA seems to give a significantly different ranking to all the three basic ones.

Table 10. Correlation between the different rankings obtained in dataset Hotels

	WA	OWAd	OWAc	Closest (>=0.95)
1 WA	1,00	0,16	0,00	
2 OWAd	0,16	1,00	0,00	
3 OWAc	0,00	0,00	1,00	
4 OWAWA.3d	1,00	0,16	0,00	IOWAd
5 OWAWA.3c	0,66	0,56	0,46	WOWAc
6 OWAWA.5d	0,71	0,77	0,00	OWAWA.7d, WOWAd
7 OWAWA.5c	0,16	0,56	0,00	-
8 OWAWA.7d	0,71	0,77	0,00	OWAWA.5c, WOWAd
9 OWAWA.7c	0,00	0,00	1,00	-
10 IOWAd	1,00	0,16	0,00	OWAWA.3d
11 IOWAc	0,06	0,81	0,44	-
12 WOWAd	0,66	0,75	0,00	OWAWA.5d,OWAWA.7d
13 WOWAc	0,66	0,56	0,46	OWAWA.3c

5.2   Generating a Ranking of Universities

The second case study comes from paper [5], with data about British universities from https://www.thecompleteuniversityguide.co.uk/league-tables/rankings   a   ranking   is built. We use the same weights and thresholds than paper [5], but we increased the number of alternatives to 20 universities. Five criteria are taken: **C01**- Academic services spend, **C02**- Completion, **C03**- Entry standards, **C04**- Facilities spend, **C05**- Good honors. Horizontal axis shows the identifier of each method given in Table 10. Again $Q(x) = \sqrt{x}$ and $Q(x) = x^2$ were used to establish the OWA weights (Fig. 1).

Aggregation with IOWA (10 & 11) and with WOWA (11 & 12) is able to change the position of some universities in this dataset. Although the ones in the best and worst positions are robust to the change of agregation operator. For example, U16 and U1 are universities that are sensible to the aggregation policy. U16 is excellent in two criteria (w = 0.2 and 0.1) and very bad in one (w = 0.3). Therefore, when using WA it appears in at rank 11/20, with OWAd it goes to upper positions (6/20). We can also observe that there are many rank reversals between universities in ranks 5 to 15 for IOWAc. A deeper analysis of this operator reveals that using the importance weights V as order inducing variable leads to strange results in some cases.

Correlations table (Table 11) shows that in this case study WA and OWA are initially highly correlated, therefore their combination also leads to high correlation values in most methods. WOWAd is the one that differentiates a bit from the rest. IOWAc is suprisingly similar to OWAd.

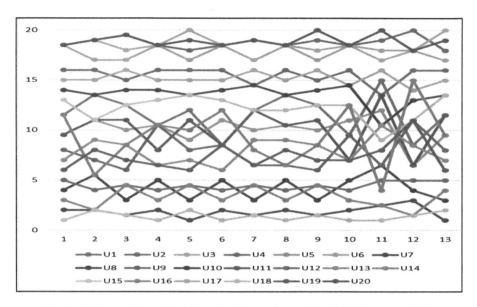

Fig. 1. Rank positions of the 20 universities values with weighted average (WA)

Table 11. Correlation between the different rankings obtained in dataset Universities

	WA	OWAd	OWAc	Closest (>=0.99)
WA	1,00	0,95	0,98	
OWAd	0,95	1,00	0,98	
OWAc	0,98	0,98	1,00	
OWAWA.5d	0,98	0,93	0,95	OWAWA.3d, IOWAd
OWAWA.5c	0,97	0,98	1,00	OWAWA.7c
IOWAd	0,99	0,92	0,95	OWAWA.3d, OWAWA.5d
IOWAc	0,88	0,97	0,93	-
WOWAd	0,93	0,86	0,88	-
WOWAc	0,99	0,96	0,98	OWAWA.5c

6 Conclusions and Future Work

This work presents a new approach to the aggregation of partial concordances in the ELECTRE outranking method. Using weighted averaging may sometimes have an undesired compensative effect between opposite values, as such we look to a family of OWA operators to avoid this effect. OWAWA, WOWA and IOWA which combine WA and OWA may substitute WA, introducing a new way of weighting values.

In the tests we observed that the results of the 3 approaches are different as they model the combination in different ways. An undesired behaviour has been seen in the IOWA conjunctive operator. If a low partial concordance is given by criteria with high importance weight, they will be placed in the first positions during aggregation, so they will receive a low *W* weight and their contribution is minimized. This seems to go in contrary to common sense. Moreover, WOWA seems to give a significantly different ranking to all the three basic ones. It may indicate that it really combines the information of the two sets of weights in a more suitable way.

Future work concerns the study of the best scenarios for each of these operators. The behaviour of other OWA policies (e.g. Olympic, Balanced) [9] will be studied. Finally, a characterisation of the properties of these operators should be investigated.

Acknowledgements. This work is supported by URV grant 2017PFR-URV-B2-60.

References

1. Figueira, J.R., Greco, S., Ehrgott, M.: Multiple Criteria Decision Analysis: State of the Art Surveys. Springer, Boston (2005). https://doi.org/10.1007/b100605
2. Figueira, J.R., Greco, S., Roy, B.: Electre methods with interaction between criteria: an extension of the concordance index. Eur. J. Oper. Res. **119**, 479–495 (2014)
3. Dong, Y., Liu, Y., Liang, H., Chiclana, F., Herrera-Viedma, E.: Strategic weight manipulation in multiple attribute decision making. Omega **75**, 154–164 (2018)
4. Govindan, K., Jepsen, M.B.: ELECTRE: a comprehensive literature review on methodologies and applications. Eur. J. Oper. Res. **250**, 1–29 (2015)
5. Ishizaka, A., Giannoulis, C.: A web-based decision support system with ELECTRE III for a personalised ranking of British universities. Decis. Support Syst. **48**, 488–497 (2010)
6. Merigo, J.M.: On the use of the OWA operator in the weighted average and its application in decision making, In: Proceedings of the World Congress on Engineering. WCE, London (2009)
7. Roy, B., Słowiński, R.: Handling effects of reinforced preference and counter-veto in credibility of outranking. Eur. J. Oper. Res. **188**(1), 185–190 (2008)
8. Torra, V.: The weighted OWA operator. Int. J. Intell. Syst. **12**, 153–166 (1997)
9. Torra, V., Narukawa, Y.: Modeling Decisions: Information Fusion and Aggregation Operators. Springer, Heidelberg (2007). https://doi.org/10.1007/978-3-540-68791-7. https://www.springer.com/gp/book/9783540687894
10. Yager, R.R.: On ordered weighted averaging aggregation operators in multi-criteria decision making. IEEE Trans. Syst. Man Cybern. **18**, 183–190 (1988)
11. Yager, R.R.: Induced ordered weighted averaging operators. IEEE Trans. Syst. Man Cybern. Part B Cybern. **29**, 141–150 (1999)

Sugeno Integrals and the Commutation Problem

Didier Dubois[1($\boxtimes$)], Hélène Fargier[1], and Agnès Rico[2]

[1] IRIT, Université Paul Sabatier, 31062 Toulouse Cedex 9, France
dubois@irit.fr
[2] ERIC, Université Claude Bernard Lyon 1, 69100 Villeurbanne, France

Abstract. In decision problems involving two dimensions (like several agents and several criteria) the properties of expected utility ensure that the result of a multicriteria multiperson evaluation does not depend on the order with which the aggregations of local evaluations are performed (agents first, criteria next, or the converse). We say that the aggregations on each dimension *commute*. Ben Amor, Essghaier and Fargier have shown that this property holds when using pessimistic possibilistic integrals on each dimension, or optimistic ones, while it fails when using a pessimistic possibilistic integral on one dimension and an optimistic one on the other. This paper studies and completely solves this problem when Sugeno integrals are used in place of possibilistic integrals, indicating that there are capacities other than possibility and necessity measures that ensure commutation of Sugeno integrals.

Keywords: Capacities · Sugeno integrals · Possibility theory
Commutation

1 Introduction

In various applications where information fusion or multifactorial evaluation is needed, an aggregation process is carried out as a two-stepped procedure whereby several local fusion operations are performed in parallel and then the results are merged into a global result. It may sometimes be natural to demand that the result does not depend on the order with which we perform the aggregation steps because there is no reason to perform either of the steps first.

For instance, in a multi-person multi-criteria decision problem, each alternative is evaluated by a matrix of ratings where the rows represent evaluations by persons and the columns represent evaluations by criteria. One may, for each row, merge the ratings according to each column with some aggregation operation and form the global rating of each person, and then merge the persons opinions using another aggregation operation. On the other hand, one may decide first to merge the ratings in each column, thus forming the collective rating according to each criterion, and then merge these evaluations across the criteria. The same considerations apply when we consider several agents under

V. Torra et al. (Eds.): MDAI 2018, LNAI 11144, pp. 48–63, 2018.
https://doi.org/10.1007/978-3-030-00202-2_5

uncertainty sharing the same knowledge. Should we average out the uncertainty for each agent prior to merging the individual evaluations (i.e., follow the so-called *ex-ante* approach), or should we average out the common uncertainty only after merging the individual evaluations for each possible state of affairs (i.e., adopt an *ex-post* approach)?

Even if it may sound natural that the two procedures should deliver the same results in any sensible approach, the problem is that this natural outcome is mathematically not obvious at all. When the two procedures yield the same results, the aggregation operations are said to *commute*. In decision under risk for instance, the *ex-ante* and *ex-post* approaches are equivalent (the aggregations commute) if and only the preferences are considered with a utilitarian view [10,13]: the expected utility of a sum is equal to the sum of the expected utilities. With an egalitarian collective utility function this is no longer the case, which leads to a timing effect: the *ex-ante* approach (minimum of the expected utilities) is not equivalent to the *ex-post* one (the expected utility of the minimum of the utilities). Some authors [10,13] proved representation theorems stating that, in a probabilistic setting, commutation occurs if and only if the two aggregations are weighted averages, i.e., the weighted average of expected utilities is the same as the expected collective utility.

More recently, Ben Amor et al. [2–4] have reconsidered the same problem in the setting of qualitative decision theory under uncertainty. They have proved that commuting alternatives to weighted average operations exist, namely qualitative possibilistic integrals [6]. Namely, Sugeno integrals with respect to possibility or necessity measures, respectively corresponding to optimistic and pessimistic possibilistic integrals. Pessimistic possibilistic integrals commute, as well as optimistic ones, but a pessimistic possibilistic integral generally does not commute with an optimistic one.

The question considered in this paper is whether there exist capacities other than possibility and necessity measures, in the qualitative setting, for which this commutation result holds, replacing pessimistic or optimistic utility functionals by Sugeno integrals with respect to general capacities.

The paper is organized as follows. After a refresher on Sugeno integrals on totally ordered sets in Sect. 2, Sect. 3 provides necessary and sufficient conditions for their commutation. Finally Sect. 4 gives the explicit format of capacities that allow for commuting Sugeno integrals.

2   A Refresher on 1D Sugeno Integral

Consider a set $\mathcal{X} = \{x_1, \cdots, x_n\}$ and L a totally ordered scale with top 1, bottom 0, and the order-reversing operation denoted by $1 - (\cdot)$ (it is involutive and such that $1 - 1 = 0$ and $1 - 0 = 1$). A decision to be evaluated is represented by a function $u : \mathcal{X} \to L$ where $u(x_i)$ is, for instance, the degree of utility of the decision in state x_i.

In the definition of Sugeno integral [14,15], the relative likelihood or importance of subsets of states is represented by a capacity (or fuzzy measure), which is a set function $\mu : 2^{\mathcal{X}} \rightarrow L$ that satisfies $\mu(\emptyset) = 0$, $\mu(\mathcal{X}) = 1$ and $A \subseteq B$ implies $\mu(A) \leq \mu(B)$.

Definition 1. *The Sugeno integral (S-integral for short) of function u with respect to a capacity μ is defined by:* $S_\mu(u) = \max_{\alpha \in L} \min(\alpha, \mu(u \geq \alpha))$, *where* $\mu(u \geq \alpha) = \mu(\{x_i \in \mathcal{X} | u(x_i) \geq \alpha\})$.

For instance, suppose that μ is a necessity measure N [5], i.e., a capacity such that $N(A \cap B) = \min(N(A), N(B))$. N is entirely defined by a function $\pi : \mathcal{X} \rightarrow L$, called the possibility distribution associated to N, namely by: $N(A) = \min_{x_i \notin A} 1 - \pi(x_i)$. The conjugate of a necessity measure is a possibility measure Π [5]: $\Pi(A) = \max_{x_i \in A} \pi(x_i)$. We have $\Pi(A \cup B) = \max(\Pi(A), \Pi(B))$ and $\Pi(A) = 1 - N(\overline{A})$ where $\overline{A}$ is the complementary of A. We thus get the following special cases of the Sugeno integral:

$$S_\Pi(u) = \max_{\alpha \in L} \min(\alpha, \Pi(u \geq \alpha)) = \max_{x_i \in \mathcal{X}} \min(\pi(x_i), u(x_i)) \tag{1}$$

$$S_N(u) = \max_{\alpha \in L} \min(\alpha, N(u \geq \alpha)) = \min_{x_i \in \mathcal{X}} \max(1 - \pi(x_i), u(x_i)). \tag{2}$$

These are the weighted maximum and minimum operations that are used in qualitative decision making under uncertainty (they are called optimistic and pessimistic qualitative utility respectively [6]). In this interpretation, $\pi(x_i)$ measures to what extent x_i is a possible state, $S_N(u)$ (resp. $S_\Pi(u)$) evaluates to what extent it is certain (resp. possible) that u is a good decision.

A Sugeno integral can be equivalently written under various forms [11,14], especially as a lattice polynomial [7] of the form $S_\mu(u) = \max_{A \subseteq \mathcal{X}} \min(\mu(A), \min_{x_i \in A} u(x_i))$. It can be expressed in a non-redundant format by means of the qualitative Möbius transform of μ [8]:

$$\mu_{\#}(T) = \begin{cases} \mu(T) \text{ if } \mu(T) > \max_{x \in T} \mu(T \setminus \{x\}) \\ 0 \text{ otherwise} \end{cases}$$

as

$$S_\mu(u) = \max_{T \subseteq \mathcal{X} : \mu_{\#}(T) > 0} \min(\mu_{\#}(T), \min_{x_i \in T} u(x_i))$$

The function $\mu_{\#}$ contains the minimal information to reconstruct the capacity μ as $\mu(A) = \max_{T \subseteq A} \mu_{\#}(T)$. Subsets T of $\mathcal{X}$ for which $\mu_{\#}(T) > 0$ are called *focal sets* of μ and the set of focal sets of μ is denoted by $\mathcal{F}(\mu)$. As a matter of fact, it is clear that the qualitative Möbius transform of a possibility measure coincides with its possibility distribution: $\Pi_{\#}(A) = \pi(s)$ if $A = \{s\}$ and 0 otherwise.

Lastly, the S-integral can be expressed in terms of Boolean capacities (i.e., of capacities that take their values in $\{0,1\}$) obtained from μ. Given a capacity μ on $\mathcal{X}$, for all $\lambda > 0, \lambda \in L$, let $\mu_\lambda : 2^{\mathcal{X}} \rightarrow \{0,1\}$ (called the λ-cut of μ) be a Boolean capacity defined by $\mu_\lambda(A) = \begin{cases} 1 & \text{if } \mu(A) \geq \lambda \\ 0 & \text{otherwise.} \end{cases}$, for all $A \subseteq \mathcal{X}$. It is clear that the capacity μ can be reconstructed from the μ_λ's as follows:

$$\mu(A) = \max_{\lambda > 0} \min(\lambda, \mu_\lambda(A)).$$

Observe that the focal sets of a Boolean capacity μ_λ form an antichain of subsets (there is no inclusion between them).

We can also express S-integrals with respect to μ by means of the cuts of μ:

Proposition 1. $S_\mu(u) = \max_{\lambda>0} \min(\lambda, S_{\mu_\lambda}(u))$

Proof:

$$S_\mu(u) = \max_{A \subseteq \mathcal{X}} \min(\max_{\lambda>0} \min(\lambda, \mu_\lambda(A)), \min_{x_i \in A} u(x_i))$$
$$= \max_{\lambda>0} \min(\lambda, \max_{A \subseteq \mathcal{X}} \min(\mu_\lambda(A), \min_{x_i \in A} u(x_i)))$$

□

Note that the expression $S_\mu(u) = \max_{\alpha \in L} \min(\alpha, \mu(u \geq \alpha))$ uses cuts of the utility function. It can be combined with Proposition 1 to yield:

$$S_\mu(u) = \max_{\alpha, \lambda \in L} \min(\alpha, \lambda, \mu_\lambda(u \geq \alpha)). \tag{3}$$

This expression can be simplified as follows

Proposition 2. $S_\mu(u) = \max_{\lambda \in L} \min(\lambda, \mu_\lambda(u \geq \lambda))$.

Proof: Note that $\mu_\lambda(u \geq \alpha)$ does not increase with α nor λ. Suppose then that $S_\mu(u) = \min(\alpha^*, \lambda^*, \mu_{\lambda^*}(u \geq \alpha^*))$. If $\mu_{\lambda^*}(u \geq \alpha^*) = 1$, and $\alpha^* > \lambda^*$, then notice that $\mu_{\lambda^*}(u \geq \lambda^*) = 1$ as well. Likewise, if $\alpha^* < \lambda^*, \mu_{\alpha^*}(u \geq \alpha^*)) = 1$. If $\mu_{\lambda^*}(u \geq \alpha^*) = 0$, this is also true for $\mu_\lambda(u \geq \alpha)$ with $\alpha > \alpha^*$ and $\lambda > \lambda^*$. So we can assume $\alpha = \lambda$ in Eq. (3). □

3   The Commutation of Sugeno Integrals

In this section, given two capacities on finite sets $\mu_{\mathcal{X}}$ on $\mathcal{X}$ and $\mu_{\mathcal{Y}}$ on $\mathcal{Y}$, we consider double Sugeno integrals of a function $u : \mathcal{X} \times \mathcal{Y} \to L$, either as $S_{\mu_{\mathcal{X}}}(S_{\mu_{\mathcal{Y}}}(u)) = S_{\mu_{\mathcal{X}}}(f)$ where $f(x) = S_{\mu_{\mathcal{Y}}}(u(x, \cdot))$ or as $S_{\mu_{\mathcal{Y}}}(S_{\mu_{\mathcal{X}}}(u)) = S_{\mu_{\mathcal{Y}}}(g)$ where $g(y) = S_{\mu_{\mathcal{X}}}(u(\cdot, y))$. In this section we look for necessary and sufficient conditions for which the two double integrals coincide, namely:

$$S_{\mu_{\mathcal{X}}}(S_{\mu_{\mathcal{Y}}}(u(x_1, \cdot)), \cdots, S_{\mu_{\mathcal{Y}}}(u(x_n, \cdot))) = S_{\mu_{\mathcal{Y}}}(S_{\mu_{\mathcal{X}}}((u(\cdot, y_1)), \ldots, S_{\mu_{\mathcal{X}}}(u(\cdot, y_p)))$$

Or for short $S_{\mu_{\mathcal{X}}}(S_{\mu_{\mathcal{Y}}}(u)) = S_{\mu_{\mathcal{Y}}}(S_{\mu_{\mathcal{X}}}(u))$. We then say that the S-integrals *commute* and write $S_{\mu_{\mathcal{Y}}} \perp S_{\mu_{\mathcal{X}}}$. This question can be considered from two points of view: for which functions u do S-integrals commute for all capacities on $\mathcal{X}$ and $\mathcal{Y}$? For which capacities do the S-integrals commute for all functions u? The first question is considered by Narukawa and Torra [12] for more general fuzzy integrals, and the second one by Behrisch et al. [1], albeit in the larger setting of distributive lattices, for general lattice polynomials. However, in our paper, we only consider a totally ordered set L. It is of interest to adapt these results for S-integrals valued on totally ordered sets, as they become more palatable.

Then an explicit description of capacities ensuring commutation is obtained. In particular the question is whether commutation holds for other pairs of capacities than possibility measures and necessity measures, a case handled in [2].

First note that Halas et al. [9] proved that any double S-integral $S_{\mu_{\mathcal{X}}}(S_{\mu_{\mathcal{Y}}}(u))$ is a 2D S-integral

$$S_{\mu_{\mathcal{X}}}(S_{\mu_{\mathcal{Y}}}(u)) = \max_{R \subseteq \mathcal{X} \times \mathcal{Y}} \min(\kappa(R), \min_{(x_i, y_j) \in R} u(x_i, y_j)) \tag{4}$$

with $\kappa(R) = S_{\mu_{\mathcal{X}}}(S_{\mu_{\mathcal{Y}}}(1_R))$ for each $R \subseteq \mathcal{X} \times \mathcal{Y}$, 1_R denoting the characteristic function of R ($1_R(x, y) = \begin{cases} 1 & \text{if } (x, y) \in R, \\ 0 & \text{otherwise} \end{cases}$). So it becomes clear that commutation holds for all functions $u : \mathcal{X} \times \mathcal{Y} \to L$ whatever the capacities if and only if commutation holds for all Boolean-valued functions $u : \mathcal{X} \times \mathcal{Y} \to \{0, 1\}$, that is, relations $R \subseteq \mathcal{X} \times \mathcal{Y}$. More precisely, $S_{\mu_{\mathcal{Y}}} \perp S_{\mu_{\mathcal{X}}}$ if and only if

$$\forall R \subseteq \mathcal{X} \times \mathcal{Y}, S_{\mu_{\mathcal{X}}}(\mu_{\mathcal{Y}}(x_1 R), \cdots, \mu_{\mathcal{Y}}(x_n R)) = S_{\mu_{\mathcal{Y}}}(\mu_{\mathcal{X}}(R y_1), \ldots, \mu_{\mathcal{X}}(R y_p)),$$

where $x_i R = \{y \in \mathcal{Y} : x_i R y\}$ is the set of images of x_i via R, and $R y_j = \{x \in \mathcal{Y} : x R y_j\}$ the set of inverse images of y_j via R.

Another result worth mentioning is a Fubini theorem for S-integrals [12]:

Proposition 3. *If $R = A \times B$, commutation always holds, i.e.,*

$$S_{\mu_{\mathcal{X}}}(S_{\mu_{\mathcal{Y}}})(1_R) = S_{\mu_{\mathcal{Y}}}(S_{\mu_{\mathcal{X}}}(1_R)) = \min(\mu_{\mathcal{X}}(A), \mu_{\mathcal{Y}}(B))$$

Proof:

$$S_{\mu_{\mathcal{X}}}(S_{\mu_{\mathcal{Y}}}(1_R)) = \max_{S \subseteq \mathcal{X}} \min(\mu_{\mathcal{X}}(S), \min_{x \in S} \mu_{\mathcal{Y}}(xR))$$

$$= \max_{S \subseteq A} \min(\mu_{\mathcal{X}}(S), \min_{x \in S} \mu_{\mathcal{Y}}(B)) = \min(\mu_{\mathcal{X}}(A), \mu_{\mathcal{Y}}(B)).$$

$\square$

Corollary 1. *If $u(x, y) = \min(u_{\mathcal{X}}(x), u_{\mathcal{Y}}(y))$, commutation holds, i.e., $S_{\mu_{\mathcal{X}}}(S_{\mu_{\mathcal{Y}}}(u)) = S_{\mu_{\mathcal{Y}}}(S_{\mu_{\mathcal{X}}}(u)) = \min(S_{\mu_{\mathcal{X}}}(u_{\mathcal{X}}), S_{\mu_{\mathcal{Y}}}(u_{\mathcal{Y}}))$.*

Proof: It follows easily noticing that λ-cuts of u, $R = \{(x, y) : u(x, y) \geq \lambda\}$ are of the form, $S_\lambda \times T_\lambda$, where $S_\Lambda = \{x : u_{\mathcal{X}}(x) \geq \lambda\}$ and $T_\Lambda = \{y : u_{\mathcal{Y}}(y) \geq \lambda\}$. $\square$

Finally we shall prove the main theorem of this section, that is

Theorem 1. $S_{\mu_{\mathcal{Y}}} \perp S_{\mu_{\mathcal{X}}}$ *if and only if $\forall A_1, A_2 \subseteq \mathcal{X}, \forall B_1, B_2 \subseteq \mathcal{Y}$:*

$$\max(\mu_{\mathcal{X}}(A_1 \cap A_2), \mu_{\mathcal{Y}}(B_1), \mu_{\mathcal{Y}}(B_2)) \geq \min(\mu_{\mathcal{X}}(A_1), \mu_{\mathcal{X}}(A_2), \mu_{\mathcal{Y}}(B_1 \cup B_2))$$
$$\max(\mu_{\mathcal{Y}}(B_1 \cap B_2), \mu_{\mathcal{X}}(A_1), \mu_{\mathcal{X}}(A_2)) \geq \min(\mu_{\mathcal{Y}}(B_1), \mu_{\mathcal{Y}}(B_2), \mu_{\mathcal{X}}(A_1 \cup A_2)).$$

Proof: The proof is inspired by a paper on the commutation of polynomials on distributive lattices [1], and requires several lemmas listed below. Our proof is easier to read and simpler, though. First we restrict to Boolean functions (relations R) on $\mathcal{X} \times \mathcal{Y}$ without loss of generality. Then we show that commutation is equivalent to a certain identity for relations R of the form $(A_1 \times B_1) \cup (A_2 \times B_2)$ (Lemma 1). We show this identity implies the two inequalities of the theorem (Lemmas 2 then 3), which proves necessity. Then we show that these inequalities can be extended to more than just pairs of sets (Lemma 4). Finally we show that these extended inequalities imply the commutation condition (Lemma 5).

In the following three lemmas, we omit the symbol min where necessary for the sake of saving space (e.g., $\mu_{\mathcal{X}}(A_1)\mu_{\mathcal{Y}}(B_1)$ stands for $\min(\mu_{\mathcal{X}}(A_1), \mu_{\mathcal{Y}}(B_1))$, etc.)

Lemma 1. $S_{\mu_{\mathcal{X}}}(S_{\mu_{\mathcal{Y}}}(1_R)) = S_{\mu_{\mathcal{Y}}}(S_{\mu_{\mathcal{X}}}(1_R))$ for $R = (A_1 \times B_1) \cup (A_2 \times B_2)$ if and only if the 2-rectangle condition holds, i.e.

$$\max(\mu_{\mathcal{X}}(A_1 \cap A_2)\mu_{\mathcal{Y}}(B_1 \cup B_2),$$
$$\mu_{\mathcal{X}}(A_1)\mu_{\mathcal{Y}}(B_1), \mu_{\mathcal{X}}(A_2)\mu_{\mathcal{Y}}(B_2), \mu_{\mathcal{X}}(A_1 \cup A_2)\mu_{\mathcal{Y}}(B_1)\mu_{\mathcal{Y}}(B_2))$$
$$= \max(\mu_{\mathcal{Y}}(B_1 \cap B_2)\mu_{\mathcal{X}}(A_1 \cup A_2),$$
$$\mu_{\mathcal{X}}(A_1)\mu_{\mathcal{Y}}(B_1), \mu_{\mathcal{X}}(A_2)\mu_{\mathcal{Y}}(B_2), \mu_{\mathcal{Y}}(B_1 \cup B_2)\mu_{\mathcal{X}}(A_1)\mu_{\mathcal{X}}(A_2))$$

Proof: The proof just spells out the various min-terms of the Sugeno integral when $R = (A_1 \times B_1) \cup (A_2 \times B_2)$. □

Lemma 2. The 2-rectangle condition of Lemma 1 implies the two following properties

$$\max(\mu_{\mathcal{X}}(A_1 \cap A_2)\mu_{\mathcal{Y}}(B_1 \cup B_2), \mu_{\mathcal{X}}(A_1)\mu_{\mathcal{X}}(A_2)\max(\mu_{\mathcal{Y}}(B_1), \mu_{\mathcal{Y}}(B_2)))$$
$$= \mu_{\mathcal{X}}(A_1)\mu_{\mathcal{X}}(A_2)\mu_{\mathcal{Y}}(B_1 \cup B_2)$$
$$\max(\mu_{\mathcal{Y}}(B_1 \cap B_2)\mu_{\mathcal{X}}(A_1 \cup A_2), \mu_{\mathcal{Y}}(B_1)\mu_{\mathcal{Y}}(B_2)\max(\mu_{\mathcal{X}}(A_1), \mu_{\mathcal{X}}(A_2)))$$
$$= \mu_{\mathcal{Y}}(B_1)\mu_{\mathcal{Y}}(B_2)\mu_{\mathcal{X}}(A_1 \cup A_2).$$

Proof: To get the first equality the idea (from [1]) is to compute the conjunction of each side of the 2-rectangle condition with $\mu_{\mathcal{X}}(A_1)\mu_{\mathcal{X}}(A_2)$ (applying distributivity). The second equality is obtained likewise, by conjunction of each side of the equality with the term $\mu_{\mathcal{Y}}(B_1)\mu_{\mathcal{Y}}(B_2)$. □

The following lemma simplifies the two obtained equalities into simpler inequalities.

Lemma 3. The two equalities in Lemma 2 are equivalent to the two respective inequalities

$$\max(\mu_{\mathcal{X}}(A_1 \cap A_2), \mu_{\mathcal{Y}}(B_1), \mu_{\mathcal{Y}}(B_2))) \geq \min(\mu_{\mathcal{X}}(A_1), \mu_{\mathcal{X}}(A_2), \mu_{\mathcal{Y}}(B_1 \cup B_2)) \tag{5}$$

$$\max(\mu_{\mathcal{Y}}(B_1 \cap B_2), \mu_{\mathcal{X}}(A_1), \mu_{\mathcal{X}}(A_2)) \geq \min(\mu_{\mathcal{Y}}(B_1), \mu_{\mathcal{Y}}(B_2), \mu_{\mathcal{X}}(A_1 \cup A_2)). \tag{6}$$

Proof: We must apply distributivity to the right-hand side of the first equality in Lemma 2: $\max(\mu_{\mathcal{X}}(A_1 \cap A_2)\mu_{\mathcal{Y}}(B_1 \cup B_2), \mu_{\mathcal{X}}(A_1)\mu_{\mathcal{X}}(A_2)\max(\mu_{\mathcal{Y}}(B_1), \mu_{\mathcal{Y}}(B_2)))$ and the first equality in Lemma 2 reduces to the equality $\mu_{\mathcal{X}}(A_1) \mu_{\mathcal{X}}(A_2)\mu_{\mathcal{Y}}(B_1 \cup B_2)\max(\mu_{\mathcal{X}}(A_1 \cap A_2), \mu_{\mathcal{Y}}(B_1), \mu_{\mathcal{Y}}(B_2)) = \mu_{\mathcal{X}}(A_1)\mu_{\mathcal{X}}(A_2)\mu_{\mathcal{Y}} (B_1 \cup B_2)$, which is equivalent to the inequality (5). The inequality (6) is proved likewise, exchanging A and B, $\mathcal{X}$ and $\mathcal{Y}$.

The two inequalities (5) and (6) extend to more than two pairs of sets, namely:

Lemma 4. *(5) and (6) imply:*

$$\max(\mu_{\mathcal{X}}(\cap_{i=1}^{k} A_i), \max_{j=1}^{\ell} \mu_{\mathcal{Y}}(B_j)) \geq \min(\min_{i=1}^{k} \mu_{\mathcal{X}}(A_i), \mu_{\mathcal{Y}}(\cup_{j=1}^{\ell} B_j)) \qquad (7)$$

$$\max(\mu_{\mathcal{Y}}(\cap_{j=1}^{\ell} B_j), \max_{i=1}^{k} \mu_{\mathcal{X}}(A_i)) \geq \min(\min_{j=1}^{\ell} \mu_{\mathcal{Y}}(B_j), \mu_{\mathcal{X}}(\cup_{i=1}^{k} A_i)). \qquad (8)$$

Proof: Inequality (7) holds for $k = \ell = 2$ (this is (5)). Suppose that inequality (7) holds for $i = 1, \ldots k - 1$ and $\ell = 2$. We can write, by assumption:

$$\max(\mu_{\mathcal{X}}(\cap_{i=1}^{k-1} A_i), \mu_{\mathcal{Y}}(B_1), \mu_{\mathcal{Y}}(B_2)) \geq \min(\min_{i=1}^{k-1} \mu_{\mathcal{X}}(A_i), \mu_{\mathcal{Y}}(B_1 \cup B_2))$$

Moreover we can write (5) for $A = \cap_{i=1}^{k-1} A_i, A_k, B_1, B_2$. Then we can write the inequality

$$\max(\mu_{\mathcal{X}}(\cap_{i=1}^{k} A_i), \mu_{\mathcal{Y}}(B_1), \mu_{\mathcal{Y}}(B_2)) \geq \min(\mu_{\mathcal{X}}(\cap_{i=1}^{k-1} A_i), \mu_{\mathcal{X}}(A_k), \mu_{\mathcal{Y}}(B_1 \cup B_2))$$

Suppose $\mu_{\mathcal{X}}(\cap_{i=1}^{k-1} A_i) \geq \max(\mu_{\mathcal{Y}}(B_1), \mu_{\mathcal{Y}}(B_2))$. So the first inequality reduces to

$$\mu_{\mathcal{X}}(\cap_{i=1}^{k-1} A_i) \geq \min(\min_{i=1}^{k-1} \mu_{\mathcal{X}}(A_i), \mu_{\mathcal{Y}}(B_1 \cup B_2)).$$

Then we can replace $\mu_{\mathcal{X}}(\cap_{i=1}^{k-1} A_i)$ by $\min(\min_{i=1}^{k-1} \mu_{\mathcal{X}}(A_i), \mu_{\mathcal{Y}}(B_1 \cup B_2))$ in the second inequality, and get (7).

Otherwise, $\mu_{\mathcal{X}}(\cap_{i=1}^{k-1} A_i) \leq \max(\mu_{\mathcal{Y}}(B_1), \mu_{\mathcal{Y}}(B_2))$, and the first inequality reads

$$\max(\mu_{\mathcal{Y}}(B_1), \mu_{\mathcal{Y}}(B_2)) \geq \min(\min_{i=1}^{k-1} \mu_{\mathcal{X}}(A_i), \mu_{\mathcal{Y}}(B_1 \cup B_2))$$

so we have $\max(\mu_{\mathcal{X}}(\cap_{i=1}^{k} A_i), \mu_{\mathcal{Y}}(B_1), \mu_{\mathcal{Y}}(B_2)) \geq \min(\min_{i=1}^{k-1} \mu_{\mathcal{X}}(A_i)), \mu_{\mathcal{X}}(A_k), \mu_{\mathcal{Y}}(B_1 \cup B_2))$, which is (7) again. Proving that the inequality (7) holds for $k = 2, \ell > 2$ is similar. So, the inequality (7) holds for any $k > 2, \ell > 2$. The inequality (8) is proved in a similar way, exchanging A and B, X and Y. $\square$

Lemma 5. *If $\mu_{\mathcal{X}}$ and $\mu_{\mathcal{Y}}$ satisfy the two inequalities (7) and (8), then $S_{\mu_{\mathcal{X}}} \perp S_{\mu_{\mathcal{Y}}}$*

Proof: Let us consider (7) written as $\max([\mu_{\mathcal{X}}(\cap_{i=1}^{k} A_i)\mu_{\mathcal{Y}}(\cup_{j=1}^{\ell} B_j)], [\min_{i=1}^{k} \mu_{\mathcal{X}}(A_i) \max_{j=1}^{\ell} \mu_{\mathcal{Y}}(B_j)]) = \min_{i=1}^{k} \mu_{\mathcal{X}}(A_i)\mu_{\mathcal{Y}}(\cup_{j=1}^{\ell} B_j)$, and prove that $\max_{S \subseteq \mathcal{X}} \min(\mu_{\mathcal{X}}(S), \min_{x \in S} \mu_{\mathcal{Y}}(xR)) \geq \max_{T \subseteq \mathcal{Y}} \min(\mu_{\mathcal{Y}}(T), \min_{y \in T} \mu_{\mathcal{X}}(Ry))$.

Consider the term $\min(\mu_{\mathcal{Y}}(T), \min_{y \in T} \mu_{\mathcal{X}}(Ry))$ that we identify with the right-hand side of (7). Denoting $S_T = \cap_{y \in T} Ry$, this equality then reads:

$$\min(\mu_{\mathcal{Y}}(T), \min_{y \in T} \mu_{\mathcal{X}}(Ry)) = \max(\min[\mu_{\mathcal{X}}(S_T), \mu_{\mathcal{Y}}(T)], \min[\min_{y \in T} \mu_{\mathcal{X}}(Ry), \max_{t \in T} \mu_{\mathcal{Y}}(\{t\})])$$

$$= \max(\min[\mu_{\mathcal{X}}(S_T), \mu_{\mathcal{Y}}(T)], \max_{t \in T}[\min(\min_{y \in T} \mu_{\mathcal{X}}(Ry), \mu_{\mathcal{Y}}(\{t\}))]).$$

We have $\mu_{\mathcal{Y}}(T) \leq \min_{x \in S_T} \mu_{\mathcal{Y}}(xR)$ because $S_T = \cap_{y \in T} Ry$ if and only if $S_T \times T \subseteq R$ if and only if $T = \cap_{x \in S_T} xR$. So, the term $\min(\mu_{\mathcal{X}}(S_T), \mu_{\mathcal{Y}}(T))$ is upper bounded by $\max_{S \subseteq \mathcal{X}} \min(\mu_{\mathcal{X}}(S), \min_{x \in S} \mu_{\mathcal{Y}}(xR))$.

The term $\min(\min_{y \in T} \mu_{\mathcal{X}}(Ry), \mu_{\mathcal{Y}}(\{t\})$ has the same upper bound since

- as $t \in T$, $\min_{y \in T} \mu_{\mathcal{X}}(Ry) \leq \mu_{\mathcal{X}}(Rt)$, choosing $y = t$;
- if $x \in Rt$, then $\mu_{\mathcal{Y}}(\{t\}) \leq \mu_{\mathcal{Y}}(xR)$ since $t \in xR$ as well.

So, $\min(\min_{y \in T} \mu_{\mathcal{X}}(Ry), \mu_{\mathcal{Y}}(\{t\})) \leq \min(\mu_{\mathcal{X}}(Rt), \mu_{\mathcal{Y}}(xR)), \forall x \in Rt$.
Hence, $\min(\min_{y \in T} \mu_{\mathcal{X}}(Ry), \mu_{\mathcal{Y}}(\{t\})) \leq \min(\mu_{\mathcal{X}}(Rt), \min_{x \in Rt} \mu_{\mathcal{Y}}(xR))$ that is also upper bounded by $\max_{S \subseteq \mathcal{X}} \min(\mu_{\mathcal{X}}(S), \min_{x \in S} \mu_{\mathcal{Y}}(xR))$. We thus get $S_{\mu_{\mathcal{X}}}(S_{\mu_{\mathcal{Y}}}(\mathbf{1}_R)) \geq S_{\mu_{\mathcal{Y}}}(S_{\mu_{\mathcal{X}}}(\mathbf{1}_R))$

The converse inequality can be proved likewise, by symmetry, using (8).   □

The proof of Theorem 1 is now complete.   □

Theorem 1 gives a necessary and sufficient condition for the commutation of two S-integrals applied to any function $u : \mathcal{X} \times \mathcal{Y} \to L$ based on capacities $\mu_{\mathcal{X}}$ and $\mu_{\mathcal{Y}}$. As these S-integrals are entirely characterized by these capacities, we shall simply say that the two capacities commute.

4   Commuting Capacities

Consider the cases when $\mu_{\mathcal{X}}$ and $\mu_{\mathcal{Y}}$ are possibility or necessity measures. In the framework of possibilistic decision under uncertainty, $\mathcal{X} = \{x_1, \cdots, x_n\}$ is a set of states, and a possibility distribution π captures the common knowledge of the agents: π_i is the possibility degree to be in state x_i. $\mathcal{Y} = \{y_1, \cdots, y_p\}$ is the set of agents. The weight vector $w = (w_1, \cdots, w_p) \in [0, 1]^p$ is modeled as a possibility distribution on $\mathcal{Y}$ where w_j is the importance of agent y_j. The attractiveness of decision u for agent y_j in the different states is captured by utility function $u(\cdot, y_j) : \mathcal{X} \to [0, 1]$. There are two possible approaches for egalitarian (min-based) aggregations of pessimistic decision-makers, and two possible approaches for egalitarian aggregations of optimistic decision-makers [2].

ex-post pessimistic
$U_{post}^{-min}(\pi, w, u) = \min_{x_i \in \mathcal{X}} \max(1 - \pi_i, \min_{y_j \in \mathcal{Y}} \max(u(x_i, y_j), 1 - w_j))$.
ex-ante pessimistic
$U_{ante}^{-min}(\pi, w, u) = \min_{y_j \in \mathcal{Y}} \max(1 - w_j, \min_{x_i \in \mathcal{X}} \max(u(x_i, y_j), 1 - \pi_i))$.
ex-post optimistic
$U_{post}^{+min}(\pi, w, u) = \max_{x_i \in \mathcal{X}} \min(\pi_i, \min_{y_j \in \mathcal{Y}} \max(u(x_i, y_j), 1 - w_j))$.

ex-ante optimistic
$$U_{ante}^{+min}(\pi, w, u) = \min_{y_j \in \mathcal{Y}} \max(1 - w_j, \max_{x_i \in \mathcal{X}} \min(u(x_i, y_j), \pi_i)).$$

It can be checked that the first two quantities are of the form $U_{post}^{-min}(\pi, w, u) = S_{N_{\mathcal{X}}}(S_{N_{\mathcal{Y}}}(u))$ and $U_{ante}^{-min}(\pi, w, u) = S_{N_{\mathcal{Y}}}(S_{N_{\mathcal{X}}}(u))$, respectively. Essghaier et al. [2] show that the two expressions are equal to $\min_{x_i \in \mathcal{X}, y_j \in \mathcal{Y}} \max(1 - \pi_i, u(x_i, y_j), 1 - w_j)$, thus $S_{N_{\mathcal{X}}}(S_{N_{\mathcal{Y}}}(u)) = S_{N_{\mathcal{Y}}}(S_{N_{\mathcal{X}}}(u))$.

In the optimistic case, qualitative decision theory [6] prescribes the use of a Sugeno integral based on a possibility measure on $\mathcal{X}$: $U_{post}^{+min}(\pi, w, u) = S_{\Pi_{\mathcal{X}}}(S_{N_{\mathcal{Y}}}(u))$ and $U_{ante}^{+min}(\pi, w, u) = S_{N_{\mathcal{Y}}}(S_{\Pi_{\mathcal{X}}}(u))$. Now the two integrals no longer coincide: Essghaier et al. [2,4] have shown that we only have the inequality $U_{ante}^{+min}(\pi, w, u) \geq U_{post}^{+min}(\pi, w, u)$ with no equality in general. The following counterexample shows that the latter inequality can be strict, when one of the capacities is a necessity measure and the other one a possibility measure, even in the Boolean case [3]:

Example 1. *Let* $\mathcal{X} = \{x_1, x_2\}$, $\pi_i = 1$, *and* $w_i = 1, \forall i = 1, 2$, $\mathcal{Y} = \{y_1, y_2\}$, $u(x_1, y_1) = u(x_2, y_2) = 1$ *and* $u(x_2, y_1) = u(x_1, y_2) = 0$. *We have* $U_{post}^{+min}(\pi, w, u)$ *as*
$$\max(\min(1, \min(\max(1 - 1, 1), \max(1 - 1, 0)), \min(1, \min(\max(1 - 1, 0), \max(1 - 1, 1)))) = 0.$$
But $U_{ante}^{+min}(\pi, w, u)$ *is computed as*
$$\min(\max(1 - 1, \max(\max(1, 1), \max(0, 1)), \max(1 - 1, \max(\max(0, 1), \max(1, 1))))) = 1.$$

In this subsection, we try to characterize all pairs of commuting capacities. Let us begin with the Boolean case. It confirms the intuitions of [2].

Proposition 4. *If one of* $\mu_{\mathcal{X}}$ *and* $\mu_{\mathcal{Y}}$ *is Boolean, S-integrals commute if and only if they are both necessity measures or possibility measures or one of them is a Dirac measure.*

Proof: Suppose $\mu_{\mathcal{X}}$ is Boolean and is not a necessity measure and $\mu_{\mathcal{Y}}$ is not a possibility measure. Then $\exists A_1, A_2 \subseteq \mathcal{X}, \mu_{\mathcal{X}}(A_1 \cap A_2) < \min(\mu_{\mathcal{X}}(A_1), \mu_{\mathcal{X}}(A_2))$, and $\exists B_1, B_2 \subseteq \mathcal{Y}, \mu_{\mathcal{Y}}(B_1 \cup B_2) > \max(\mu_{\mathcal{Y}}(B_1), \mu_{\mathcal{X}}(B_2))$. For $\mu_{\mathcal{X}}$, it reads $\mu_{\mathcal{X}}(A_1 \cap A_2) = 0, \mu_{\mathcal{X}}(A_1) = \mu_{\mathcal{X}}(A_2) = 1$. Then the 2-rectangle condition (5) fails since it reads $\max(0, \mu_{\mathcal{Y}}(B_1), \mu_{\mathcal{X}}(B_2), \min(\mu_{\mathcal{Y}}(B_1), \mu_{\mathcal{X}}(B_2))) = \max(\mu_{\mathcal{X}}(B_2), \mu_{\mathcal{Y}}(B_1)) < \max(\mu_{\mathcal{Y}}(B_1 \cap B_2), \mu_{\mathcal{Y}}(B_1), \mu_{\mathcal{X}}(B_2), \mu_{\mathcal{Y}}(B_1 \cup B_2)) = \mu_{\mathcal{Y}}(B_1 \cup B_2)$.

The second inequality (6) is violated by choosing $A_1, A_2 \subseteq \mathcal{X}, B_1, B_2 \subseteq \mathcal{Y}$, such that $\mu_{\mathcal{Y}}(B_1 \cap B_2) = 0, \mu_{\mathcal{Y}}(B_1) = \mu_{\mathcal{Y}}(B_2) = 1, \mu_{\mathcal{X}}(A_1 \cup A_2) = 1, \mu_{\mathcal{X}}(A_1) = \mu_{\mathcal{X}}(A_2) = 0$, assuming $\mu_{\mathcal{Y}}$ is not a necessity measure and $\mu_{\mathcal{X}}$ is not a possibility measure. Obeying the two inequalities (5) and (6) enforces the following constraints in the Boolean case

$$\mu_{\mathcal{Y}} \text{ possibility measure or } \mu_{\mathcal{X}} \text{ necessity measure}$$
$$\text{and}$$
$$\mu_{\mathcal{Y}} \text{ necessity measure or } \mu_{\mathcal{X}} \text{ possibility measure}$$

It leads to possibility measures on both sets $\mathcal{X}$ and $\mathcal{Y}$, or necessity measures (known cases where commuting occurs). Alternatively, if we enforce $\mu_{\mathcal{Y}}$ to be a possibility measure and a necessity measure, it is a Dirac function on $\mathcal{Y}$, and any capacity on the other space. □

Corollary 2. *S-integrals w.r.t. Boolean capacities $\mu_{\mathcal{X}}$ and $\mu_{\mathcal{Y}}$ commute if and only if they are both necessity measures or possibility measures or one of them is a Dirac measure.*

Note that, to violate the necessary condition for commutation (5), it is enough that neither $\mu_{\mathcal{X}}$ nor $\mu_{\mathcal{Y}}$ are possibility and necessity measures, and moreover for A_1, A_2, B_1, B_2 where, say $\mu_{\mathcal{X}}$ violates the axiom of necessities and $\mu_{\mathcal{Y}}$ violates the axiom of possibilities, we have that $\mu_{\mathcal{X}}(A_1)$ and $\mu_{\mathcal{X}}(A_2)$ are both greater than each of $\mu_{\mathcal{Y}}(B_1), \mu_{\mathcal{Y}}(B_2)$ and moreover $\mu_{\mathcal{Y}}(B_1 \cup B_2) > \mu_{\mathcal{Y}}(A_1 \cap A_2)$. Then the integrals will not commute.

In the following we solve the commutation problem for non-Boolean capacities. We can give examples of commuting capacities that are neither only possibility measures, nor only necessity measures nor a Dirac function contrary to the Boolean case of Corollary 2.

Example 2. *Let $\mathcal{X} = \{x_1, x_2\}; \mathcal{Y} = \{y_1, y_2\}$. Then let $\mu_{\mathcal{X}}(x_1) = \alpha, \mu_{\mathcal{X}}(x_2) = \alpha$, $\mu_{\mathcal{Y}}(y_1) = 1, \mu_{\mathcal{Y}}(y_2) = \alpha$, so a constant capacity and a possibility measure.*
We have $\max(\mu_{\mathcal{X}}(A_1 \cap A_2), \mu_{\mathcal{Y}}(B_1), \mu_{\mathcal{Y}}(B_2)) \geq \min(\mu_{\mathcal{X}}(A_1), \mu_{\mathcal{X}}(A_2), \mu_{\mathcal{Y}}(B_1 \cup B_2))$ because the possible values are α or 1. The right-hand side is equal to 1 if and only if $A_1 = A_2 = \mathcal{X}$; in this case $\mu_{\mathcal{X}}(A_1 \cap A_2) = 1$.
We have $\max(\mu_{\mathcal{Y}}(B_1 \cap B_2), \mu_{\mathcal{X}}(A_1), \mu_{\mathcal{X}}(A_2)) \geq \min(\mu_{\mathcal{Y}}(B_1), \mu_{\mathcal{Y}}(B_2), \mu_{\mathcal{X}}(A_1 \cup A_2))$ because the possible values are α or 1. The right-hand side is equal to 1 if and only if $y_1 \in B_1$ and $y_2 \in B_2 = \mathcal{X}$; in this case $\mu_{\mathcal{Y}}(B_1 \cap B_2) = 1$.
So $S_{\mu_{\mathcal{X}}} \perp S_{\mu_{\mathcal{Y}}}$.

In the following, we lay bare the pairs of capacities that commute by applying the result of Corollary 2 to cuts of the capacities. We first prove that for Boolean functions on $\mathcal{X} \times \mathcal{Y}$, the double S-integrals are completely defined by the cuts of the involved capacities, thus generalizing Proposition 1 to double S-integrals.

Proposition 5. $S_{\mu_{\mathcal{X}}}(S_{\mu_{\mathcal{Y}}}(u)) = \max_{\lambda > 0} \min(\lambda, S_{\mu_{\mathcal{X}\lambda}}(S_{\mu_{\mathcal{Y}\lambda}}(u)))$ *when $u = 1_R$.*

Proof: For simplicity we denote $\mu_{\mathcal{X}}$ by μ and $\mu_{\mathcal{Y}}$ by ν

$$S_\mu(S_\nu(u)) = \max_{A \subseteq \mathcal{X}} \min(\mu(A), \min_{x \in A} S_\nu(u(x, \cdot)))$$

$$= \max_{A \subseteq \mathcal{X}} \min(\max_{\lambda > 0} \min(\lambda, \mu_\lambda(A)), \min_{x \in A} \max_{\alpha > 0} \min(\alpha, S_{\nu_\alpha}(u(x, \cdot))))$$

Note that $\min_{x \in A} \max_{\alpha > 0} \min(\alpha, S_{\nu_\alpha}(u(x, \cdot))) \geq \max_{\alpha > 0} \min_{x \in A} \min(\alpha, S_{\nu_\alpha}(u(x, \cdot)))$. Let us prove the converse inequality when $u = 1_R$. Let $\alpha^*, \hat{x}$ be optima for $\min(\alpha, \nu_\alpha(xR))$ on the right hand side, that is, $\max_{\alpha > 0} \min_{x \in A} \min(\alpha, \nu_\alpha(xR)) = \min(\alpha^*, \nu_{\alpha^*}(\hat{x}R))$. Note that $\min(\alpha^*, \nu_{\alpha^*}(\hat{x}R))$ takes the values 0 or α^*.

- If $\min(\alpha^*, \nu_{\alpha^*}(\hat{x}R)) = 0$ then forall α there exists x such that $\nu_\alpha(xR) = 0$; so the left side is also equal to 0.
- If $\min(\alpha^*, \nu_{\alpha^*}(\hat{x}R)) = \alpha^*$ then for all α, there exists x such that $\min(\alpha, \nu_\alpha(xR)) \leq \alpha^*$. Hence if $\alpha > \alpha^*$ then there exists x such that $\nu_\alpha(xR) = 0$ and $\min(\alpha, \nu_\alpha(xR)) = 0$. If $\alpha \leq \alpha^*$ then $\min(\alpha, \nu_\alpha(xR)) \leq \alpha^*$.

So the left side is less than the right side. We get the equality as follows:

$$\begin{aligned}
S_\mu(S_\nu(1_R)) &= \max_{A \subseteq \mathcal{X}} \min(\max_{\lambda > 0} \min(\lambda, \mu_\lambda(A)), \max_{\alpha > 0} \min_{x \in A} \min(\alpha, \nu_\alpha(xR))) \\
&= \max_{\lambda > 0} \max_{A \subseteq \mathcal{X}} \min(\min(\lambda, \mu_\lambda(A)), \max_{\alpha > 0} \min_{x \in A} \min(\alpha, \nu_\alpha(xR))) \\
&= \max_{\lambda > 0} \min(\lambda, \max_{A \subseteq \mathcal{X}} \min(\mu_\lambda(A)), \max_{\alpha > 0} \min(\alpha, \min_{x \in A} \nu_\alpha(xR))) \\
&= \max_{\lambda > 0, \alpha > 0} \min(\lambda, \alpha, \max_{A \subseteq \mathcal{X}} \min(\mu_\lambda(A), \min_{x \in A} \nu_\alpha(xR))) \\
&= \max_{\lambda > 0, \alpha > 0} \min(\lambda, \alpha, S_{\mu_\lambda}(S_{\nu_\alpha}(1_R)))
\end{aligned}$$

Due to the monotonicity of the Sugeno integral and due to the use of minimum, the maximum is attained for $\alpha = \lambda$.   □

We know that commutation between integrals holds for functions $u(x, y)$ if it holds for relations. The above result shows that commutation between capacities will hold if and only if it will hold for their cuts, to which we can apply Corollary 2.

Corollary 3. *Capacities $\mu_\mathcal{X}$ and $\mu_\mathcal{Y}$ commute if and only if their cuts $\mu_{\mathcal{X}\lambda}$ and $\mu_{\mathcal{Y}\lambda}$ commute for all $\lambda \in L$.*

Proof: Suppose $\mu_\mathcal{X}$ and $\mu_\mathcal{Y}$ commute. It means that $S_{\mu_\mathcal{X}}(S_{\mu_\mathcal{Y}}(1_R))$ and $S_{\mu_\mathcal{Y}}(S_{\mu_\mathcal{X}}(1_R))$ are the same 2D capacity κ on $\mathcal{X} \times \mathcal{Y}$, namely $S_{\mu_\mathcal{X}}(S_{\mu_\mathcal{Y}}(1_R)) = S_{\mu_\mathcal{Y}}(S_{\mu_\mathcal{X}}(1_R)) = \kappa(R)$. It is then clear that using Proposition 2:

$$\begin{aligned}
\kappa_\lambda(R) &= S_{\mu_\mathcal{X}}(S_{\mu_\mathcal{Y}}(1_R))_\lambda = S_{\mu_\mathcal{Y}}(S_{\mu_\mathcal{X}}(1_R))_\lambda \\
&= S_{\mu_{\mathcal{X}\lambda}}(1_{[\mu_\mathcal{Y}(R(x_1, \cdot)) \geq \lambda]}, \cdots, 1_{[\mu_\mathcal{Y}(R(x_n, \cdot)) \geq \lambda]}) \\
&= S_{\mu_{\mathcal{Y}\lambda}}(1_{[\mu_\mathcal{X}(R(\cdot, y_1)) \geq \lambda]}, \cdots, 1_{[\mu_\mathcal{X}(R(\cdot, y_n)) \geq \lambda]}) \\
&= S_{\mu_{\mathcal{X}\lambda}}(S_{\mu_{\mathcal{Y}\lambda}}(R(x_1, \cdot)), \ldots, S_{\mu_{\mathcal{Y}\lambda}}(R(x_n, \cdot))) \\
&= S_{\mu_{\mathcal{Y}\lambda}}(S_{\mu_{\mathcal{X}\lambda}}(R(\cdot, y_1)), \ldots, S_{\mu_{\mathcal{Y}\lambda}}(R(\cdot, y_n))) \\
&= S_{\mu_{\mathcal{X}\lambda}}(S_{\mu_{\mathcal{Y}\lambda}}(1_R)) = S_{\mu_{\mathcal{Y}\lambda}}(S_{\mu_{\mathcal{X}\lambda}}(1_R))
\end{aligned}$$

Conversely, using Proposition 5 if $S_{\mu_{\mathcal{X}\lambda}}(S_{\mu_{\mathcal{Y}\lambda}}(1_R)) = S_{\mu_{\mathcal{Y}\lambda}}(S_{\mu_{\mathcal{X}\lambda}}(1_R))$ for all $\lambda \in L$ and $R \subseteq \mathcal{X} \times \mathcal{Y}$ it implies $S_{\mu_\mathcal{X}}(S_{\mu_\mathcal{Y}}(1_R)) = S_{\mu_\mathcal{Y}}(S_{\mu_\mathcal{X}}(1_R))$ for all $R \subseteq \mathcal{X} \times \mathcal{Y}$, which is equivalent to commutation of S-integrals w.r.t. $\mu_\mathcal{X}$ and $\mu_\mathcal{Y}$ for all 2-place functions u.   □

In the above Example 2, the commutation becomes obvious because the λ-cut of $\mu_\mathcal{X}$ is a necessity (with focal set $\mathcal{X}$) and $\mu_\mathcal{Y}$ is a Dirac function on y_1 for $\lambda > 1$. And the λ-cut of $\mu_\mathcal{X}$ is the vacuous possibility, as well as the λ-cut of $\mu_\mathcal{Y}$ for $\lambda \leq \alpha$. More generally we can claim:

Corollary 4. *Capacities* $\mu_{\mathcal{X}}$ *and* $\mu_{\mathcal{Y}}$ *commute if and only if for each* $\lambda \in L$, *their cuts* $\mu_{\mathcal{X}\lambda}$ *and* $\mu_{\mathcal{Y}\lambda}$ *are two possibility measures, two necessity measures, or one of them is a Dirac measure.*

To check commutation using Corollary 4, one must compute the focal sets of the cuts of a capacity.

Lemma 6. *The focal sets of* μ_λ *form the family* $\mathcal{F}(\mu_\lambda) = \min_{\subseteq} \{E \subseteq \mathcal{X} : \mu_\#(E) \geq \lambda\}$, *containing the smallest sets for inclusion in the family* $\mathcal{F}(\mu)$ *of focal sets of* μ *with weights at least* λ.

Indeed the focal sets of a Boolean capacity form an antichain, that is, they are not nested, and if $\mu_\#(E) > \mu_\#(F) \geq \lambda$, while $F \subset E$, then E is not focal for μ_λ. The above results lead us to conclude as follows:

Proposition 6. *For any capacity* μ *on* $\mathcal{X}$,

1. μ_λ *is a necessity measure if and only if there is a single focal set* E *with* $\mu_\#(E) \geq \lambda$ *such that for all focal sets* F *in* $\mathcal{F}(\mu)$ *with weights* $\mu_\#(F) \geq \lambda$, *we have* $E \subset F$.
2. μ_λ *is a possibility measure if and only if there is a set* S *of singletons* $\{x_i\}$ *with* $\mu_\#(\{x_i\}) \geq \lambda$ *such that for all focal sets* F *in* $\mathcal{F}(\mu)$ *with weights* $\mu_\#(F) \geq \lambda$, *we have* $S \cap F \neq \emptyset$.
3. μ_λ *is a Dirac measure if and only if there is a focal singleton* $\{x\}$ *with* $\mu_\#(\{x\}) \geq \lambda$ *such that for all focal sets* F *in* $\mathcal{F}(\mu)$ *with weights* $\mu_\#(F) \geq \lambda$, *we have* $x \in F$.

Proof: We apply Lemma 6.

1. The condition does ensure that E is the only focal set of μ_λ hence it is a necessity measure. If the condition does not hold it is clear that μ_λ has more than one focal set, hence is a not a necessity measure.
2. The condition does ensure that the focal sets of μ_λ are the singletons in S, hence it is a possibility measure. If the condition does not hold it is clear that μ_λ has a focal set that is not a singleton, hence is not a possibility measure.
3. The condition implies that μ_λ is both a possibility and a necessity measure, hence a Dirac measure. If it is not satisfied, either μ_λ has more than one focal set or its focal set is not a singleton. □

Note that if μ_λ is a possibility measure with focal sets that are the singletons of S and $\alpha < \lambda$ then μ_α cannot be a necessity measure, since if a set E is focal for μ_α, it must be disjoint from S so that $\mathcal{F}(\mu_\lambda)$ contains all singletons of S and E at least. So we have the following claim: if $\forall\lambda, \in L, \mu_\lambda$ is either a possibility measure or a necessity measure, there is a threshold value θ such that $\forall\lambda \leq \theta \ \mu_\lambda$ is a possibility measure (possibly a Dirac measure), and $\forall\lambda > \theta, \mu_\lambda$ is a necessity measure. We are then in a position to state the main result of this section, as pictured on Fig. 1.

Theorem 2. *Two capacities* $\mu_{\mathcal{X}}$ *and* $\mu_{\mathcal{Y}}$ *commute if and only if there exist at most two thresholds* $\theta_N \leq \theta_\Pi \in L$ *such that*

λ	μ_X^λ	μ_Y^λ
1 θ_N	necessity	necessity
θ_Π	any capacity or Dirac	Dirac any capacity
0	possibility	possibility

Fig. 1. Commuting capacities

- For $1 \geq \lambda > \theta_N$, the λ-cuts of μ_X and μ_Y are necessity measures.
- For $\theta_N \geq \lambda > \theta_\Pi$, the λ-cut of one of μ_X, μ_Y is a Dirac measure, the other one being any Boolean capacity.
- For $\theta_\Pi \geq \lambda$, the λ-cuts of μ_X and μ_Y are possibility measures.

Proof: We just apply Corollary 4, noticing that if the λ-cut of μ_X is a possibility measure, its λ'-cuts for $\lambda' < \lambda$ cannot be necessity measures. □

Example 3. *We can apply Theorem 2 to find the condition for commutation on $\{x_1, x_2\} \times \{y_1, y_2\}$ where in general $\mu_X(x_1) = \alpha_1, \mu_X(x_2) = \alpha_2, \mu_Y(y_1) = \beta_1, \mu_Y(y_2) = \beta_2$. Note that cuts of capacity on two-element sets can only be Boolean possibility or necessity measures. So the capacities will commute except if there is $\lambda \in L$ such that the cut of μ_X is a possibility measure and the cut of μ_Y is a necessity measure. So commutation will hold in any one of the following situations and only for them:*

- *μ_X is a possibility measure with $\alpha_1 > \alpha_2$ and μ_Y is a necessity measure with mass $\beta_1 > \beta_2 = 0$ with $\beta_1 > \alpha_2$.*
- *μ_X is a capacity $(1 > \alpha_1 \geq \alpha_2)$ and μ_Y a possibility measure with $\beta_1 = 1 > \beta_2$, where $\alpha_1 > \beta_2$.*
- *μ_X is a capacity $(1 > \alpha_1 \geq \alpha_2)$ then μ_Y is a necessity measure with mass $\beta_1 > \beta_2 = 0$ with $\beta_1 \geq \alpha_2$.*
- *μ_X and μ_Y are genuine capacities $(1 > \alpha_1 \geq \alpha_2; 1 > \beta_1 \geq \beta_2)$, then $\max(\alpha_1, \alpha_2) \geq \min(\beta_1, \beta_2)$ and $\max(\beta_1, \beta_2) \geq \min(\alpha_1, \alpha_2)$.*

The latter condition $\max(\alpha_1, \alpha_2) \geq \min(\beta_1, \beta_2)$ and $\max(\beta_1, \beta_2) \geq \min(\alpha_1, \alpha_2)$ covers all 4 cases. To check that this is correct, note that the only cases when the cuts are a possibility vs. a necessity measure are when $\max(\alpha_1, \alpha_2) < \min(\beta_1, \beta_2)$ or $\max(\beta_1, \beta_2) < \min(\alpha_1, \alpha_2)$ (take λ in the interval). Note that this is the case in Example 1 since then $\alpha_1 = \alpha_2 = 1$ and $\beta_1 = \beta_2 = 0$. However the commutation condition is clearly satisfied in Example 2.

Finally we shall express commuting capacities in closed form. Without loss of generality, and up to a permutation between $\mathcal{X}$ and $\mathcal{Y}$, if μ_X and μ_Y commute, the set of focal sets $\mathcal{F}(\mu_X)$ is partitioned in $\mathcal{F}_N(\mu_X) \cup \mathcal{F}_\Pi(\mu_X)$, where

- $\mathcal{F}_N(\mu_X) = \{E \in \mathcal{F}(\mu_X) : \mu_{X\#}(E) > \theta_N\}$ is nested, say $E_p \subset \cdots \subset E_1$.

- $\mathcal{F}_\Pi(\mu_\mathcal{X}) = \{E \in \mathcal{F}(\mu_\mathcal{X}) : \mu_{\mathcal{X}\#}(E) \leq \theta_\Pi\}$ contains only singletons.
- $\exists x \in E_p, \mu_{\mathcal{X}\#}(\{x\}) = \theta_N$ (no set in $\mathcal{F}_N(\mu_\mathcal{X})$ is focal for the λ-cut of $\mu_\mathcal{X}$ when $\lambda \leq \theta_N$).

while the set of focal sets $\mathcal{F}(\mu_\mathcal{Y})$ is partitioned in $\mathcal{F}_N(\mu_\mathcal{Y}) \cup \mathcal{F}_D(\mu_\mathcal{Y}) \cup \mathcal{F}_\Pi(\mu_\mathcal{Y})$, where

- $\mathcal{F}_N(\mu_\mathcal{Y}) = \{F \in \mathcal{F}(\mu_\mathcal{Y}) : \mu_{\mathcal{Y}\#}(F) > \theta_N\}$ is nested.
- $\mathcal{F}_\Pi(\mu_\mathcal{Y}) = \{F \in \mathcal{F}(\mu_\mathcal{Y}) : \mu_{\mathcal{Y}\#}(F) \leq \theta_\Pi\}$ contains only singletons.
- $\forall F \in \mathcal{F}(\mu) \setminus \mathcal{F}_\Pi(\mu_\mathcal{Y}), \exists y \in \mathcal{Y}$ such that $\mu_{\mathcal{Y}\#}(\{y\}) = \theta_\Pi$ and $y \in F$ (so that no focal set of μ outside of $\mathcal{F}_\Pi(\mu_\mathcal{Y})$ is focal for the λ-cut of $\mu_\mathcal{Y}$ when $\lambda \leq \theta_\Pi$).
- the focal sets in $\mathcal{F}_D(\mu_\mathcal{Y}) = \{F \in \mathcal{F}(\mu_\mathcal{Y}) : \theta_\Pi < \mu_{\mathcal{Y}\#}(F) \leq \theta_N\}$ are not constrained otherwise.

We can exchange $\mu_\mathcal{X}$ and $\mu_\mathcal{Y}$ above. Moreover, $\mathcal{F}_D(\mu_\mathcal{X}) = \mathcal{F}_D(\mu_\mathcal{Y}) = \emptyset$ if $\theta_\Pi = \theta_N$.

Let $N_\mathcal{X}^\mu$ be the necessity measure such that $N_{\mathcal{X}\#}^\mu(E) = \mu_{\mathcal{X}\#}(E), E \in \mathcal{F}_N(\mu_\mathcal{X})$ (likewise for $N_\mathcal{Y}^\mu$), $\Pi_\mathcal{X}^\mu$ be the possibility measure such that

$$\pi_\mathcal{X}^\mu(x) = \begin{cases} 1 & \text{if } \mu_{\mathcal{X}\#}(\{x\}) = \theta_N, \\ \mu_{\mathcal{X}\#}(\{x\}) & \text{if } \mu_{\mathcal{X}\#}(\{x\}) < \theta_N. \end{cases}$$

We have that $\theta_N = \max\{\mu_{\mathcal{Y}\#}(F) : F \in \mathcal{F}_D(\mu_\mathcal{X})\}$. Let $\kappa_\mathcal{Y}^\mu$ be the capacity with qualitative Möbius transform defined by

$$\kappa_\#^\mu(F) = \begin{cases} 1 & \text{if } \mu_{\mathcal{Y}\#}(F) = \theta_N, F \in \mathcal{F}_D(\mu_\mathcal{Y}), \\ \mu_{\mathcal{Y}\#}(F) & \text{if } \mu_{\mathcal{Y}\#}(F) < \theta_D, F \in \mathcal{F}_D(\mu_\mathcal{Y}), \\ 0 & \text{otherwise.} \end{cases}$$ Finally let $\Pi_\mathcal{Y}$ be the

possibility measure such that $\pi_\mathcal{Y}^\mu(y) = \begin{cases} 1 & \text{if } \mu_{\mathcal{Y}\#}(\{y\}) = \theta_\Pi, \\ \mu_{\mathcal{Y}\#}(\{y\}) & \text{if } \mu_{\mathcal{Y}\#}(\{y\}) < \theta_\Pi \end{cases}$. Concluding:

Corollary 5. *Up to exchanging $\mathcal{X}$ and $\mathcal{Y}$, $\mu_\mathcal{X}$ and $\mu_\mathcal{Y}$ commute if and if they are of the form $\mu_\mathcal{X}(A) = \max(N_\mathcal{X}^\mu(A), \min(\theta_N, \Pi_\mathcal{X}^\mu(A)))$; $\mu_\mathcal{Y}(B) = \max(N_\mathcal{Y}^\mu(B), \min(\theta_N, \kappa_\mathcal{Y}^\mu(B)), \min(\theta_\Pi, \Pi_\mathcal{Y}^\mu(B)))$.*

These expressions provide a convenient tool for explicitly constructing commuting capacities.

5   Conclusion

In this paper we have provided a characterization of capacities such that the Sugeno integrals induced for them commute, based on the Boolean capacities obtained as their cuts. We can see that the cut-worthy property of min and max is instrumental for obtaining this result. Hence it cannot be simply extended to more general integrals [12], involving operations other than min and max. Contrary to the numerical case where only regular expectations commute (in the setting of decision under risk), the commutation of Sugeno integrals is not ensured only by possibility measures, nor by necessity measures: other, rather special, capacities (their cuts must be Boolean possibility measures, necessity measures or Dirac functions) ensure commutation. In the future, we should find a decision-theoretic setting with axioms implying that uncertainty and agent importance can be represented by commuting capacities, which would highlight the practical significance of our results. Finally, at the theoretical level, one should study conditions for which a standard Sugeno integral on the 2D space $\mathcal{X} \times \mathcal{Y}$ is equal to one of, or both, double integrals with respect to the projections of the 2D capacity.

Acknowledgements. This work is supported by ANR-11-LABX-0040-CIMI (Centre International de Mathématiques et d'Informatique) within the program ANR-11-IDEX-0002-02, project ISIPA.

References

1. Behrisch, M., Couceiro, M., Kearnes, K.A., Lehtonen, E., Szendrei, A.: Commuting polynomial operations of distributive lattices. Order **29**(2), 245–269 (2012)
2. Ben Amor, N., Essghaier, F., Fargier, H.: Solving multi-criteria decision problems under possibilistic uncertainty using optimistic and pessimistic utilities. In: Laurent, A., Strauss, O., Bouchon-Meunier, B., Yager, R.R. (eds.) IPMU 2014. CCIS, vol. 444, pp. 269–279. Springer, Cham (2014). https://doi.org/10.1007/978-3-319-08852-5_28
3. Ben Amor, N., Essghaier, F., Fargier, H.: Décision collective sous incertitude possibiliste. Principes et axiomatisation. Revue d'Intelligence Artificielle **29**(5), 515–542 (2015)
4. Ben Amor, N., Essghaier, F., Fargier, H.: Egalitarian collective decision making under qualitative possibilistic uncertainty: principles and characterization. In: Proceedings of AAAI 2015, pp. 3482–3488. AAAI Press (2015)
5. Dubois, D., Prade, H.: Possibility Theory. Plenum Press, New York (1988)
6. Dubois, D., Prade, H., Sabbadin, R.: Decision theoretic foundations of qualitative possibility theory. Eur. J. Oper. Res. **128**, 459–478 (2001)
7. Goodstein, R.L.: The solution of equations in a lattice. Proc. Roy. Soc. Edinb. Sect. A **67**, 231–242 (1967)
8. Grabisch, M.: The Möbius transform on symmetric ordered structures and its application to capacities on finite sets. Discrete Math. **287**, 17–34 (2004)
9. Halas, R., Mesiar, R., Pocs, J., Torra, V.: A note on discrete Sugeno integrals: composition and associativity. Fuzzy Sets Syst. (2018, to appear)
10. Harsanyi, J.: Cardinal welfare, individualistic ethics, and interpersonal comparisons of utility. J Polit. Econ. **63**, 309–321 (1955)
11. Marichal, J.-L.: On Sugeno integrals as an aggregation function. Fuzzy Sets Syst. **114**(3), 347–365 (2000)
12. Narukawa, Y., Torra, V.: Multidimensional generalized fuzzy integral. Fuzzy Sets Syst. **160**(6), 802–815 (2009)
13. Myerson, R.: Utilitarianism, egalitarianism, and the timing effect in social choice problems. Econometrica **49**, 883–97 (1981)

14. Sugeno, M.: Theory of fuzzy integrals and its applications. Ph.D. thesis. Tokyo Institute of Technology, Tokyo (1974)
15. Sugeno,M.: Fuzzy measures and fuzzy integrals: a survey. In: Gupta, M.M., Saridis, G.N., Gaines, B.R. (eds.) Fuzzy Automata and Decision Processes, North-Holland, pp. 89–102 (1977)

Characterization of k-Choquet Integrals

Ľubomíra Horanská and Zdenko Takáč[(✉)]

Institute of Information Engineering, Automation and Mathematics,
Faculty of Chemical and Food Technology, Slovak University of Technology
in Bratislava, Radlinského 9, 812 37 Bratislava, Slovak Republic
{lubomira.horanska,zdenko.takac}@stuba.sk

Abstract. In the present paper we characterize the class of all n-ary k-Choquet integrals and we find a minimal subset of points in the unit hypercube, the values on which fully determine the k-Choquet integral.

Keywords: Aggregation function · Choquet integral · k-additivity
Multicriteria decision making · Comonotone additivity

1 Introduction

The problem of finding an appropriate model of aggregation is crucial for multicriteria decision making. It is desirable to handle aggregated data properly, but with a model which is as simple as possible. One of the simplest models of aggregation on the unit hypercube $[0,1]^n$ is an additive aggregation function where the values of aggregation at $n+1$ points fully determine the aggregation function on the whole domain. Unfortunately, additive aggregation functions are insufficient to model even quite simple preferences, although some of these situations can be treated with the Choquet integrals [3, Example 7]. For the Choquet integral defined on the unit hypercube $[0,1]^n$ full information is contained in the set of values at points $\{0,1\}^n$ corresponding to its related capacity. On the other hand, k-additive aggregation functions recently introduced by Kolesárová et al. [5] yield more complex models (still fully determined by a relatively small set of values of the aggregation function) which can handle many situations where both the additive aggregation functions and the Choquet integrals fail. The authors have constructed the class of k-additive aggregation functions as a natural extension of the class of k-additive capacities [2,6].

The Choquet integrals can be regarded as a generalization of additive aggregation functions replacing the requirement of additivity by that of comonotone additivity. Kolesárová et al. [5] have defined (but not studied) the k-Choquet integrals as an analogous generalization of the k-additive aggregation functions.

The aim of the present paper is to characterize the class of all n-ary k-Choquet integrals and to find a minimal subset of points in the unit hypercube, the values at which fully determine the k-Choquet integral.

The paper is organized as follows. In Sect. 2, we recall basic notions needed throughout the paper. In Sect. 3, we start with definition of the n-ary k-Choquet

© Springer Nature Switzerland AG 2018
V. Torra et al. (Eds.): MDAI 2018, LNAI 11144, pp. 64–76, 2018.
https://doi.org/10.1007/978-3-030-00202-2_6

integrals denoting class of all n-ary k-Choquet integrals by $\mathcal{C}h_{k,n}$. Then we characterize classes $\mathcal{C}h_{k,1}, \mathcal{C}h_{2,2}, \mathcal{C}h_{2,n}$ and we finish with $\mathcal{C}h_{k,n}$. We also compare the minimal sets of points in the unit hypercube fully determining n-ary k-additive aggregation functions and k-Choquet integrals. Finally, some concluding remarks are provided.

2   Preliminaries

In this section we recall some definitions and results which will be used in the sequel. We also fix the notation, mostly according to [4], wherein more information concerning the theory of aggregation functions can be found.

Let $n \in \mathbb{N}$. A finite set $\{1, \ldots, n\}$ will be denoted by $[n]$. We will denote vectors $(x_1, \ldots, x_n) \in [0,1]^n$ by bold symbols $\mathbf{x}$, in particular, vectors $(0, \ldots, 0)$ and $(1, \ldots, 1)$ by $\mathbf{0}$ and $\mathbf{1}$ respectively. For any $K \subseteq [n]$ and any $\mathbf{x}, \mathbf{y} \in [0,1]^n$, we will denote by $\mathbf{x}_K\mathbf{y}$ the vector whose ith coordinate is x_i, if $i \in K$ and y_i otherwise. For any $k \in \mathbb{N}$ and $x \in [0,1]$, we set $(k \odot x) := (x, \ldots, x)$ (k times).

Definition 1. *An n-ary aggregation function is a function $F \colon [0,1]^n \to [0,1]$ satisfying*

(i) $F(\mathbf{0}) = 0$, $F(\mathbf{1}) = 1$,
(ii) F is monotone in each variable.

Definition 2 ([5]). *An n-ary aggregation function $F \colon [0,1]^n \to [0,1]$ is defined to be k-additive if and only if it holds*

$$\sum_{i=1}^{k+1}(-1)^{k+1-i}\left(\sum_{\substack{I \subseteq \{1,\ldots,k+1\} \\ |I|=i}} F\left(\sum_{j \in I}\mathbf{x}_j\right)\right) = 0 \tag{1}$$

for all $(k+1)$-tuples of vectors $\mathbf{x}_1, \ldots, \mathbf{x}_{k+1} \in [0,1]^n$ with $\sum_{i=1}^{k+1}\mathbf{x}_i \in [0,1]^n$.

The class of all n-ary k-additive aggregation functions is denoted by $\mathcal{A}_{k,n}$. The following characterization of the class $\mathcal{A}_{k,n}$ was given in [5].

Theorem 1 ([5]). *A function $F \colon [0,1]^n \to [0,1]$ is a k-additive aggregation function, i.e., $F \in \mathcal{A}_{k,n}$, if and only if there are appropriate constants (ensuring the boundary condition $F(\mathbf{1}) = 1$ and the monotonicity of F) such that for all $(x_1, \ldots, x_n) \in [0,1]^n$,*

$$F(x_1, \ldots, x_n) = \sum_{i=1}^{k}\left(\sum_{1 \le j_{1,i} \le \ldots \le j_{i,i} \le n} a_{j_{1,i},\ldots,j_{i,i}}\left(\prod_{p=1}^{i} x_{j_{p,i}}\right)\right),$$

i.e., F is a polynomial with degree not exceeding k.

Denote $\mathfrak{S}_{[n]}$ the set of all permutations of the set $[n]$.

Definition 3. *Vectors* $\mathbf{x}_1 = (x_{1,1}, \ldots, x_{1,n}), \ldots, \mathbf{x}_k = (x_{k,1}, \ldots, x_{k,n}) \in [0, 1]^n$ *are comonotone if and only if there exists a permutation* $\sigma \in \mathfrak{S}_{[n]}$ *such that* $x_{i,\sigma(1)} \leq \ldots \leq x_{i,\sigma(n)}$ *for all* $i \in [k]$.

Let $n \geq 2$ and let $\sigma \in \mathfrak{S}_{[n]}$ be a fixed permutation. We set

$$S_\sigma = \{\mathbf{x} \in [0, 1]^n \mid x_{\sigma(1)} \leq \ldots \leq x_{\sigma(n)}\}.$$

A subset S_σ of the unit hypercube $[0, 1]^n$ will be called a simplex. Clearly, all vectors in a simplex S_σ are pairwise comonotone and $\bigcup_{\sigma \in \mathfrak{S}_{[n]}} S_\sigma = [0, 1]^n$.

3   Characterization of k-Choquet Integrals

In this section we describe the notion of k-Choquet integrals and characterize the classes of n-ary k-Choquet integrals starting with particular values of n and k and finishing with general n-ary k-Choquet integrals. Note that the k-Choquet integrals must not be confused with the Choquet integrals based on the k-additive capacities.

3.1 k-Choquet Integrals

Definition 4. *An* n-ary *aggregation function* $F\colon [0, 1]^n \to [0, 1]$ *is defined to be a* k-Choquet integral *if and only if it is comonotone* k-additive, *i.e., Eq. (1) holds for all* $(k + 1)$-tuples *of comonotone vectors* $\mathbf{x}_1, \ldots, \mathbf{x}_{k+1} \in [0, 1]^n$ *with* $\sum_{i=1}^{k+1} \mathbf{x}_i \in [0, 1]^n$.

The class of all n-ary k-Choquet integrals will be denoted by $\mathcal{C}h_{k,n}$.

It can be checked that if an n-ary aggregation function is a k-Choquet integral then it is also p-Choquet integral for any integer $p > k$, i.e., it holds

$$\mathcal{C}h_{1,n} \subseteq \ldots \subseteq \mathcal{C}h_{k,n} \subseteq \mathcal{C}h_{k+1,n} \subseteq \ldots.$$

Clearly, the standard n-ary Choquet integrals form a class $\mathcal{C}h_{1,n}$.

In what follows, we are going to characterize k-Choquet integrals and to study a minimal set of points from $[0, 1]^n$ needed for determining a k-Choquet integral. Since the general characterization is based on an induction, we proceed stepwise, that is, we begin with the class $\mathcal{C}h_{k,1}$, then $\mathcal{C}h_{2,2}$, $\mathcal{C}h_{2,n}$ and finally $\mathcal{C}h_{k,n}$.

3.2   The Class $\mathcal{C}h_{k,1}$

Since for $n = 1$ the comonotone k-additivity coincides with the k-additivity, the class of all unary k-Choquet integrals $\mathcal{C}h_{k,1}$ coincides with the class of all unary k-additive functions $A_{k,1}$. Using the characterization of the class given by Theorem 1 we obtain the following assertion.

Proposition 1. *Let k be a positive integer. A function $F\colon [0,1] \to [0,1]$ is a k-Choquet integral, i.e., $F \in Ch_{k,1}$ if and only if F is a non-decreasing function on $[0,1]$ given by:*

$$F(x) = a_1 x + a_2 x^2 + \cdots + a_k x^k,$$

where $\sum_{i=1}^{k} a_i = 1$. Moreover, for $k \geq 2$ each coefficient a_i can be expressed by means of the values of F at the points of the set $\{\frac{1}{k}, \frac{2}{k}, \ldots, \frac{k-1}{k}\}$.

Proof. We only need to prove that all coefficients a_i can be expressed by means of the values of F at the points $\frac{1}{k}, \frac{2}{k}, \ldots, \frac{k-1}{k}$. That can be obtained by solving the system of linear equations $F(\frac{i}{k}) = \sum_{j=1}^{k} a_j \left(\frac{i}{k}\right)^j$, $i = 1, \ldots, k-1$ together with

$$\sum_{i=1}^{k} a_i = 1.$$

Remark 1. It is clear, that the set of points fully determining an operator F in the previous Proposition is not unique. A different choice of the set (preserving number of points) can be done according to a problem to be solved.

Example 1. (i) Let us consider $F \in Ch_{2,1}$, i.e., $F(x) = ax^2 + (1-a)x$. Then the coefficient a can be expressed by the value $F(1/2)$, in particular, $F(1/2) = 1/2 - 1/4a$, thus $a = 2 - 4F(1/2)$. Since the monotonicity of F is satisfied if and only if $F'(0) \geq 0$ and $F'(1) \geq 0$, we have $a \in [-1,1]$ and consequently $F(1/2) \in [1/4, 3/4]$. Conclusion: each unary 2-Choquet integral F is fully determined by a single value $F(1/2)$:

$$F(x) = (2 - 4F(1/2))x^2 + (4F(1/2) - 1)x \quad \text{where} \quad F(1/2) \in [1/4, 3/4].$$

(ii) For $Ch_{3,1}$ we have $F(x) = ax^3 + bx^2 + (1-a-b)x$ and a, b can be expressed by the values $F(1/3)$ and $F(2/3)$. Since

$$F(1/3) = \frac{1}{27}a + \frac{1}{9}b + \frac{1}{3}(1-a-b), \quad F(2/3) = \frac{8}{27}a + \frac{4}{9}b + \frac{2}{3}(1-a-b),$$

we have

$$a = \frac{27}{2}F(1/3) - \frac{27}{2}F(2/3) + \frac{9}{2}, \quad b = -\frac{45}{2}F(1/3) + 18F(2/3) - \frac{9}{2}.$$

Moreover, F is non-decreasing on $[0,1]$ if and only if one of the following three conditions is satisfied:

1. $a \leq 0$ and $F'(0) \geq 0$ and $F'(1) \geq 0$;
2. $a > 0$ and F' has at most one real root (i.e. $4b^2 - 12a(1-a-b) \leq 0$);
3. $a > 0$ and F' has two different real roots (i.e. $4b^2 - 12a(1-a-b) > 0$) and $F'(0) \geq 0$ and $F'(1) \geq 0$ and $F''(0) \cdot F''(1) \geq 0$.

From the three items we obtain the following restrictions for $F(1/3)$ and $F(2/3)$:

1. $F(2/3) \geq F(1/3)+1/3$, $F(2/3) \leq 2F(1/3)+2/9$ and $F(2/3) \leq (1/2)F(1/3)+ 11/18$ (see the blue area in Fig. 1);
2. $e_1 \leq F(2/3) \leq e_2$ (see the green area, ellipse, in Fig. 1), where

$$e_{1,2} = (1/42)\left(9 + 39F(1/3) \mp \sqrt{3}\sqrt{66F(1/3) - 81F^2(1/3) - 1}\right)$$

form a "lower and upper boundary line" of the ellipse;
3. $F(2/3) < F(1/3) + 1/3$ and $F(2/3) \geq e_2$ and $F(2/3) \leq 2F(1/3) + 2/9$ and $F(2/3) \leq (1/2)F(1/3)+11/18$ and $-81(2+4F(1/3)-5F(2/3))(1+5F(1/3)- 4F(2/3)) \geq 0$ (see the red area in Fig. 1).

Conclusion: each unary 3-Choquet integral F is fully determined by the pair of values $F(1/3), F(2/3)$:

$$F(x) = \left(\frac{27}{2}F\left(\frac{1}{3}\right) - \frac{27}{2}F\left(\frac{2}{3}\right) + \frac{9}{2}\right)x^3 + \left(-\frac{45}{2}F\left(\frac{1}{3}\right) + 18F\left(\frac{2}{3}\right) - \frac{9}{2}\right)x^2$$
$$+ \left(\frac{18}{2}F\left(\frac{1}{3}\right) + \frac{9}{2}F\left(\frac{2}{3}\right) + 1\right)x$$

where $F(1/3), F(2/3)$ satisfy

$$\begin{cases} F(2/3) \in [e_1, e_2], & \text{if } F(1/3) \in \left[\frac{11-4\sqrt{7}}{27}, \frac{1}{27}\right] \cup \left[\frac{19}{27}, \frac{11+4\sqrt{7}}{27}\right], \\ F(2/3) \in [e_2, 2F(1/3) + 2/9], & \text{if } F(1/3) \in \left[\frac{1}{27}, \frac{7}{27}\right], \\ F(2/3) \in [e_2, 1/2F(1/3) + 11/18], & \text{if } F(1/3) \in \left[\frac{7}{27}, \frac{19}{27}\right]. \end{cases}$$

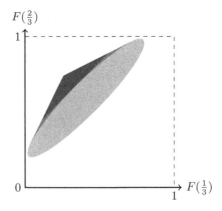

Fig. 1. The points of colored areas correspond to all admissible pairs $(F(1/3), F(2/3))$ determining an unary 3-Choquet integral, see Example 1 (ii). (Color figure online)

3.3   The Class $\mathcal{C}h_{2,2}$

Theorem 2. *A function $F\colon [0,1]^2 \to [0,1]$ is a 2-Choquet integral, i.e., $F \in \mathcal{C}h_{2,2}$, if and only if F is given by:*

$$F(x,y) = \begin{cases} F_{21}(x,y), & \text{if } (x,y) \in S_{21}; \\ F_{12}(x,y), & \text{if } (x,y) \in S_{12}; \end{cases} \tag{2}$$

where $S_{21} = \{(x,y)\,|\,x \geq y\}$, $S_{12} = \{(x,y)\,|\,x \leq y\}$ and

- $F_{21}(x,y) = a_1 x + b_1 y + c_1 x^2 + d_1 y^2 + e_1 xy$
  where
  $a_1 + b_1 + c_1 + d_1 + e_1 = 1, \qquad a_1 \geq 0, \qquad b_1 \geq 0, \qquad a_1 + 2c_1 \geq 0,$
  $b_1 + e_1 \geq 0, \qquad a_1 + 2c_1 + e_1 \geq 0, \qquad b_1 + 2d_1 + e_1 \geq 0;$
- $F_{12}(x,y) = a_2 x + b_2 y + c_2 x^2 + d_2 y^2 + e_2 xy$
  where
  $a_2 + b_2 + c_2 + d_2 + e_2 = 1, \qquad a_2 \geq 0, \qquad b_2 \geq 0, \qquad a_2 + e_2 \geq 0,$
  $b_2 + 2d_2 \geq 0, \qquad a_2 + 2c_2 + e_2 \geq 0, \qquad b_2 + 2d_2 + e_2 \geq 0;$
- $F_{21}(x,x) = F_{12}(x,x)$ *for all $x \in [0,1]$.*

Proof. ($\Leftarrow$) It is easy to check that each function F given by (2) is a 2-Choquet integral.

($\Rightarrow$) We will only deal with F_{21}, the case of F_{12} is similar. Let us define a function $U : S_{21} \to [0,1]$ as $U(x,y) = F_{21}(x,y) - F_{21}(x-y,0) - F_{21}(y,y)$.

Let $\mathbf{x} = \mathbf{y} = \left(\frac{x}{2} - \frac{y}{2}, 0\right)$, $\mathbf{z} = (y,y)$. From 2-additivity of F_{21} we have

$$F_{21}(x,y) - 2F_{21}\left(\frac{x}{2} + \frac{y}{2}, y\right) - F_{21}(x-y,0) + 2F_{21}\left(\frac{x}{2} - \frac{y}{2}, 0\right) + F_{21}(y,y) = 0,$$

hence

$$U(x,y) = -2F_{21}(y,y) + 2F_{21}\left(\frac{x}{2} + \frac{y}{2}, y\right) - 2F_{21}\left(\frac{x}{2} - \frac{y}{2}, 0\right) = 2U\left(\frac{x}{2} + \frac{y}{2}, y\right)$$

and consequently

$$U\left(\frac{x+y}{2}, y\right) = \frac{1}{2}U(x,y), \qquad \text{for all } (x,y) \in S_{21}. \tag{3}$$

Now, let $q \in [0,1]$ be fixed and let $h(x) = U(x+q,q)$ for all $x \in [0,1-q]$. Then, by (3), we have

$$h\left(\frac{x-q}{2}\right) = U\left(\frac{x+q}{2}, q\right) = \frac{1}{2}U(x,q) = \frac{1}{2}h(x-q),$$

for all $x \in [q,1]$, which means that h is linear. Moreover, $h(1-q) = U(1,q)$ and $h(0) = 0$, thus $h(x) = \frac{U(1,q)}{1-q}x$ for all $x \in [0,1-q]$ and $U(x,q) = \frac{U(1,q)}{1-q}(x-q)$, for all $x \in [q,1]$. Hence

$$U(x_2,q) = \frac{x_2 - q}{x_1 - q}U(x_1,q), \qquad \text{for all } x_1, x_2 \in [q,1]. \tag{4}$$

Similarly, for $\mathbf{x} = \mathbf{y} = \left(\frac{y}{2}, \frac{y}{2}\right)$, $\mathbf{z} = (x - y, 0)$ we obtain

$$U\left(x - \frac{y}{2}, \frac{y}{2}\right) = \frac{1}{2}U(x, y), \qquad \text{for all } (x, y) \in S_{21}. \tag{5}$$

Let $g(y) = U(y + 1 - p, y)$, for some fixed $p \in [0, 1]$ and every $y \in [0, p]$. Then, by (5), $g\left(\frac{y}{2}\right) = \frac{1}{2}g(y)$ for all $y \in [0, p]$ and therefore g is linear. Moreover, $g(0) = 0$ and $g(p) = U(1, p)$, thus $g(y) = U(1, p)y$ for all $y \in [0, p]$ which implies $g(y_2) = \frac{y_2}{y_1}g(y_1)$ and finally

$$U(y_2 + 1 - p, y_2) = \frac{y_2}{y_1}U(y_1 + 1 - p, y_1), \qquad \text{for all } y_1, y_2 \in [0, p]. \tag{6}$$

Let $x, y \in S_{21}$ be such that $y \geq x - \frac{1}{2}$ and $p = y - x + 1$. Then

$$U(x, y) = U(y - p + 1, y) = 2yU\left(\frac{1}{2} + 1 - p, \frac{1}{2}\right)$$

$$= 2yU\left(x - y + \frac{1}{2}, \frac{1}{2}\right) = 4y(x - y)U\left(1, \frac{1}{2}\right), \tag{7}$$

where the second equality follows from (6) for $y_2 = y$, $y_1 = \frac{1}{2}$ and the last from (4) for $x_1 = 1$, $x_2 = x - y + \frac{1}{2}$ and $q = \frac{1}{2}$.

Let $x, y \in S_{21}$ be such that $y \leq x - \frac{1}{2}$ and $q = y$. Then

$$U(x, y) = 2(x - y)U\left(\frac{1}{2} + y, y\right) = 4y(x - y)U\left(1, \frac{1}{2}\right), \tag{8}$$

where the first equality follows from (4) for $x_2 = x$, $x_1 = \frac{1}{2} + q$ and the second from (6) for $y_2 = y$, $y_1 = p = \frac{1}{2}$ and $x_1 = 1$.

So, by (7) and (8), the following holds for all $x, y \in S_{21}$:

$$U(x, y) = 4y(x - y)U\left(1, \frac{1}{2}\right). \tag{9}$$

Since unary functions $F_{21}(x, 0)$ and $F_{21}(x, x)$ are 2-additive, using Example 1 we can show that F_{21} is a polynomial of degree not exceeding 2:

$$F_{21}(x, y) = F_{21}(x - y, 0) + F_{21}(y, y) + U(x, y)$$
$$= F_{21}(1, 0)\left((1 - \alpha)(x - y) + \alpha(x - y)^2\right) + (1 - \beta)y + \beta y^2$$
$$+ 4y(x - y)\left(F_{21}\left(1, \frac{1}{2}\right) - F_{21}\left(\frac{1}{2}, 0\right) - F_{21}\left(\frac{1}{2}, \frac{1}{2}\right)\right)$$

$$= a_1 x + b_1 y + c_1 x^2 + d_1 y^2 + e_1 xy. \tag{10}$$

The conditions imposed on $a_1, \ldots, e_1$ follow from the boundary condition $F_{21}(1, 1) = 1$ and monotonicity of F_{21}, that is the partial derivatives $\frac{\partial F_{21}}{\partial x}(0, 0)$, $\frac{\partial F_{21}}{\partial x}(1, 0)$, $\frac{\partial F_{21}}{\partial x}(1, 1)$, $\frac{\partial F_{21}}{\partial y}(0, 0)$, $\frac{\partial F_{21}}{\partial y}(1, 0)$ and $\frac{\partial F_{21}}{\partial y}(1, 1)$ are non-negative.

Remark 2. According to the proof of Theorem 2, a binary 2-Choquet integral $F\colon [0,1]^2 \to [0,1]$ is in S_{21} fully characterized by 4 values $F(1/2,0)$, $F(1,0)$, $F(1/2,1/2)$, $F(1,1/2)$ and in S_{12} by 4 values $F(0,1/2)$, $F(0,1)$, $F(1/2,1/2)$, $F(1/2,1)$, i.e., F is fully characterized by values at points $(x,y) \in \{0,1/2,1\}^2$, see Fig. 2 (recall that $F(0,0) = 0$ and $F(1,1) = 1$). From (10) it follows:

$$a_1 = (1-\alpha)F_{21}(1,0),$$
$$b_1 = -(1-\alpha)F_{21}(1,0) + (1-\beta),$$
$$c_1 = \alpha F_{21}(1,0),$$
$$d_1 = \alpha F_{21}(1,0) + \beta - 4\left(F_{21}\left(1,\frac{1}{2}\right) - F_{21}\left(\frac{1}{2},0\right) - F_{21}\left(\frac{1}{2},\frac{1}{2}\right)\right),$$
$$e_1 = -2\alpha F_{21}(1,0) + 4\left(F_{21}\left(1,\frac{1}{2}\right) - F_{21}\left(\frac{1}{2},0\right) - F_{21}\left(\frac{1}{2},\frac{1}{2}\right)\right),$$

where α depends on $F_{21}\left(\frac{1}{2},0\right)$ and β on $F_{21}\left(\frac{1}{2},\frac{1}{2}\right)$, see Example 1 and recall that unary functions $F_{21}(x,0)$ and $F_{21}(x,x)$ are 2-additive.

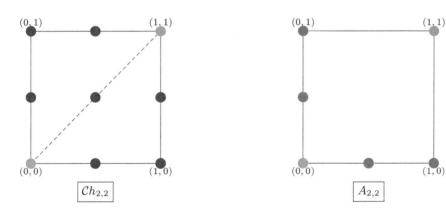

Fig. 2. On the left: Binary 2-Choquet integral F is fully determined by the values at the seven blue points apart from the points **0**, **1**, see Remark 2. Compare with the situation for $A_{2,2}$ on the right where we only need values at four red points, see [5, Proof of Proposition 2]. (Color figure online)

3.4   The Class $\mathcal{Ch}_{2,n}$

Theorem 3. *Let n be a positive integer. A function $F\colon [0,1]^n \to [0,1]$ is a 2-Choquet integral, i.e., $F \in \mathcal{Ch}_{2,n}$, if and only if there exist $n!$ functions F_{σ_l}, $l = 1,\ldots,n!$ corresponding to all permutations $\sigma_l \in \mathfrak{S}_{[n]}$ such that the following conditions are satisfied*

(i) $F(\mathbf{x}) = F_{\sigma_l}(\mathbf{x})$, *for all* $\mathbf{x} \in S_{\sigma_l}$,
(ii) $F_{\sigma_i}(\mathbf{x}) = F_{\sigma_j}(\mathbf{x})$ *for all* $\mathbf{x} \in S_{\sigma_i} \cap S_{\sigma_j}$ *and* $i,j \in \{1,\ldots,n!\}$;

(iii) $F_{\sigma_l}(x_1, \ldots, x_n) = \sum\limits_{1 \leq i \leq j \leq n} a_{ij}^{\sigma_l} x_i x_j + \sum\limits_{i=1}^{n} b_i^{\sigma_l} x_i,$ where coefficients $a_{ij}^{\sigma_l}, b_i^{\sigma_l}$
satisfy for each $l \in \{1, \ldots, n!\}$ the following conditions:

- $\sum\limits_{1 \leq i \leq j \leq n} a_{ij}^{\sigma_l} + \sum\limits_{i=1}^{n} b_i^{\sigma_l} = 1$

- $b_r^{\sigma_l} + \sum\limits_{i=1}^{s} c_{\sigma_l(i),r}^{\sigma_l} \geq 0,$ where

$$c_{ir}^{\sigma_l} = \begin{cases} a_{ir}^{\sigma_l}, & \text{if } i < r, \\ 2a_{rr}^{\sigma_l}, & \text{if } i = r, \\ a_{ri}^{\sigma_l}, & \text{if } i > r, \end{cases}$$

for all $r, s \in [n]$,
- $b_r^{\sigma_l} \geq 0$ for all $r \in [n]$.

Proof. ($\Leftarrow$) It is the matter of simple computation to check that each function F fullfiling (i)-(iii) is a 2-Choquet integral.

($\Rightarrow$) The proof is done by induction on n. We have proved the cases $n = 1$ and $n = 2$ in Proposition 1 and Theorem 2, respectively. Now, suppose that $n > 2$ and the assertion holds for all $k < n$. Let $\sigma_l \in \mathfrak{S}_{[n]}$, and let $\mathbf{a} = (a_1, \ldots, a_n) \in S_{\sigma_l}$, i.e., $a_{\sigma_l(1)} \leq \ldots \leq a_{\sigma_l(n)}$. For simplicity, we will suppose that σ_l is the identity, the same can be done for any permutation. Let

$$\mathbf{x} = (n \odot a_1),$$
$$\mathbf{y} = (0, (n-1) \odot (a_2 - a_1)),$$
$$\mathbf{z} = (0, 0, a_3 - a_2, \ldots, a_n - a_2).$$

Then $\mathbf{x} + \mathbf{y} + \mathbf{z} = \mathbf{a}$, and

$$\mathbf{x} + \mathbf{y} = (a_1, (n-1) \odot a_2),$$
$$\mathbf{x} + \mathbf{z} = (2 \odot a_1, a_3 - a_2 + a_1, \ldots a_n - a_2 + a_1),$$
$$\mathbf{y} + \mathbf{z} = (0, a_2 - a_1, a_3 - a_1, \ldots, a_n - a_1).$$

Since $\mathbf{x}, \mathbf{y}, \mathbf{z} \in S_{\sigma_l}$, also $\mathbf{x} + \mathbf{y} + \mathbf{z}, \mathbf{x} + \mathbf{y}, \mathbf{x} + \mathbf{z}, \mathbf{y} + \mathbf{z} \in S_{\sigma_l}$, hence, from comonotone 2-additivity of F it follows:

$$F_{\sigma_l}(\mathbf{a}) = F_{\sigma_l}(\mathbf{x} + \mathbf{y}) + F_{\sigma_l}(\mathbf{x} + \mathbf{z}) + F_{\sigma_l}(\mathbf{y} + \mathbf{z}) - F_{\sigma_l}(\mathbf{x}) - F_{\sigma_l}(\mathbf{y}) - F_{\sigma_l}(\mathbf{z}).$$

Now the fact that F_{σ_l} is a polynomial not exceeding degree 2 follows from the observations:

- $F_{\sigma_l}(\mathbf{x} + \mathbf{y})$ is a binary 2-additive aggregation function,
- $F_{\sigma_l}(\mathbf{x} + \mathbf{z})$ and $F_{\sigma_l}(\mathbf{y} + \mathbf{z})$ are $(n-1)$-ary 2-additive aggregation functions,
- $F_{\sigma_l}(\mathbf{z})$ is $(n-2)$-ary 2-additive aggregation function,
- $F_{\sigma_l}(\mathbf{x})$ and $F_{\sigma_l}(\mathbf{y})$ are unary 2-additive aggregation functions,

The partial derivatives of F_{σ_l} are following:

$$\frac{\partial F_{\sigma_l}}{\partial x_r} = b_r^{\sigma_l} + \sum_{i=1}^{n} c_{ir}^{\sigma_l} x_i, \qquad \text{where} \qquad c_{ir}^{\sigma_l} = \begin{cases} a_{ir}^{\sigma_l}, & \text{if } i < r, \\ 2a_{rr}^{\sigma_l}, & \text{if } i = r, \\ a_{ri}^{\sigma_l}, & \text{if } i > r, \end{cases}$$

The conditions imposed on coefficients $a_{ij}^{\sigma_l}, b_i^{\sigma_l}$, for each σ_l, follow from the boundary condition $F_{\sigma_l}(\mathbf{1}) = 1$ and the monotonicity of F_{σ_l}, that is, all partial derivatives in the all vertices of the considered simplex S_{σ_l} are non-negative.

Remark 3. (i) A ternary 2-Choquet integral $F\colon [0,1]^3 \to [0,1]$ is in arbitrary simplex S_σ fully determined by the values $F(x_1, x_2, x_3)$ where $x_1, x_2, x_3 \in \{0, 1/2, 1\}$ and $x_{\sigma(1)} \le x_{\sigma(2)} \le x_{\sigma(3)}$. Hence, F on the whole domain is fully characterized by the values at the points $(x_1, x_2, x_3) \in \{0, 1/2, 1\}^3$, see Fig. 3.

(ii) In general, an n-ary 2-Choquet integral $F\colon [0,1]^n \to [0,1]$ is fully characterized by 3^n values at the points of the set $\{0, 1/2, 1\}^n$.

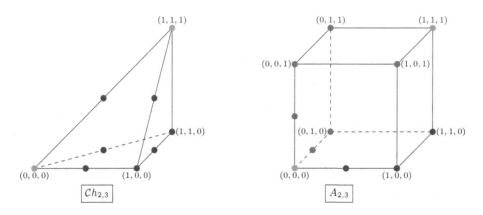

Fig. 3. On the left: Ternary 2-Choquet integral F is in the simplex S_{321} fully determined by values in the seven blue points apart from the points $\mathbf{0}, \mathbf{1}$, see Remark 3. Compare with the situation for $A_{2,n}$ on the right where we only need values at the three blue points from the simplex, see [5, Proof of Theorem 2]. (Color figure online)

3.5   The Class $\mathcal{C}h_{k,n}$

Theorem 4. *Let k, n be positive integers. A function $F\colon [0,1]^n \to [0,1]$ is a k-Choquet integral, i.e., $F \in \mathcal{C}h_{k,n}$,*

if and only if there exist $n!$ functions F_{σ_l}, $l = 1, \ldots, n!$ corresponding to all permutations $\sigma_l \in \mathfrak{S}_{[n]}$ such that the following conditions are satisfied

(i) $F(\mathbf{x}) = F_{\sigma_l}(\mathbf{x})$, for all $\mathbf{x} \in S_{\sigma_l}$,
(ii) $F_{\sigma_i}(\mathbf{x}) = F_{\sigma_j}(\mathbf{x})$, for all $\mathbf{x} \in S_{\sigma_i} \cap S_{\sigma_j}$ and $i, j \in \{1, \ldots, n!\}$

(iii) $F_{\sigma_l}(x_1, \ldots, x_n) = \sum\limits_{i=1}^{k} \left(\sum\limits_{1 \le j_{1,i} \le \cdots \le j_{i,i} \le n} a_{j_{1,i},\ldots,j_{i,i}}^{\sigma_l} \left(\prod\limits_{p=1}^{i} x_{j_{p,i}} \right) \right),$

where the coefficients $a_{j_{i,1},\ldots,j_{i,i}}^{\sigma_l}$ satisfy conditions ensuring the monotonicity and

$$\sum\limits_{i=1}^{k} \left(\sum\limits_{1 \le j_{1,i} \le \cdots \le j_{i,i} \le n} a_{j_{i,1},\ldots,j_{i,i}}^{\sigma_l} \right) = 1, \quad \text{for all } l \in [n!].$$

Proof. ($\Leftarrow$) It is the matter of computation to check that each function F given by (i)-(iii) is a k-Choquet integral.

($\Rightarrow$) The proof for $k < n$ is similar to that of Theorem 3. For simplicity, let σ_l be the identity and let $\mathbf{a} = (a_1, \ldots, a_n) \in S_{\sigma_l}$. The same can be done for any permutation $\sigma_l \in \mathfrak{S}_{[n]}$. We define $k+1$ vectors $\mathbf{x}_1, \ldots, \mathbf{x}_{k+1}$, such that $\mathbf{x}_1 + \ldots + \mathbf{x}_{k+1} = \mathbf{a}$, as follows:

$$\begin{aligned}
\mathbf{x}_1 &= (n \odot a_1), \\
\mathbf{x}_2 &= (0, (n-1) \odot (a_2 - a_1)), \\
\mathbf{x}_3 &= (0, 0, (n-2) \odot (a_3 - a_2)), \\
&\vdots \\
\mathbf{x}_k &= ((k-1) \odot 0, (n-k+1) \odot (a_k - a_{k-1})), \\
\mathbf{x}_{k+1} &= (k \odot 0, a_{k+1} - a_k, \ldots, a_n - a_k).
\end{aligned}$$

Then $F(\sum\limits_{j \in K} \mathbf{x}_j)$ for all $K \subsetneq [k+1]$ can be regarded as operators belonging to $Ch_{k,l}$ for some $l < n$. So, using the assumption that the claim holds for any $l < n$ we obtain the assertion.

The proof for $k \ge n$ is similar to that of Theorem 2.

Remark 4. An n-ary k-Choquet integral $F: [0,1]^n \to [0,1]$ is fully characterized by $(k+1)^n$ values $F(x_1, \ldots, x_n)$ at the points $(x_1, \ldots, x_n) \in \{0, \frac{1}{k}, \ldots, \frac{k-1}{k}, 1\}^n$, see Table 1. In other words, each coefficient $a_{j_{i,1},\ldots,j_{i,i}}^{\sigma_l}$ can be expressed by means of the values of F in these points. Note that values at the points $\mathbf{0}, \mathbf{1}$ are given by the boundary conditions.

Table 1. The sets of points and the numbers of points needed to determine functions from $A_{k,n}$ and $Ch_{k,n}$. In the last row, the points and their number for each simplex are given.

	Points	Number of points
$A_{k,n}$	$\{0,1\}^n \cup \{((j/k)_i \mathbf{0}) \mid i \in [n], j \in [k-1]\}$	$2^n + n(k-1)$
$Ch_{k,n}$	$\{0, 1/k, \ldots, (k-1)/k, 1\}^n$	$(k+1)^n$
S_σ	$\{0, 1/k, \ldots, (k-1)/k, 1\}^n \cap S_\sigma$	$\binom{n+k}{k}$

Remark 5. Recall that the k-OWA operators defined in [5] are symmetrized k-additive aggregation functions, i.e., $G : [0,1]^n \to [0,1]$ is a k-OWA operator if and only if there exists a k-additive aggregation function $F : [0,1]^n \to [0,1]$ such that $G(x_1, \ldots, x_n) = F(x_{\sigma(1)}, \ldots, x_{\sigma(n)})$, where σ is a permutation of $[n]$ such that $x_{\sigma(1)} \leq \ldots \leq x_{\sigma(n)}$. It means that the k-OWA operator G is fully determined by its values on the simplex $S_{12\ldots n}$ and can be regarded as a symmetric k-Choquet integral. Therefore, the number of points fully determining k-OWA operator is the same as for a unique simplex in the last row of Table 1.

Example 2. (i) Consider four competitors C_1, C_2, C_3, C_4 and their respective score vectors $(1,0)$, $(0.75, 0.75)$, $(0.75, 0)$, $(0.5, 0.5)$. We want to find such aggregation function that $C_1 \prec C_2$ and $C_4 \prec C_3$. Obviously, in this case neither additive nor comonotone additive aggregation function can be used. But we can use an symmetric 2-additive aggregation function, e.g., $J(x,y) = \frac{1}{2}(x^2 + y^2)$.

(ii) Consider four competitors C_5, C_6, C_7, C_8 and their respective score vectors $(0,1)$, $(0.25, 0.75)$, $(0.25, 1)$, $(0.5, 0.75)$. We want to find such aggregation function that $C_6 \prec C_5$ and $C_7 \prec C_8$. Again, in this case neither additive nor comonotone additive aggregation function can be used. But we can use an 2-additive aggregation function, e.g., $K(x,y) = \frac{1}{3}(2x^2 + y)$.

(iii) Consider competitors C_1-C_8 from (i) and (ii) and preferences from therein. One can use the following 2-Choquet integral as suitable aggregation function

$$M(x,y) = \begin{cases} \frac{1}{2}(x^2 + y^2), & \text{if } x \geq y, \\ \frac{1}{3}(2x^2 + y), & \text{if } x < y. \end{cases}$$

4   Conclusion

We have proved that the class of n-ary k-Choquet integrals consists of all aggregation functions the restriction of which on each simplex is a polynomial with degree not exceeding k and the polynomials agree on the common surfaces of simplices. The k-Choquet integrals can be useful in handling multicriteria decision problems where the standard Choquet integrals or k-additive aggregation functions fail.

Acknowledgement. The authors gratefully acknowledge support of the project VEGA 1/0614/18.

References

1. Choquet, G.: Theory of capacities. Annales de l'institut Fourier **5**, 131–295 (1953–1954)
2. Grabisch, M.: k-order additive discrete fuzzy measures and their representation. Fuzzy Sets Syst. **92**, 167–189 (1997)
3. Grabisch, M., Labreuche, C.: A decade of applications of the Choquet and Sugeno integrals in multicriteria decision aid. Ann. Oper. Res. **175**, 247–290 (2010)

4. Grabisch, M., Marichal, J.-L., Mesiar, R., Pap, E.: Aggregation Functions. Cambridge University Press, Cambridge (2009)
5. Kolesárová, A., Li, J., Mesiar, R.: k-additive aggregation functions and their characterization. Eur. J. Oper. Res. **265**, 985–992 (2018)
6. Mesiar, R.: k-order additivity and maxitivity. Atti del Seminario Matematico e Fisico dell'Universita di Modena **23**, 179–189 (2003)

Event-Based Transformations of Set Functions and the Consensus Requirement

Andrey G. Bronevich[1(✉)] and Igor N. Rozenberg[2]

[1] National Research University Higher School of Economics,
Myasnitskaya 20, 101000 Moscow, Russia
brone@mail.ru
[2] JSC "Research and Design Institute for Information Technology,
Signalling and Telecommunications on Railway Transport",
Orlikov per.5, building 1, 107996 Moscow, Russia
I.Rozenberg@gismps.ru

Abstract. Non-additive measures, capacities or generally set functions are widely used in decision models, data processing and game theory. In these applications we can find many structures identified as linear transformations or linear operators. The most remarkable of them are Choquet integral, Möbius transform, interaction transform, Shapley value. The main goal of the presented paper is to study some of them recently called event-based linear transformations. We describe them considering the set of all possible linear operators as a linear space w.r.t. their linear combinations and compute the dimensions of its some subspaces. We also study the consensus requirement, i.e. we analyze the condition when the linear operator maps one family of non-additive measures to other family.

Keywords: Set functions · Event-based transformations
Linear operators · Consensus requirement

1 Introduction

The theory of non-additive measures has many applications in decision theory [1], cooperative game theory [2] and in various models of uncertainty [3]. In terms of classical algebra useful structures, based on non-additive measures, like Choquet integral[1] [4], Möbius transform [5,6], interaction transform [6], and Shapley value [6] are linear transformations or linear operators. The aim of this paper to investigate various ways for defining linear transformations, and investigate their properties, which can be useful in applications. We study the conditions, under which the generated set function preserves properties of the

---

[1] To see Choquet integral as a linear transformation of measures, one need to fix the integrand.

© Springer Nature Switzerland AG 2018
V. Torra et al. (Eds.): MDAI 2018, LNAI 11144, pp. 77–88, 2018.
https://doi.org/10.1007/978-3-030-00202-2_7

original set function; in other words, when a chosen family of set functions has the image within the same family. This property is usually called the consensus requirement. In addition, we study the properties of linear operators recently called event-based transformations of capacities [7–9]. We consider linear spaces of linear combinations of such operators, and compute the dimensions of them.

The paper has the following structure. At first we introduce some definitions, which will be used later, and families of non-additive set functions. After that we investigate the connection between linear operators and set functions with two arguments, and prove some results concerning the consensus requirement. The rest of the paper is devoted to class of event-based transformations of set functions and connected to them linear spaces of transformations.

2   Preliminaries

Let H be the set of all set functions defined on the powerset 2^X of a finite set $X = \{x_1, ..., x_N\}$. The set H is as a linear space w.r.t. usual sum of set functions and product of a set function and a real number. According to the definition, $\varphi : H \to H$ is a linear operator if $\varphi(a\mu_1 + b\mu_2) = a\varphi(\mu_1) + b\varphi(\mu_2)$ for arbitrary $a, b \in \mathbb{R}$ and $\mu_1, \mu_2 \in H$. For brevity, we denote $\varphi(\mu)$ by μ^φ. In the paper we consider the following families of set functions:

- $H_0 = \{\mu \in H | \forall A \in 2^X : \mu(A) \geq 0\}$ is the set of all non-negative set functions on 2^X;
- H_1 is the set of all non-negative monotone set functions on 2^X, i.e. $\mu \in H_1$ if $\mu \in H_0$ and $A \subseteq B$, $A, B \in 2^X$ implies that $\mu(A) \leq \mu(B)$;
- H_k, $k = 2, 3, ...$, is the set of all k-monotone set functions on 2^X. Let us remind that $\mu \in H_k$ if $\mu \in H_1$ and for any system of sets $C_1, ..., C_m \in 2^X$, $m \leq k$,

$$\mu\left(\bigcup_{i=1}^{m} C_i\right) + \sum_{B \subseteq \{1,...,m\}} (-1)^{|B|} \mu\left(\bigcap_{i \in B} C_i\right) \geq 0;$$

- H_∞ is the set of all totally monotone set functions on 2^X, i.e. $\mu \in H_\infty$ if $\mu \in H_k$ for any $k = 1, 2, ...$

The analysis of linear transformations is greatly simplified by choosing some suitable basis of H. We consider two bases of H. One of them consists of the set functions of the type:

$$\chi_B(A) = \begin{cases} 1, \ A = B, \\ 0, \ A \neq B, \end{cases}$$

and another one

$$\eta_{\langle B \rangle}(A) = \begin{cases} 1, \ B \subseteq A, \\ 0, \ \text{otherwise.} \end{cases}$$

It is clear that any set function has the unique representation by

$$\mu = \sum_{B \in 2^X} \mu(B)\chi_B;$$

and there is the representation of the type

$$\mu = \sum_{B \in 2^X} \mu^m(B)\eta_{\langle B \rangle}.$$

The set function μ^m is called the Möbius transform of μ computed by

$$\mu^m(B) = \sum_{A \subseteq B} (-1)^{|B \setminus A|}\mu(A).$$

It is clear that the Möbius transform m is a linear transformation uniquely defined by $\eta_{\langle B \rangle}^m = \chi_B$. The inverse of it is computed by

$$\mu(B) = \sum_{B \subseteq A} \mu^m(A).$$

One can show that introduced families of set functions H_k, $k = 0, 1, ...$, are convex cones with finite sets of generating elements. If a given family, say $\mathcal{M}$, is a cone with the finite system $\{\mu_1, ..., \mu_n\}$ of generating elements $\mu_i \in \mathcal{M}$, $i = 1, ..., n$, then any $\mu \in \mathcal{M}$ is represented by $\mu = \sum_{i=1}^n a_i\mu_i$, where $a_i \in \mathbb{R}_+$, $i = 1, ..., n$ ($\mathbb{R}_+$ is the set of all non-negative real numbers). As a rule it is a hard problem to get a description of generating elements for a given convex cone $\mathcal{M}$. In some cases it is simply solved. For example, $\{\chi_B\}_{B \in 2^X}$ is a set of generating elements for H_0; $\{\eta_{\langle B \rangle}\}_{B \in 2^X}$ is the set of generating elements for H_∞.

We can also describe the system of generating elements for monotone set functions. It consists of so-called $\{0, 1\}$-valued monotone set functions. Such monotone set functions can be described by semi-filters (upper sets) of the algebra 2^X. In this case we consider the algebra 2^X as a partially ordered set w.r.t. $\subseteq$. A subset $\mathbf{f} \subseteq 2^X$ is called a semi-filter (an upper set) if $A \in \mathbf{f}$ and $A \subseteq B$ implies $B \in \mathbf{f}$. Any $\{0, 1\}$-valued monotone set function μ is a characteristic function of some semi-filter $\mathbf{f}$, i.e.

$$\mu(A) = \begin{cases} 1, \ A \in \mathbf{f}, \\ 0, \ A \notin \mathbf{f}. \end{cases}$$

Each semi-filter contains a finite set of minimal elements $\{A_1, A_2, ..., A_k\}$, i.e. mutually incomparable elements w.r.t. $\subseteq$ such that $\mathbf{f} = \{A \in 2^X | \exists A_i \subseteq A\}$. Clearly minimal elements generate any semi-filter $\mathbf{f}$. This fact is denoted by $\mathbf{f} = \langle A_1, A_2, ..., A_k \rangle$. The semi-filter $\mathbf{f}$ is principal if it is generated by one minimal element, i.e. $\mathbf{f} = \langle A \rangle$. Further we will use notation $\eta_{\mathbf{f}}$ if $\{0, 1\}$-valued monotone set function is associated with the semi-filter $\mathbf{f}$.

3 Linear Operators and Consensus Requirement

We begin this section, giving the following straightforward proposition.

Proposition 1. *Let $\mathcal{M}_{first}, \mathcal{M}_{second}$ be cones in H, and let $\{\mu_1, ..., \mu_n\}$ be a finite set of generating elements for $\mathcal{M}_{first}$, then $\varphi : \mathcal{M}_{first} \to \mathcal{M}_{second}$ for a linear operator φ iff $\mu_k^\varphi \in \mathcal{M}_{second}$ for $k = 1, ..., n$.*

Proof. Let $\mathcal{M}_{first}, \mathcal{M}_{second}$ be cones in H as in the proposition. Necessity is obvious. For proving sufficiency, consider an arbitrary $\mu = \sum_{i=1}^{n} a_i \mu_i$, $a_i \geq 0$, $i = 1, ..., n$, in $\mathcal{M}_{first}$. Then $\mu^{\varphi} = \sum_{i=1}^{n} a_i \mu_i^{\varphi}$, i.e. $\mu^{\varphi} = \sum_{i=1}^{n} a_i \mu_i^{\varphi}$, i.e. $\mu^{\varphi} \in \mathcal{M}_{second}$.

Proposition 1 gives the solution of the stated problem, when we suppose that $\mathcal{M}_{first}$ is equal to H_0 or H_{∞}. If $\mathcal{M}_{first} = H_0$, then any linear operator φ with $\chi_B^{\varphi} \in \mathcal{M}_{second}$, $B \in 2^X$, provides the mapping $\varphi : H_0 \to \mathcal{M}_{second}$. Since $\{\chi_B\}_{B \in 2^X}$ is a basis of H, set functions χ_B^{φ} can be chosen arbitrarily in $\mathcal{M}_{second}$. We can come to the same conclusions assuming that $\mathcal{M}_{first} = H_{\infty}$ and taking the basis $\{\eta_{\langle B \rangle}\}_{B \in 2^X}$ of H.

If $\mathcal{M}_{first} = H_1$, then the solution of the stated problem is based on the following proposition.

Proposition 2. *Any $\{0,1\}$-valued monotone set function $\eta_{\langle B_1,...,B_k \rangle}$ has the following representation through the basis $\{\eta_{\langle B \rangle}\}_{B \in 2^X}$*

$$\eta_{\langle B_1,...,B_k \rangle} = - \sum_{C \subseteq \{1,...,k\}} (-1)^{|B|} \eta_{\langle \bigcup_{i \in C} C_i \rangle}.$$

Proof. The proof of the proposition is based on the equalities $\eta_{\mathbf{f}} = \bigcup_{i=1}^{k} \langle B_i \rangle$, $\langle A \cup B \rangle = \langle A \rangle \cap \langle B \rangle$ for any $A, B \in 2^X$, and on the inclusion-exclusion formula.

We immediately have the following consequence from Proposition 2.

Proposition 3. *The linear operator φ induces the mapping $\varphi : H_1 \to \mathcal{M}_{second}$, where $\mathcal{M}_{second}$ is some cone in H, iff for any system of sets $\{B_1, B_2, ..., B_k\}$ the set function*

$$\eta_{\mathbf{f}}^{\varphi} = - \sum_{C \subseteq \{1,...,k\}} (-1)^{|B|} \eta_{\langle \bigcup_{i \in C} B_i \rangle}^{\varphi}$$

is in $\mathcal{M}_{second}$.

Now we will try to interpret obtained results with the help of set functions with two arguments. Let $\varphi : H \to H$ be a linear operator. We introduce into consideration the set function $\varphi(B, A) = \chi_B^{\varphi}(A)$, $A, B \in 2^X$. The set function φ can be considered as a matrix representation of the linear operator φ in the basis $\{\chi_B\}_{B \in 2^X}$. The image of μ is expressed by

$$\mu^{\varphi}(A) = \sum_{B \in 2^X} \mu(B)\varphi(B, A).$$

We can also consider representations of set functions through the basis $\{\eta_{\langle B \rangle}\}_{B \in 2^X}$. Then we use the set function $\varphi^m(B, A) = \eta_{\langle B \rangle}^{\varphi}(A)$. One can show that

$$\varphi^m(B, A) = \sum_{C \in 2^X | C \supseteq B} \varphi(C, A).$$

and we can compute the inverse transform by

$$\varphi(B, A) = \sum_{C \in 2^X | C \supseteq B} (-1)^{|C \backslash B|} \varphi^m(C, A).$$

The image of μ is expressed by

$$\mu^{\varphi}(A) = \sum_{B \in 2^X} \mu^m(B) \varphi^m(B, A).$$

Now we can express statements of Propositions 1 and 3 through set functions φ and φ^m. According to Proposition 1 $\varphi : H_0 \rightarrow \mathcal{M}_{second}$ iff for any fixed $B \in 2^X$ the set function $\varphi(B, A)$ as a function of A is in $\mathcal{M}_{second}$, and according to Proposition 3 for any finite system of sets $\{B_1, B_2, ..., B_k\}$ the set function

$$\nu(A) = - \sum_{C \subseteq \{1, ..., k\}} (-1)^{|B|} \varphi^m \left(\bigcup_{i \in C} B_i, A \right)$$

is in $\mathcal{M}_{second}$.

4   Linear Operators and Event-Based Transformations

Further we will consider the set of all linear operators on H as a linear space denoted by LH. We write $\varphi = a\varphi_1 + b\varphi_2$, $a, b \in \mathbb{R}$, if $\varphi(\mu)$, $\mu \in H$, is expressed by $\varphi(\mu) = a\varphi_1(\mu) + b\varphi_2(\mu)$, where $\varphi_1, \varphi_2 \in LH$. As we will see the next linear operators give a rich structure for representing event-based transformations of set functions. Let $\mu \in H$ and introduce the following linear operators:

1. $\mu(A) \xrightarrow{\varphi \cap B} \mu(A \cap B)$ for $A \in 2^X$ and a fixed $B \in 2^X$.
2. $\mu(A) \xrightarrow{\varphi \cup B} \mu(A \cup B)$ for $A \in 2^X$ and a fixed $B \in 2^X$.
3. $\mu(A) \xrightarrow{\varphi_\neg} \mu(\neg A)$, where $\neg A$ is the complement of $A \in 2^X$.
4. $\mu(A) \xrightarrow{\varphi B} \mu(B)$ for $A \in 2^X$ and a fixed $B \in 2^X$.
5. $\mu(A) \xrightarrow{\varphi \equiv} \mu(A)$ for $A \in 2^X$.

As the following examples show, the introduced linear operators enable to express event-based linear transformations of set functions (capacities) [7–9]:

1. Let $\varphi_1 = \varphi_X - \varphi_\neg$, then $\mu(A) \xrightarrow{\varphi_1} \mu(X) - \mu(\neg A)$, $A \in 2^X$.
2. Let $\varphi_2 = 0.5(\varphi_X + \varphi_\equiv - \varphi_\neg)$, then $\mu(A) \xrightarrow{\varphi_2} 0.5 (\mu(X) + \mu(A) - \mu(\neg A))$, $A \in 2^X$.
3. Let $\varphi_3 = \varphi_{\cup B} - \varphi_B$, then $\mu(A) \xrightarrow{\varphi_3} \mu(A \cup B) - \mu(B)$, $A \in 2^X$.
4. Let $\varphi_4 = \varphi_{\cup B} + \varphi_{\cap B} - \varphi_B$, then $\mu(A) \xrightarrow{\varphi_4} \mu(A \cup B) + \mu(A \cap B) - \mu(B)$, $A \in 2^X$.
5. Let $\varphi_5 = \varphi_{\cup B} - \varphi_{\cap B} \circ \varphi_\neg$, then $\mu(A) \xrightarrow{\varphi_5} \mu(A \cup B) - \mu(\neg A \cap B)$, $A \in 2^X$.
6. Let $\varphi_6 = \varphi_X - \varphi_{\cup B} \circ \varphi_\neg - \varphi_{\cap B}$, then $\mu(A) \xrightarrow{\varphi_5} \mu(X) - \mu(\neg A \cup B) - \mu(A \cap B)$, $A \in 2^X$.

7. Let $\varphi_7 = \varphi_X - \varphi_{\cup B} \circ \varphi_\neg + \varphi_B - \varphi_{\cap B} \circ \varphi_\neg$, then $\mu(A) \xrightarrow{\varphi_6} \mu(X) - \mu(\neg A \cup B) + \mu(B) - \mu(\neg A \cap B)$, $A \in 2^X$.

Proposition 4. *Let* $\varphi \in LH$, *then* $\varphi \circ \varphi_{\cap C} = \varphi_{\cap C} \circ \varphi$ *for all* $C \in 2^X$ *iff there is a* $g \in H$ *such that* $\varphi^m(B, A) = g(B)\eta_{\langle B \rangle}(A)$.

Proof. It is clear that $\varphi \circ \varphi_{\cap C} = \varphi_{\cap C} \circ \varphi$ for all $C \in 2^X$ iff $\eta_{\langle B \rangle}^{\varphi \circ \varphi_{\cap C}} = \eta_{\langle B \rangle}^{\varphi_{\cap C} \circ \varphi}$ for all $B, C \in 2^X$. Next we find $\eta_{\langle B \rangle}^{\varphi \circ \varphi_{\cap C}}(A) = \varphi^m(B, A \cap C)$ and

$$\eta_{\langle B \rangle}^{\varphi_{\cap C} \circ \varphi}(A) = \eta_{\langle B \rangle}^{\varphi}(A \cap C) = \begin{cases} \varphi^m(B, A), & B \subseteq C, \\ 0, & B \nsubseteq C \end{cases} = \varphi^m(B, A)\eta_{\langle B \rangle}(C),$$

since

$$\eta_{\langle B \rangle}(A \cap C) = \begin{cases} 1, & B \subseteq A \cap C, \\ 0, & B \nsubseteq A \cap C \end{cases} = \begin{cases} \eta_{\langle B \rangle}(A), & B \subseteq C, \\ 0, & B \nsubseteq C. \end{cases}$$

We derive from the equality $\varphi^m(B, A \cap C) = \varphi^m(B, A)\eta_{\langle B \rangle}(C)$ that $\varphi^m(B, A) = \varphi^m(B, A)\eta_{\langle B \rangle}(A)$ and $\varphi^m(B, B) = \varphi^m(B, A)$ if $B \subseteq A$. So we get the required representation denoting $g(B) = \varphi^m(B, B)$.

Let us check that our representation satisfies all necessary conditions. We see that

$$\eta_{\langle B \rangle}^{\varphi \circ \varphi_{\cap C}}(A) = g(B)\eta_{\langle B \rangle}(A \cap C) = g(B)\eta_{\langle B \rangle}(A)\eta_{\langle B \rangle}(C)$$

and

$$\eta_{\langle B \rangle}^{\varphi_{\cap C} \circ \varphi}(A) = \eta_{\langle B \rangle}^{\varphi}(A \cap C) = \eta_{\langle B \rangle}^{\varphi}(A)\eta_{\langle B \rangle}(C) = g(B)\eta_{\langle B \rangle}(A)\eta_{\langle B \rangle}(C).$$

The proposition is proved.

We denote by $LH_\cap$ the set of all linear operators, which are commutative w.r.t. $\varphi_{\cap C}$, $C \in 2^X$, i.e. $LH_\cap = \{\varphi \in LH | \varphi \circ \varphi_{\cap C} = \varphi_{\cap C} \circ \varphi \text{ for all } C \in 2^X\}$.

Proposition 5. *The set* $LH_\cap$ *is a linear subspace of* LH *and* $\{\varphi_{\cap C}\}_{C \in 2^X}$ *is a basis of* $LH_\cap$.

Proof. Let us check that $LH_\cap$ is a linear subspace of LH. Let $\varphi_1, \varphi_2 \in LH_\cap$ and $\varphi = a\varphi_1 + b\varphi_2$, then $\varphi \circ \varphi_{\cap C} = (a\varphi_1 + b\varphi_2) \circ \varphi_{\cap C} = a\varphi_1 \circ \varphi_{\cap C} + b\varphi_2 \circ \varphi_{\cap C} = a\varphi_{\cap C} \circ \varphi_1 + b\varphi_{\cap C} \circ \varphi_2 = \varphi_{\cap C} \circ (a\varphi_1 + b\varphi_2) = \varphi_{\cap C} \circ \varphi$, i.e. $LH_\cap$ is a linear subspace of LH. Then we find that $\varphi_{\cap C}^m(B, A) = \eta_{\langle B \rangle}^{\varphi_{\cap C}}(A) = \eta_{\langle B \rangle}(A \cap C) = \eta_{\langle B \rangle}(C)\eta_{\langle B \rangle}(A)$.

Therefore, $\varphi_{\cap C}^m(B, A) = g(B)\eta_{\langle B \rangle}(A)$, where $g(B) = \eta_{\langle B \rangle}(C) = \begin{cases} 1, & B \subseteq C, \\ 0, & B \nsubseteq C. \end{cases}$

i.e. $\varphi_{\cap C} \in LH_\cap$. It is clear that set functions $\eta_{\langle B \rangle}(C)$ of $B \in 2^X$ for a fixed $C \in 2^X$, are linear independent and form a basis of H. This finishes the proof of the proposition.

Proposition 6. *Let* $\varphi \in LH_\cap$, *then* $\varphi : H_1 \to H_1$ *iff* $\varphi = \sum_{C \in 2^X} m(C)\varphi_{\cap C}$,

where $m \in H_0$.

Proof. Sufficiency is obvious, because any $\varphi_{\cap C}$ induces the mapping $\varphi_{\cap C} : H_1 \to H_1$.

Necessity. Assume to the contrary that $m \notin H_0$, i.e. there is a $B \in 2^X$ such that $m(B) < 0$ and φ induces the mapping $H_1 \to H_1$. Consider the semi-filter $\mathbf{f} = \{A \in 2^X | A \supset B \backslash \{x\}\}$, where $x \in B$. We see that

$$\eta_{\mathbf{f}}^{\varphi}(X) - \eta_{\mathbf{f}}^{\varphi}(X \backslash \{x\}) = \sum_{C \in 2^X} m(C)\,[\eta_{\mathbf{f}}(C) - \eta_{\mathbf{f}}(C \backslash \{x\})]$$

$$= \sum_{C \in 2^X | x \in C} m(C)\,[\eta_{\mathbf{f}}(C) - \eta_{\mathbf{f}}(C \backslash \{x\})]$$

$$= \sum_{C \in 2^X | x \in C} m(C)\eta_{\mathbf{f}}(C) - \sum_{C \in 2^X | x \in C} m(C)\eta_{\mathbf{f}}(C \backslash \{x\})$$

$$= \sum_{C \in 2^X | B \subseteq C} m(C) - \sum_{C \in 2^X | B \subset C} m(C) = m(B) < 0.$$

Therefore, $\eta_{\mathbf{f}}^{\varphi}$ is not monotone, but this contradicts to our assumption.

Corollary 1. *Let* $\varphi \in LH_{\cap}$, *i.e.* $\varphi^m(B, A) = g(B)\eta_{\langle B \rangle}(A)$. *Then* $\varphi : H_1 \to H_1$ *iff the set function* $g(\neg B)$ *of* $B \in 2^X$ *is totally monotone.*

Proof. Let us show first necessity. Let $\varphi \in LH_{\cap}$ and $\varphi : H_1 \to H_1$, then by Proposition 6

$$\varphi = \sum_{C \in 2^X} m(C)\varphi_{\cap C}, \text{ where } m \in H_0.$$

We find next that

$$\varphi_{\cap C}^m(B, A) = \eta_{\langle B \rangle}^{\varphi_{\cap C}}(A) = \eta_{\langle B \rangle}(A \cap C)$$

$$= \eta_{\langle B \rangle}(C)\,\eta_{\langle B \rangle}(A) = \eta_{\langle \neg C \rangle}(\neg B)\,\eta_{\langle B \rangle}(A),$$

and

$$\varphi^m(B, A) = \eta_{\langle B \rangle}(A) \sum_{C \in 2^X} m(C)\eta_{\langle \neg C \rangle}(\neg B),$$

i.e. $g(B) = \sum_{C \in 2^X} m(C)\eta_{\langle \neg C \rangle}(\neg B)$ or $g(\neg B) = \sum_{C \in 2^X} m(C)\eta_{\langle \neg C \rangle}(B)$. It is clear that $g(\neg B)$ of $B \in 2^X$ is totally monotone.

Sufficiency is proved by using the fact that any totally monotone set function $g(\neg B)$ of $B \in 2^X$ can be represented in the form $g(\neg B) = \sum_{C \in 2^X} m(C)\eta_{\langle \neg C \rangle}(B)$, $m \in H_0$.

Proposition 7. *Let* $\varphi \in LH_{\cap}$ *and* $\varphi : H_1 \to H_1$, *then* $\varphi : H_k \to H_k$, $k = 2, 3, ..., \infty$.

Proof. We use the following necessary and sufficient feature of k-monotonicity [10]: $\mu \in H_0$ is k-monotone if the set function $\nu(A) = \mu(A) - \mu(A \cap \neg\{x\})$ is $(k-1)$-monotone for any $x \in X$. Assume at first that $\mu \in H_2$, then we see that $\mu^{\varphi_{\cap\neg\{x\}}} \in H_1$ for any $x \in X$. According to our supposition $\varphi : H_1 \to H_1$, therefore $\mu^{\varphi_{\cap\neg\{x\}} \circ \varphi} \in H_1$. Since $\varphi_{\cap\{x\}} \circ \varphi = \varphi \circ \varphi_{\cap\neg\{x\}}$, we have $\mu^{\varphi \circ \varphi_{\cap\neg\{x\}}} \in H_1$ for any $x \in X$, i.e. $\mu^{\varphi} \in H_2$.

The general case is proved by induction. We assume that $\varphi : H_i \to H_i$, $i = 1, 2, ..., k-1$, and show that $\varphi : H_k \to H_k$. Following this scheme, let $\mu \in H_k$, then we see that $\mu^{\varphi_{\cap\neg\{x\}}} \in H_{k-1}$ for any $x \in X$. According to our supposition $\varphi : H_{k-1} \to H_{k-1}$, therefore $\mu^{\varphi_{\cap\neg\{x\}} \circ \varphi} \in H_{k-1}$. Since $\varphi_{\cap\{x\}} \circ \varphi = \varphi \circ \varphi_{\cap\neg\{x\}}$, we have $\mu^{\varphi \circ \varphi_{\cap\neg\{x\}}} \in H_{k-1}$ for any $x \in X$, i.e. $\mu^{\varphi} \in H_k$.

Example 1. Consider the Choquet integral $g(A) = (Ch) \int_A f d\mu$ w.r.t. a set function μ and an non-negative valued function f. Obviously, for a fixed f the Choquet integral defines the linear transformation φ, in which the set function μ is mapped to the set function g. Let us show that $\varphi \in LH_\cap$. Every function f can be represented as a linear combination $f = \sum_{i=1}^{m} b_i 1_{B_i}$ of simple functions
$$1_{B_i}(x) = \begin{cases} 1, & x \in B_i, \\ 0, & x \notin B_i, \end{cases}$$
such that $B_1 \subset B_2 \subset ... \subset B_m$, $B_i \in 2^X$, $b_i \geq 0$, $i = 1, ..., m$. Because such functions are comonotone, we have
$$g(A) = (Ch) \int_A \left(\sum_{i=1}^{m} b_i 1_{B_i} \right) d\mu = \sum_{i=1}^{m} b_i (Ch) \int_A 1_{B_i} d\mu = \sum_{i=1}^{m} b_i \mu(A \cap B_i).$$

Proposition 8. *Let $\varphi \in LH$, then $\varphi \circ \varphi_{\cup C} = \varphi_{\cup C} \circ \varphi$ for all $C \in 2^X$ iff there is a $g \in H$ such that $\varphi^m(B, A) = g(\neg B \cup A)$.*

Proof. It is clear that $\varphi \circ \varphi_{\cup C} = \varphi_{\cup C} \circ \varphi$ for all $C \in 2^X$ iff $\eta_{\langle B \rangle}^{\varphi \circ \varphi_{\cup C}} = \eta_{\langle B \rangle}^{\varphi_{\cup C} \circ \varphi}$ for all $B, C \in 2^X$. Next we find
$$\eta_{\langle B \rangle}^{\varphi \circ \varphi_{\cup C}}(A) = \varphi^m(B, A \cup C)$$
and
$$\eta_{\langle B \rangle}^{\varphi_{\cup C} \circ \varphi}(A) = \eta_{\langle B \backslash C \rangle}^{\varphi}(A) = \varphi^m(B \backslash C, A),$$
since
$$\eta_{\langle B \rangle}(A \cup C) = \begin{cases} 1, & B \subseteq A \cup C, \\ 0, & B \not\subseteq A \cup C \end{cases} = \begin{cases} 1, & B \backslash C \subseteq A, \\ 0, & B \backslash C \not\subseteq A \end{cases} = \eta_{\langle B \backslash C \rangle}(A).$$
We derive from the equality
$$\varphi^m(B, A \cup C)\, \eta_{\langle B \rangle}^{\varphi_{\cup C} \circ \varphi}(A) = \varphi^m(B \backslash C, A)$$
that $\varphi^m(B_1, A_1) = \varphi^m(B_2, A_2)$ if $\neg B_1 \cup A_1 = \neg B_2 \cup A_2$, i.e. we get the required representation, defining $g(\neg B \cup A) = \varphi^m(B, A)$. Let us check that our representation satisfies all necessary conditions. We see that
$$\eta_{\langle B \rangle}^{\varphi \circ \varphi_{\cup C}}(A) = g(\neg B \cup A \cup C)$$

and

$$\eta_{\langle B\rangle}^{\varphi_{\cup C}\circ\varphi}(A) = \eta_{\langle B\backslash C\rangle}^{\varphi}(A) = g\left(\neg(B\backslash C)\cup A\right) = g(\neg B\cup A\cup C),$$

i.e. the proposition is proved.

We denote by $LH_{\cup}$ the set of all linear operators, which are commutative w.r.t. $\varphi_{\cup C}$, $C \in 2^X$, i.e. $LH_{\cup} = \{\varphi \in LH | \varphi \circ \varphi_{\cup C} = \varphi_{\cup C} \circ \varphi \text{ for all } C \in 2^X\}$.

Proposition 9. *The set $LH_{\cup}$ is a linear subspace of LH and $\{\varphi_{\cup C}\}_{C\in2^X}$ is a basis of $LH_{\cup}$.*

Proof. Let us check that $LH_{\cup}$ is a linear subspace of LH. Let $\varphi_1, \varphi_2 \in LH_{\cap}$ and $\varphi = a\varphi_1 + b\varphi_2$, then $\varphi \circ \varphi_{\cup C} = (a\varphi_1 + b\varphi_2) \circ \varphi_{\cup C} = a\varphi_1 \circ \varphi_{\cup C} + b\varphi_2 \circ \varphi_{\cup C} = a\varphi_{\cup C} \circ \varphi_1 + b\varphi_{\cup C} \circ \varphi_2 = \varphi_{\cup C} \circ (a\varphi_1 + b\varphi_2) = \varphi_{\cup C} \circ \varphi$, i.e. $LH_{\cup}$ is a linear subspace of LH. Then we find that

$$\varphi_{\cup C}^m(B, A) = \eta_{\langle B\rangle}^{\varphi_{\cup C}}(A) = \eta_{\langle B\rangle}(A\cup C) = \begin{cases} 1, & B \subseteq A\cup C, \\ 0, & B \subseteq A\cup C \end{cases}$$

$$= \begin{cases} 1, & \neg B\cup A\cup C = X, \\ 0, & \neg B\cup A\cup C \neq X \end{cases} = \begin{cases} 1, & \neg C \subseteq \neg B\cup A, \\ 0, & \neg C \nsubseteq \neg B\cup A \end{cases} = \eta_{\langle\neg C\rangle}(\neg B\cup A).$$

Therefore, $\varphi^m(B, A) = g(\neg B\cup A)$, where $g = \eta_{\langle\neg C\rangle}$. It is clear that set functions $\eta_{\langle\neg C\rangle}$, $C \in 2^X$, are linear independent and form a basis of H. This finishes the proof of the proposition.

Proposition 10. *Let $\varphi \in LH_{\cup}$, then $\varphi : H_1 \rightarrow H_1$ iff $\varphi = \sum\limits_{C\in2^X} m(C)\varphi_{\cup C}$, where $m \in H_0$.*

Proof. Sufficiency is obvious, because any $\varphi_{\cup C}$ induces the mapping $\varphi_{\cup C} : H_1 \rightarrow H_1$.

Necessity. Assume to the contrary that $m \notin H_0$, i.e. there is a $B \in 2^X$ such that $m(B) < 0$, however, φ induces the mapping $H_1 \rightarrow H_1$. Let $x \notin B$, and consider the semi-filter $\mathbf{f} = \{A \in 2^X | A \supset B\}$. We see that

$$\eta_{\mathbf{f}}^{\varphi}(\emptyset) = \sum_{C\in2^X} m(C)\eta_{\mathbf{f}}(C) = \sum_{C\supset B} m(C)$$

and

$$\eta_{\mathbf{f}}^{\varphi}(\{x\}) = \sum_{C\in2^X} m(C)\eta_{\mathbf{f}}(\{x\}\cup C) = \sum_{C\supseteq B} m(C).$$

Therefore, $\eta_{\mathbf{f}}^{\varphi}(\{x\}) - \eta_{\mathbf{f}}^{\varphi}(\emptyset) = m(B) < 0$, i.e. $\eta_{\mathbf{f}}^{\varphi}$ is not monotone, but this contradicts to our assumption.

Corollary 2. *Let $\varphi \in LH_{\cup}$, i.e. $\varphi^m(B, A) = g(\neg B\cup A)$. Then $\varphi : H_1 \rightarrow H_1$ iff the set function g is totally monotone.*

Proof. Necessity. Let $\varphi \in LH_\cup$ and $\varphi : H_1 \to H_1$, then by Proposition 10

$$\varphi = \sum_{C \in 2^X} m(C)\varphi_{\cup C}, \text{ where } m \in H_0.$$

We find that $\varphi_{\cup C}^m(B, A) = \eta_{\langle \neg C \rangle}(\neg B \cup A)$, $\varphi^m(B, A) = \sum_{C \in 2^X} m(C)\eta_{\langle \neg C \rangle}$ $(\neg B \cup A)$, i.e. $g = \sum_{C \in 2^X} m(C)\eta_{\langle \neg C \rangle}$, i.e. g is totally monotone.

Sufficiency is proved by using the fact that every totally monotone set function g can be represented as $g = \sum_{C \in 2^X} m(C)\eta_{\langle \neg C \rangle}$, where $m \in H_0$.

Proposition 11. *Let $\varphi \in LH_\cup$ and $\varphi : H_1 \to H_1$, then $\varphi : H_k \to H_k$, $k = 2, 3, ..., \infty$.*

Proof. We use the following necessary and sufficient feature of k-monotonicity [10]: $\mu \in H_0$ is k-monotone if the set function $\nu(A) = \mu(A \cup \{x\}) - \mu(A)$ is $(k-1)$-monotone for any $x \in X$. Assume at first that $\mu \in H_2$, then we see that $\mu^{\varphi_{\cup\{x\}}} \in H_1$ for any $x \in X$. According to our assumption $\varphi : H_1 \to H_1$, therefore $\mu^{\varphi_{\cup\{x\}} \circ \varphi} \in H_1$. Since $\varphi_{\cup\{x\}} \circ \varphi = \varphi \circ \varphi_{\cup\{x\}}$, we have $\mu^{\varphi \circ \varphi_{\cup\{x\}}} \in H_1$ for any $x \in X$, i.e. $\mu^\varphi \in H_2$.

The general case is proved by induction. Assume that $\varphi : H_i \to H_i$, $i = 1, 2, ..., k-1$, and show that $\varphi : H_k \to H_k$. Following this scheme, let $\mu \in H_k$, then we see that $\mu^{\varphi_{\cup\{x\}}} \in H_{k-1}$ for any $x \in X$. According to our assumption $\varphi : H_{k-1} \to H_{k-1}$, therefore $\mu^{\varphi_{\cup\{x\}} \circ \varphi} \in H_{k-1}$. Since $\varphi_{\cup\{x\}} \circ \varphi = \varphi \circ \varphi_{\cup\{x\}}$, we have $\mu^{\varphi \circ \varphi_{\cup\{x\}}} \in H_{k-1}$ for any $x \in X$, i.e. $\mu^\varphi \in H_k$.

Proposition 12. *Let $LH_{\cup\cap}$ be the minimal linear space, closed under composition of linear operators and containing $LH_\cup$ and $LH_\cap$. Then $\{\varphi_{\cup C} \circ \varphi_{\cap D}\}_{C \subseteq D}$ is the basis of $LH_{\cup\cap}$ and the dimension of $LH_{\cup\cap}$ is $3^{|X|}$.*

Proof. Consider the composition of operators

$$\mu^{\varphi_{\cup C} \circ \varphi_{\cap D}}(A) = \mu((A \cup C) \cap D) = \mu((A \cup C_1) \cap D),$$

where $C_1 = C \cap D$, i.e. $\varphi_{\cup C} \circ \varphi_{\cap D} = \varphi_{\cup C_1} \circ \varphi_{\cap D}$ and $C_1 \subseteq D$. Analogously,

$$\mu^{\varphi_{\cap D} \circ \varphi_{\cup C}}(A) = \mu((A \cap D) \cup C) = \mu((A \cup C) \cap D_1),$$

where $D_1 = C \cup D$, i.e. $\varphi_{\cap D} \circ \varphi_{\cup C} = \varphi_{\cup C} \circ \varphi_{\cap D_1}$ and $C \subseteq D_1$. We see that

$$\varphi_{\cup B} \circ \varphi_{\cup C} \circ \varphi_{\cap D} = \varphi_{\cup(B \cup C)} \circ \varphi_{\cap D},$$

$$\varphi_{\cap B} \circ \varphi_{\cup C} \circ \varphi_{\cap D} = \varphi_{\cup C} \circ \varphi_{\cap((B \cup C) \cap D)},$$

$$\varphi_{\cup C} \circ \varphi_{\cap D} \circ \varphi_{\cap B} = \varphi_{\cup C} \circ \varphi_{\cap(B \cap D)},$$

$$\varphi_{\cup C} \circ \varphi_{\cap D} \circ \varphi_{\cup B} = \varphi_{\cup((C \cap D) \cup B)} \circ \varphi_{\cap(B \cup D)}.$$

Thus, every operator from $LH_{\cup\cap}$ can be represented as a linear combination of operators from $\{\varphi_{\cup C} \circ \varphi_{\cap D}\}_{C \subseteq D}$.

Let us show that $\{\varphi_{\cup C} \circ \varphi_{\cap D}\}_{C \subseteq D}$ is the basis of $LH_{\cup \cap}$. At first we will simplify the expression

$$\eta_{\langle B \rangle}^{\varphi_{\cup C} \circ \varphi_{\cap D}} (A) = \eta_{\langle B \rangle} ((A \cup C) \cap D) = \eta_{\langle X \rangle} ((A \cup C) \cap D \cup \neg B)$$

$$= \eta_{\langle X \rangle} ((A \cap D) \cup C \cup \neg B) = \eta_{\langle B \backslash C \rangle} (A \cap D)$$

$$= \eta_{\langle B \backslash C \rangle} (A) \eta_{\langle B \backslash C \rangle} (D) = \eta_{\langle B \backslash C \rangle} (A) \eta_{\langle B \rangle} (D).$$

Consider next an arbitrary linear operator L in $LH_{\cup \cap}$ given by

$$L = \sum_{(C,D)|C \subseteq D} \alpha(C, D) \varphi_{\cup C} \circ \varphi_{\cap D}, \tag{1}$$

where numbers $\alpha(C, D)$ are defined for pairs (C, D), in which $C \subseteq D$. Then

$$L \left(\eta_{\langle B \rangle} \right) = \sum_{(C,D)|C \subseteq B \subseteq D} \alpha(C, D) \eta_{\langle B \backslash C \rangle}.$$

Now we will prove that the system of linear operators $\{\varphi_{\cup C} \circ \varphi_{\cap D}\}_{C \subseteq D}$ are linear independent. Let us assume the inverse. Then there is a representation (1) of the linear operator L, that is identical to zero, in which there is a coefficient $\alpha(C, D) \neq 0$ for some $C \subseteq D \subseteq X$. Because this linear operator is identical to zero, $L \left(\eta_{\langle B \rangle} \right) = 0$ for all $B \in 2^X$. Thus, there is $C \in 2^X$ such that $\{D | \alpha(C, D) \neq 0\} \neq \emptyset$. Let us take an arbitrary maximal element D of $\{D | \alpha(C, D) \neq 0\}$. Then

$$L \left(\eta_{\langle D \rangle} \right) = \sum_{B \in 2^X} m(B) \eta_{\langle B \rangle},$$

and $m(D \backslash C) = \alpha(C, D) \neq 0$. Therefore, we see that $L \left(\eta_{\langle D \rangle} \right) \neq 0$, i.e. our assumption is wrong, and $\{\varphi_{\cup C} \circ \varphi_{\cap D}\}_{C \subseteq D}$ is the basis of $LH_{\cup \cap}$.

Let compute the cardinality of $\{\varphi_{\cup C} \circ \varphi_{\cap D}\}_{C \subseteq D}$. Assume that $|X| = N$. For every C with cardinality k there are 2^{N-k} sets D with $C \subseteq D$. Therefore, the total number of pairs (C, D) with $C \subseteq D$ is

$$\sum_{k=0}^{N} \binom{N}{k} 2^{N-k} = 3^N.$$

Thus, the dimension of $LH_{\cup \cap}$ is 3^N. The proposition is proved.

5   Conclusion

We can define many linear operators on set functions and in this paper we have studied the linear space of operators connected with event-based transformations of capacities. The next problems are connected with studying

(a) the consensus requirement in $LH_{\cup\cap}$;
(b) the linear space containing all possible event-based transformations of set functions;
(c) the matrix description of event-based transformations;
(d) other linear transformations, for example, $\varphi^\gamma : H \to H$, defined through one-to-one mappings $\gamma : X \to X$ by $\mu^{\varphi^\gamma}(A) = \mu(\gamma(A))$, $A \in 2^X$.

Acknowledgment. This work has been supported by the grant 18-01-00877 of RFBR (Russian Foundation for Basic Research).

References

1. Grabisch, M., Roubens, M.: Application of the Choquet integral in multicriteria decision making. In: Grabisch, M., Murofushi, T., Sugeno, M. (eds.) Fuzzy Measures and Integrals: Theory and Applications. Studies on Fuzziness and Soft Computing, pp. 415–434. Physica-Verlag, Heidelberg (2000)
2. Grabisch, M.: Set Functions, Games and Capacities in Decision Making. Springer, Cham (2016). https://doi.org/10.1007/978-3-319-30690-2
3. Klir, G.J.: Uncertainty and Information: Foundations of Generalized Information Theory. Wiley-Interscience, Hoboken (2006)
4. Denneberg, D.: Non-additive Measure and Integral. Kluwer, Dordrecht (1997)
5. Chateauneuf, A., Jaffray, J.Y.: Some characterizations of lower probabilities and other monotone capacities through the use of Möbius inversion. Math. Soc. Sci. **17**, 263–283 (1989)
6. Grabisch, M.: The interaction and Möbius representations of fuzzy measures on finite spaces, k-additive measures: a survey. In: Grabisch, M., Murofushi, T., Sugeno, M. (eds.) Fuzzy Measures and Integrals: Theory and Applications. Studies on Fuzziness and Soft Computing, pp. 70–93. Physica-Verlag, Heidelberg (2000)
7. Borkotokey, S., Mesiar, R., Li, J.: Event-based transformations of capacities. In: Torra, V., Narukawa, Y., Honda, A., Inoue, S. (eds.) MDAI 2017. LNCS (LNAI), vol. 10571, pp. 33–39. Springer, Cham (2017). https://doi.org/10.1007/978-3-319-67422-3_4
8. Mesiar, R., Borkotokey, S., Jin, L., Kalina, M.: Aggregation functions and capacities, Fuzzy Sets and Systems, 24 August 2017. https://doi.org/10.1016/j.fss.2017.08.007
9. Kouchakinejad, F., Šipošová, A.: On some transformations of fuzzy measures. Tatra Mt. Math. Publ. **69**, 75–86 (2017)
10. Bronevich, A.G.: On the closure of families of fuzzy measures under eventwise aggregations. Fuzzy Sets Syst. **153**(1), 45–70 (2005)

Association Analysis on Interval-Valued Fuzzy Sets

Petra Murinová[(✉)], Viktor Pavliska, and Michal Burda

Institute for Research and Applications of Fuzzy Modeling,
Centre of Excellence IT4Innovations, Division University of Ostrava,
30. dubna 22, 701 03 Ostrava, Czech Republic
{petra.murinova,viktor.pavliska,michal.burda}@osu.cz

Abstract. The aim of this paper is to generalize the concept of association rules for interval-valued fuzzy sets. Interval-valued fuzzy sets allow for intervals of membership degrees to be assigned to each element of the universe. These intervals may be interpreted as partial information where the exact membership degree is not known. The paper provides a generalized definition of support and confidence, which are the most commonly known measures of quality of a rule.

Keywords: Association rules · Missing values
Interval-valued fuzzy sets · Support · Confidence

1 Introduction

Searching for association rules is a tool for explanatory analysis of large data sets. Association rule is a formula of the form $A \rightharpoonup C$, where A is called an antecedent and C is a consequent, and which denotes some interesting relationship between A and C. There exist many different types of association rules. In this paper, we focus on implicative rules.

Association rules were firstly introduced by Hájek et al. in the late 1960s [1] by formulating the GUHA (General Unary Hypotheses Automaton) method [2]. Independently on them, a similar framework was developed by Agrawal [3] in 1993. Many different authors extended the association rules framework for fuzzy data, see [4] for a recent survey. A framework for a construction of linguistic summaries is also very closely related to fuzzy association rules. It was proposed by Yager in [5] and later further developed by Kacprzyk [6]. Another approach [7] introduces intermediate quantifiers to interpret association rules in natural language.

In real-world applications, data being analyzed are sometimes missing. Non-availability comes very often from the fact that some values are unknown or concealed. Handling of missing values is very common in data processing. There were developed many techniques for missing values imputation, and many existing methods were extended to directly work with unknown values. Hájek et al.

© Springer Nature Switzerland AG 2018
V. Torra et al. (Eds.): MDAI 2018, LNAI 11144, pp. 89–100, 2018.
https://doi.org/10.1007/978-3-030-00202-2_8

Table 1. Sobociński's $\wedge_S$, Bochvar's internal $\wedge_B$ and Kleene's (strong) $\wedge_K$ variants for handling of the third truth value $*$ in conjunction

$\wedge_S$	0	$*$	1		$\wedge_B$	0	$*$	1		$\wedge_K$	0	$*$	1
0	0	0	0		0	0	$*$	0		0	0	0	0
$*$	0	$*$	1		$*$	$*$	$*$	$*$		$*$	0	$*$	$*$
1	0	1	1		1	0	$*$	1		1	0	$*$	1

proposed within their GUHA method [2] an extension capable of searching for association rules on data with missing values. However, their approach is applicable on *binary* (or categorical) data only.

In practical data analytical tasks, missing data are preferably *imputed*, if the missingness is not dependent on unobserved or the missing value itself [8]. Imputation is done by substituting missing values with average or a value obtained from regression or from application of some of the machine learning techniques [9].

The objective of handling missing or undefined values is not new also in mathematical logic. The fundamental grounds were established by Kleene, Bochvar, Sobociński and others, who studied the properties of three-valued logics 0/1/$*$, which was also studied by Łukasiewicz in 1920 in [10]. These authors showed that the third value $*$ may represent an unknown, undefined or indeterminate truth value. An overview of main contributions can be found e.g. in [11].

The logic called *Bochvar's internal* [12] (also known as Kleene's weak) defines the third truth value $*$ as an annihilator. *Sobociński's* variant handles $*$ as an ignorable non-sense, so that $*$ is treated like 1 (resp. 0) in conjunction (resp. disjunction). *Kleene's* (strong) logic's indeterminate value $*$ preserve absorbing elements 0 (in conjunction) and 1 (in disjunction), while annihilating in other cases. See Table 1.

In [13], authors present three-valued logics where the third-value means *borderline* or *unknown*. Deeper interpretations of the third-truth value (*possible, undefined, half-true, inconsistent*) can be found in [14]. A generalization of Kleene's, Bochvar's, Sobociński's approach to predicate fuzzy logic was introduced in in [15,16]. In [17], the author proposed a study of fuzzy type theory (FTT) with partial functions which are used for a characterization of the undefined values. In [18], several truth values representing different kinds of unavailability together with a single type of fuzzy logical connectives were introduced. A previous work of the authors comprises association rules on undefined values (i.e. on fuzzy sets with *non-existent* membership degrees) [19].

The aim of this paper is to generalize the concept of association rules for interval-valued fuzzy sets. Interval-valued fuzzy sets allow for intervals of membership degrees to be assigned to each element of the universe. These intervals may be interpreted as partial information where the exact membership degree is not known. Interval-valued fuzzy sets were studied in [20,21].

For instance, a measurement tool has often some limits of quantification, which means that some values below or above a threshold are undistinguishable.

E.g., a tool for measuring the concentration of alcohol in blood is unable to distinguish values below 0.2‰. Such concentrations may be represented by the interval $[0, 0.2]$ as the exact value is unknown. Moreover, each measurement tool has defined a measure error e so that each measured value m may differ from exact value by error e, and so the exact value is only known to be in an interval $[m - e, m + e]$.

In this paper, we discuss handling of missing values in crisp data (with respect to association rules) and we show that transition to interval-valued fuzzy data brings some difficulties, which we analyze and solve at the end of the paper.

The rest of the paper is organized as follows. Section 2 recalls basic definitions from interval-valued fuzzy set theory. Section 3 introduces the reader into association rules and Sect. 4 generalizes association rules for interval-valued fuzzy sets. Section 5 concludes the paper and by drawing some directions for future work.

2   Operations on Interval-Valued Fuzzy Sets

The main goal of this section is to introduce a mathematical background which will be used for fuzzy association rules.

2.1   Basic Logical Operations

A fuzzy set [22] is defined as a mapping from universe of discourse U to a real interval $[0, 1]$, i.e. $F: U \to [0, 1]$. Unlike crisp sets, where an object fully belongs or does not belong to a set, fuzzy sets enable an object $u \in U$ to belong partially to a set F in a degree $F(u)$. A fuzzy set X is a subset of a fuzzy set Y, $X \subseteq Y$, if $X(u) \le Y(u)$, for all $u \in U$. A size of a fuzzy set X is $|X| = \sum_{u \in U} X(u)$.

Triangular norms, t-norms, are binary operations $\otimes : [0, 1]^2 \to [0, 1]$, which fulfils commutativity, associativity, monotonicity and boundary condition and which have been mainly studied by Klement, Mesiar and Pap in [23] and later elaborated by many others. A concept associated with t-norm is the triangular conorm (t-conorm) $\oplus : [0, 1]^2 \to [0, 1]$. A *generalized implication* is a binary operation $\leadsto : [0, 1]^2 \to [0, 1]$ that is monotone decreasing in the first argument and monotone increasing in the second argument and that satisfies the boundary conditions. The precise definition can be found in [24]. Finally, the *negation* is a non-increasing operation $\neg : [0, 1] \to [0, 1]$ such that $\neg(0) = 1$ and $\neg(1) = 0$. We say that it is *involutive* if $\neg(\neg(a)) = a$ holds for every $a \in [0, 1]$.

Convention. In the sequel we will work with intervals $[x_1, x_2] \subseteq [0, 1]$ and will be for simplicity denoted by capital letter X. By I we will denote the corresponding support which consists of all intervals on $[0, 1]$. We will put $I = \{X \mid [x_1, x_2] \subseteq [0, 1]\}$.

First of all, we start with a definition of the ordering $\le_I$ on I.

Definition 1. *Let* $X, Y \in I$, $X = [x_1, x_2]$ *and* $Y = [y_1, y_2]$ *then*

$$X \le_I Y \text{ iff } X \vee_I Y = Y \text{ (equivalently } X \wedge_I Y = X),$$

where

$$X \vee_I Y = [\max(x_1, y_1), \max(x_2, y_2)],$$
$$X \wedge_I Y = [\min(x_1, y_1), \min(x_2, y_2)].$$

We continue with definitions of basic operations on I. We generalize this approach to interval fuzzy logic in definitions below which were introduced in [20, 21].

Definition 2 (T-norm on I). *A t-norm on I is a binary operation $\otimes_I \colon I \times I \to I$ such that the following axioms are satisfied for all $X, Y, Z \in I$:*

(a) commutativity: $X \otimes_I Y = Y \otimes_I X$,
(b) associativity: $X \otimes_I (Y \otimes_I Z) = (X \otimes_I Y) \otimes_I Z$,
(c) monotonicity: $X \leq_I Y$ implies $(X \otimes_I Z) \leq_I (Y \otimes_I Z)$,
(d) boundary condition: $1_I \otimes_I X = X$.

Definition 3 (T-conorm on I). *A t-conorm on I is a binary operation $\oplus_I \colon I \times I \to I$ such that the following axioms are satisfied for all $X, Y, Z \in I$:*

(a) commutativity: $X \oplus_I Y = Y \oplus_I X$,
(b) associativity: $X \oplus_I (Y \oplus_I Z) = (X \oplus_I Y) \oplus_I Z$,
(c) monotonicity: $X \leq_I Y$ implies $(X \oplus_I Z) \leq_I (Y \oplus_I Z)$,
(d) boundary condition: $0_I \oplus_I X = X$.

Theorem 1 (t-representability). *Let $\otimes$ be a t-norm on $[0,1]$ and let $\oplus$ be a t-conorm. For $X, Y \in I$, $X = [x_1, x_2]$, $Y = [y_1, y_2]$, we put*

(a) $X \otimes_I Y = [x_1 \otimes y_1, x_2 \otimes y_2]$,
(b) $X \oplus_I Y = [x_1 \oplus y_1, x_2 \oplus y_2]$.

Then $\otimes_I$ is a t-norm on I and $\oplus_I$ is a t-conorm on I.

Proof. The proof can be found in [25].

Definition 4. *A binary operation $\leadsto_I \colon I \times I \to I$ is said to be a generalized implication function on I (similarly as in [24]) if for each $X, Y \in I$ the following holds:*

(a) $X \leadsto_I [1,1] = [1,1]$,
(b) $[0,0] \leadsto_I Y = [1,1]$,
(c) $[1,1] \leadsto_I Y = Y$.

Moreover, we require $\leadsto_I$ to be decreasing in its first, and increasing in its second argument.

Theorem 2. *Let $\leadsto \colon [0,1] \times [0,1] \to [0,1]$ be a function that it is monotone decreasing in the first and monotone increasing in the second argument. Let $X = [x_1, x_2]$ and $Y = [y_1, y_2]$. We put*

$$X \leadsto_I Y = [x_2 \leadsto y_1, x_1 \leadsto y_2]. \tag{1}$$

Then $\leadsto_I$ is an implication function on I.

Proof. The proof can be found in [25].

Definition 5 (Negation on I). *The* negation on I *is a non-increasing operation* $\neg_I \colon I \to I$ *such that* $\neg_I([0,0]) = [1,1]$ *and* $\neg_I([1,1]) = [0,0]$. *The negation is* involutive *if* $\neg_I(\neg_I(X)) = X$ *holds for every* $X \in I$.

Theorem 3. *Let* $\neg$ *be a negation function on* $[0,1]$ *and* $X = [x_1, x_2] \in I$. *We put*

$$\neg_I([x_1, x_2]) = [\neg x_2, \neg x_1] \tag{2}$$

Then $\neg_I$ *is a negation function on* I.

Convention. Whenever we will need to perform an operation between an element $a \in [0,1]$ and an interval $X \in I$ (for example $a \rightsquigarrow_I X$) then we can promote the element a to the interval $[a,a]$ so we can use the operation defined for intervals.

Definition 6. *Interval-valued fuzzy set is a mapping* $F \colon U \to I$ *assigning an interval from* I *to any element from the universe* U. *Cardinality* $|F|$ *of an interval-valued fuzzy set* F *is defined as*

$$|F| = \sum_{u \in U} F(u),$$

where $A + B = [a_1 + b_1, a_2 + b_2]$ for any $A, B \in I$, $A = [a_1, a_2]$ and $B = [b_1, b_2]$.

3   Introduction to Association Rules

Let $\mathcal{O} = \{o_1, o_2, \ldots, o_N\}$, $N > 0$, be a finite set of abstract elements called objects and $\mathcal{A} = \{a_1, a_2, \ldots, a_M\}$, $M > 0$, be a finite set of attributes. Within the association rules framework, a dataset D is a mapping that assigns to each object $o \in \mathcal{O}$ and attribute $a \in \mathcal{A}$ a truth degree $D(a, o) \in [0, 1]$, which represents the intensity of assignment of attribute a to object o.

For fixed D, we can treat the attribute a as a predicate, which assigns a truth value $a(o) \in [0, 1]$ to each object $o \in \mathcal{O}$. Similarly, for each subset $X \subseteq \mathcal{A}$ of attributes, we define a predicate $X(o)$ for a selected t-norm $\otimes$ as follows:

$$X(o) = \bigotimes_{a \in X} a(o). \tag{3}$$

Association rule is a formula $A \to C$, where $A \subseteq \mathcal{A}$ is the *antecedent* and $C \subseteq \mathcal{A}$ is the *consequent*. It is natural to assume $A \cap C = \emptyset$ and also $|C| = 1$.

As each combination of predicates in antecedent and consequent form a well-formed association rules, an important problem is to identify such rules that are relevant to the given dataset D. So far, there exist a large number of measures of such relevance. An overview can be found in [26].

Perhaps the most commonly known indicators of a rule quality are the *support* and *confidence*. Dubois et al. [24] define them on the basis of a partition of $\mathcal{O}$:

Table 2. Admissible operators induced by (4) and (5) accordingly to [24]: for Gödel, Goguen, Łukasiewicz t-norms the following implications are induced: Łukasiewicz, Reichenbach, Kleene-Dienes

$\otimes$	$\rightsquigarrow$
$\min(a,b)$	$\min(1, 1 - a + b)$
$a \cdot b$	$1 - a(1 - b)$
$\max(a + b - 1, 0)$	$\max(1 - a, b)$

they argue that a rule $A \rightharpoonup C$ is a three valued entity, which partitions the objects from $\mathcal{O}$ into three (fuzzy) subsets, namely, into a set of *positive examples* S_+ that verify the rule, *negative examples* S_- that falsify the rule, and *irrelevant examples* $S_\pm$ that do not contribute in either direction. For any $o \in \mathcal{O}$, Dubois et al. [24] provide the following formal definitions for a fixed t-norm $\otimes$ and a generalized implication $\rightsquigarrow$:

$$
\begin{aligned}
S_+(o) &= A(o) \otimes C(o); \\
S_-(o) &= \neg\big(A(o) \rightsquigarrow C(o)\big); \\
S_\pm(o) &= \neg A(o).
\end{aligned}
\tag{4}
$$

They argue that for $\langle S_+, S_-, S_\pm \rangle$ to be a proper fuzzy partition, all o should satisfy Ruspini condition:

$$
S_+(o) + S_-(o) + S_\pm(o) = 1.
\tag{5}
$$

As noted in [24], Eqs. (4) and (5) lead to the admissible operator problem. [24] identifies three pairs of $\otimes$ and $\rightsquigarrow$, which together satisfy both (4) and (5), see Table 2. Dubois et al. [24] assumes that $\neg(a) = 1 - a$, which together with conditions (4) and (5) results in

$$
a \rightsquigarrow c = \neg a \oplus (a \otimes b).
\tag{6}
$$

Based on (4), the *support* and *confidence* of a fuzzy association rule $A \rightharpoonup C$ may be defined as follows [24].

Definition 7. *Let $R = A \rightharpoonup C$ be a rule and $S = \langle S_+, S_-, S_\pm \rangle$ be a partition of $\mathcal{O}$ with respect to R. Then*

$$
\mathrm{supp}(A \rightharpoonup C) = |S_+|,
\tag{7}
$$

$$
\mathrm{conf}(A \rightharpoonup C) = \frac{|S_+|}{|S_+| + |S_-|}.
\tag{8}
$$

4 Association Rules on Interval-Valued Fuzzy Sets

In order to extend the association rules framework for data containing *interval-valued* membership degrees, one has to switch the range of membership degrees from $[0,1]$ to I. In other words, dataset D becomes a mapping such that $D(a,o) \in I$ for each $a \in \mathcal{A}$ and each $o \in \mathcal{O}$. Similarly to (3), an attribute a may be treated as a predicate with truth value $a(o) \in I$ and each subset $X \subseteq \mathcal{A}$ may be used to define a predicate $X(o) \in I$ by applying an extended t-norm $\otimes_I$ in (3):

$$X(o) = x_1(o) \otimes_I x_2(o) \otimes_I \ldots \otimes_I x_k(o), \tag{9}$$

where $x_1, \ldots, x_k \in X$.

Dubois [24] definitions of S_+, S_-, $S_\pm$ may be extended for interval-valued degrees as follows:

$$\begin{aligned} S_+(o) &= A(o) \otimes_I C(o); \\ S_-(o) &= \neg_I\big(A(o) \rightsquigarrow_I C(o)\big); \\ S_\pm(o) &= \neg_I A(o). \end{aligned} \tag{10}$$

Note that the degrees of S_+, S_- and $S_\pm$ become interval-valued too so that the sum $S_+(o) + S_-(o) + S_\pm(o)$ no longer results in $[1,1]$. It is because intervals only capture a limited knowledge; if all values were known exactly, the Ruspini condition (5) should hold again. Based on that, it is better to compute the confidence as

$$\mathrm{conf}(A \rightharpoonup C) = \frac{|S_+|}{1 - |S_\pm|} = \frac{\sum_{o \in \mathcal{O}}(A(o) \otimes_I C(o))}{\sum_{o \in \mathcal{O}} A(o)}. \tag{11}$$

4.1   Motivational Example

There are many attempts for searching for association rules on crisp data containing unknown values. In order to obtain intervals of support and confidence, it lasts to consider their most optimistic and pessimistic variants for each $o \in \mathcal{O}$. For instance, for a rule $A \rightharpoonup C$ and $o \in \mathcal{O}$ such that $A(o) = unknown$ and $C(o) = 0$, a pessimistic (resp. optimistic) variant is to take $A(o) = 1$ (resp. $A(o) = 0$).

The computation of support and confidence on (interval-valued) fuzzy sets depends on the selection of a t-norm as $\otimes_I$ and an implication function as $\rightsquigarrow_I$. In this paper, we fix

$$\begin{aligned} X \otimes_I Y &= [\min(x_1, y_1), \min(x_2, y_2)], \\ X \rightsquigarrow_I Y &= [\min(1, 1 - x_2 + y_1), \min(1, 1 - x_1 + y_2)]. \end{aligned}$$

Obtaining support based on interval-valued fuzzy sets is as easy as determination of $|S_+|$. Unfortunately, the computation of confidence is not so straightforward. Before providing a detailed theorem about confidence on interval-valued fuzzy sets, let us focus on a toy example that demonstrates some difficulties with that computation. Let us denote $\mathrm{supp}_D(A \rightharpoonup C) = [ls_D, us_D]$ and $\mathrm{conf}_D(A \rightharpoonup C) = [lc_D, uc_D]$, which are support and confidence of a rule $A \rightharpoonup C$ in dataset D.

Table 3. Example datasets

D_1:	A	C		D_2:	A	C		D_3:	A	C
o_1	1	0.3		o_1	1	0.1		o_1	1	0.3
o_2	$[0.2, 0.8]$	$[0.4, 0.5]$		o_2	$[0.2, 0.8]$	$[0.4, 0.5]$		o_2	$[0.2, 0.8]$	$[0.4, 0.5]$
								o_3	$[0.9, 1]$	$[0, 0.1]$

Example 1. Consider datasets D_1, D_2 and D_3 from Table 3.

- Supports can be obtained quite easily. $ls_{D_1} = \min(1, 0.3) + \min(0.2, 0.4) = 0.5$ and $us_{D_1} = \min(1, 0.3) + \min(0.8, 0.5) = 0.8$. Analogously for D_2 and D_3, we obtain $ls_{D_2} = 0.3$, $us_{D_2} = 0.6$, $ls_{D_3} = 0.5$, and $us_{D_3} = 0.9$.
- For both D_1 and D_2, the upper bound of the confidence is evidently reached if $A(o_2) = C(o_2) = 0.5$: $uc_{D_1} = \frac{0.3+0.5}{1+0.5} \doteq 0.53$, $uc_{D_2} = \frac{0.1+0.5}{1+0.5} = 0.4$. For other variants of $A(o_2)$ and $C(o_2)$, we obtain lower values of confidences. The similar holds for D_3, where the upper bound is reached for $A(o_2) = C(o_2) = 0.5$, $A(o_3) = 0.9$ and $C(o_3) = 0.1$.
- Unfortunately, the choice of such $A(o_2)$ and $C(o_2)$ that the confidence reaches its lower bound, does not depend only on the intervals of A and C for o_2, but also on the other values on other rows of the dataset. For D_1, the lower bound of confidence is reached if $A(o_2) = 0.8$ and $C(o_2) = 0.4$: $lc_{D_1} = \frac{0.3+0.4}{1+0.8} \doteq 0.39$. On the other hand, a lower bound of confidence for D_2 is obtained for $A(o_2) = 0.2$ and any $C(o_2) \in [0.4, 0.5]$: $lc_{D_2} = \frac{0.1+0.2}{1+0.2} = 0.25$. Similarly, a lower bound of confidence for D_3 is obtained for $A(o_2) = 0.2$, any $C(o_2) \in [0.4, 0.5]$, $A(o_3) = 1$, and $C(o_3) = 0$: $lc_{D_3} = \frac{0.3+0.2+0}{1+0.2+1} \doteq 0.23$.

Note that although o_2 is identical in all datasets D_1, D_2 and D_3, the minimal confidence is obtained from different values of $A(o_2)$ and $C(o_2)$. That is, the choice of the most pessimistic case (with respect to $\mathrm{conf}_{D_i}(A \rightharpoonup C)$) of $A(o_2)$ and $C(o_2)$ within the intervals *depends* not only on the intervals of $A(o_2)$ and $C(o_2)$, but also *on all other values in the dataset* D_i. This fundamental observation indicates that computing confidence threshold from interval-valued fuzzy sets must be more complex than in the case of crisp sets.

4.2 Computing Support and Confidence on Interval-Valued Data

Let D be a dataset that assigns an interval of truth values to each pair $\langle a, o \rangle$ of attribute $a \in \mathcal{A}$ and object $o \in \mathcal{O}$, i.e. $D(a, o) \in I$. Let us have a rule $A \rightharpoonup C$ such that $A, C \subseteq \mathcal{A}$ and $A \cap C = \emptyset$ and in the sense of (9) we have $A(o), C(o) \in I$ for all $o \in \mathcal{O}$.

Consider a mapping $w \colon \mathcal{O} \to [0, 1] \times [0, 1]$ such that $w(o) = \langle w_A(o), w_C(o) \rangle$ where $w_A(o) \in A(o)$ and $w_C(o) \in C(o)$ for any $o \in \mathcal{O}$. A class of all such mappings will be denoted with W. Then

$$\mathrm{supp}_w(A \rightharpoonup C) = \sum_{o \in \mathcal{O}} \min(w_A(o), w_C(o)) \in [0, |\mathcal{O}|], \qquad (12)$$

$$\mathrm{conf}_w(A \rightharpoonup C) = \frac{\sum_{o \in \mathcal{O}} \min(w_A(o), w_C(o))}{\sum_{o \in \mathcal{O}} w_A(o)} \in [0, 1]. \qquad (13)$$

The problem of finding $\text{supp}_D(A \rightharpoonup C) = [ls_D, us_D]$ and $\text{conf}_D(A \rightharpoonup C) = [lc_D, uc_D]$ is in finding such maximal ls_D, lc_D and minimal us_D, uc_D that

$$\forall w \in W : ls_D \leq \text{supp}_w(A \rightharpoonup C) \leq us_D,$$

$$\forall w \in W : lc_D \leq \text{conf}_w(A \rightharpoonup C) \leq uc_D.$$

Theorem 4. *Let $A(o) = [a_1(o), a_2(o)]$, $C(o) = [c_1(o), c_2(o)]$ and let $lw, uw \in W$ such that*

$$lw_A(o) = a_1(o), \qquad\qquad lw_C(o) = c_1(o),$$
$$uw_A(o) = a_2(o), \qquad\qquad uw_C(o) = c_2(o).$$

Then $\text{supp}_{lw}(A \rightharpoonup C) = ls_D$ and $\text{supp}_{uw}(A \rightharpoonup C) = us_D$.

Proof. Evident.

Theorem 5. *Let $A(o) = [a_1(o), a_2(o)]$, $C(o) = [c_1(o), c_2(o)]$ and let $lw, uw \in W$ such that*

$$lw_A(o) = \begin{cases} a_2(o), & \text{for } c_1(o) < a_1(o), \\ a_1(o), & \text{for } a_1(o) \leq c_1(o) \leq a_2(o) \text{ and } \dfrac{c_1(o) - a_1(o)}{a_2(o) - a_1(o)} > lc_D, \\ a_2(o), & \text{for } a_1(o) \leq c_1(o) \leq a_2(o) \text{ and } \dfrac{c_1(o) - a_1(o)}{a_2(o) - a_1(o)} \leq lc_D, \\ a_1(o), & \text{for } a_2(o) < c_1(o), \end{cases}$$

$$lw_C(o) = c_1(o),$$

$$uw_A(o) = \begin{cases} a_1(o), & \text{for } c_2(o) < a_1(o), \\ a_2(o), & \text{for } a_2(o) < c_2(o), \\ c_2(o), & \text{otherwise}, \end{cases}$$

$$uw_C(o) = c_2(o).$$

Then $\text{conf}_{lw}(A \rightharpoonup C) = lc_D$ and $\text{conf}_{uw}(A \rightharpoonup C) = uc_D$.

Sketch of the proof. For $w \in W$, let

$$M_w = \sum_{o' \in (\mathcal{O} \setminus \{o\})} \min(w_A(o'), w_C(o')),$$

$$N_w = \sum_{o' \in (\mathcal{O} \setminus \{o\})} w_A(o'),$$

then

$$\text{conf}_w(A \rightharpoonup C) = \frac{M_w + \min(w_A(o), w_C(o))}{N_w + w_A(o)}. \tag{14}$$

Let us examine the influence of a single fixed object $o \in \mathcal{O}$ on the resulting confidence, if M_w and N_w are fixed.

In order to maximize (resp. minimize) confidence (14), one has to select the greatest (resp. the lowest) $w_C(o)$. Hence evidently, $lw_C(o) = c_1(o)$ and $uw_C(o) = c_2(o)$ and we can assume $w_C(o)$ to be fixed from now on.

If varying only $w_A(o)$, the confidence (14) is non-decreasing for $w_A(o) \leq w_C(o)$ and non-increasing for $w_A(o) \geq w_C(o)$. Hence (14) is maximized for $w_C(o) = c_2(o)$ and such $w_A(o) \in [a_1(o), a_2(o)]$ that is closest to $w_C(o)$. A maximum of confidence (14) is evidently obtained for:

- $w_A(o) = a_1(o)$ if $c_2(o) < a_1(o)$,
- $w_A(o) = a_2(o)$ if $a_2(o) < c_2(o)$, and
- $w_A(o) = c_2(o)$ otherwise.

A minimum of confidence (14) is obtained for $w_C(o) = c_1(o)$ and for either $w_A(o) = a_1(o)$ or $w_A(o) = a_2(o)$. If $c_1(o) < a_1(o)$ then evidently the minimum appears for $w_A(o) = a_2(o)$. On the other hand, if $a_2(o) < c_1(o)$, we obtain minimum of (14) for $w_A(o) = a_1(o)$.

The most complicated situation appears if $a_1(o) \leq c_1(o) \leq a_2(o)$. We can rewrite (14) (while searching for the minimum confidence) as follows:

$$\mathrm{conf}_w(A \rightharpoonup C) = lc_D = \frac{M_w + a_1(o) + k(o) \cdot (c_1(o) - a_1(o))}{N_w + a_1(o) + k(o) \cdot (a_2(o) - a_1(o))}, \qquad (15)$$

where $k(o) = 0$ corresponds to $w_A(o) = a_1$ and $k(o) = 1$ corresponds to $w_A(o) = a_2$. To continue with the proof, the following lemma would be helpful:

Lemma 1. *Let $p, P \geq 0$, $q, Q > 0$ and $\frac{p}{q} \leq \frac{P}{Q}$ then*

$$\frac{p}{q} \leq \frac{p + P}{q + Q} \leq \frac{P}{Q}.$$

Proof of the lemma is evident.

If

$$lc_D < \frac{c_1(o) - a_1(o)}{a_2(o) - a_1(o)}$$

then

$$\frac{M_w + a_1(o) + k(o) \cdot (c_1(o) - a_1(o))}{N_w + a_1(o) + k(o) \cdot (a_2(o) - a_1(o))} < \frac{c_1(o) - a_1(o)}{a_2(o) - a_1(o)}$$

and accordingly to Lemma 1 we obtain

$$\frac{M_w + a_1(o)}{N_w + a_1(o)} \leq \frac{M_w + a_1(o) + k(o) \cdot (c_1(o) - a_1(o))}{N_w + a_1(o) + k(o) \cdot (a_2(o) - a_1(o))} < \frac{c_1(o) - a_1(o)}{a_2(o) - a_1(o)}.$$

Since lc_D is a minimal possible confidence, we immediately see that $\frac{M_w + a_1(o)}{N_w + a_1(o)} = lc_D$ and therefore $k(o) = 0$, which corresponds to $w_A(o) = a_1$.

The opposite case,

$$lc_D \geq \frac{c_1(o) - a_1(o)}{a_2(o) - a_1(o)},$$

leads using similar construction to $k(o) = 1$, which corresponds to $w_A(o) = a_2$, and the proof is finished.

Although the theorem is mathematically correct, it does not provide an exact recipe for finding such $lw \in W$ that $\text{conf}_{lw}(A \rightharpoonup C)$ is minimal among all $w \in W$, because for finding lw, we need to know the lower bound of the confidence, in advance. However, the theorem would help to formulate the efficient algorithm, which can be based on iterative traversal through $\mathcal{O}$, computing the estimates of lc_D and updating $k(o)$. Details of the algorithm are left for the future.

5 Conclusion and Future Work

In this paper, an association analysis framework was developed that allows to process data with missing values, i.e. values that are unknown or known only partially. The approach of Dubois et al. [24] for association rules was extended to handle interval-valued fuzzy sets. Based on that, an extended definition of association rule's support and confidence was proposed. It was shown that generalization of the computation of confidence interval is not straightforward. Some observations were discussed that would allow to develop an efficient computational algorithm in the future.

Acknowledgements. Authors acknowledge support by project "LQ1602 IT4Innovations excellence in science" and by GAČR 16-19170S.

References

1. Hájek, P.: The question of a general concept of the GUHA method. Kybernetika **4**, 505–515 (1968)
2. Hájek, P., Havránek, T.: Mechanizing Hypothesis Formation (Mathematical Foundations for a General Theory). Springer, Heidelberg (1978). https://doi.org/10.1007/978-3-642-66943-9
3. Agrawal, R., Srikant, R.: Fast algorithms for mining association rules. In: Proceedings of 20th International Conference on Very Large Databases, Chile, pp. 487–499. AAAI Press (1994)
4. Ralbovský, M.: Fuzzy GUHA. Ph.D. thesis, University of Economics, Prague (2009)
5. Yager, R.R.: A new approach to the summarization of data. Inf. Sci. **28**(1), 69–86 (1982)
6. Kacprzyk, J., Yager, R.R., Zadrożny, S.: A fuzzy logic based approach to linguistic summaries of databases. Int. J. Appl. Math. Comput. Sci. **10**(4), 813–834 (2000)
7. Murinová, P., Burda, M., Pavliska, V.: An algorithm for intermediate quantifiers and the graded square of opposition towards linguistic description of data. In: Kacprzyk, J., Szmidt, E., Zadrożny, S., Atanassov, K.T., Krawczak, M. (eds.) IWIFSGN/EUSFLAT-2017. AISC, vol. 642, pp. 592–603. Springer, Cham (2018). https://doi.org/10.1007/978-3-319-66824-6_52
8. Gelman, A., Hill, J.: Data Analysis Using Regression and Multilevel/hierarchical Models. Analytical methods for social research. Cambridge University Press, New York (2007)

9. Liu, Y., Gopalakrishnan, V.: An overview and evaluation of recent machine learning imputation methods using cardiac imaging data. Data **2**(1), 8 (2017)
10. Lukasiewicz, J.: O logice trojwartosciowej. Ruch filozoficzny **5**, 170–171 (1920)
11. Malinowski, G.: The Many Valued and Nonmonotonic Turn in Logic. North-Holand, Amsterdam (2007)
12. Bergmann, M.: An Introduction To Many-Valued and Fuzzy Logic: Semantics, Algebras, and Derivation Systems. Cambridge University Press, Cambridge New York (2008)
13. Ciucci, D., Dubois, D., Lawry, J.: Borderline vs. unknown: comparing three-valued representations of imperfect information. Int. J. Approx. Reason. **55**, 1866–1889 (2014)
14. Ciucci, D., Dubois, D.: Three-valued logics, uncertainty management and rough set. Int. J. Approx. Reason. **55**, 1866–1889 (2014)
15. Běhounek, L., Novák, V.: Towards fuzzy partial logic. In: Proceedings of the IEEE 45th International Symposium on Multiple-Valued Logics (ISMVL 2015), pp. 139–144 (2015)
16. Běhounek, L., Daňková, M.: Towards fuzzy partial set theory. In: Carvalho, J.P., Lesot, M.-J., Kaymak, U., Vieira, S., Bouchon-Meunier, B., Yager, R.R. (eds.) IPMU 2016. CCIS, vol. 611, pp. 482–494. Springer, Cham (2016). https://doi.org/10.1007/978-3-319-40581-0_39
17. Novák, V.: Towards fuzzy type theory with partial functions. In: Kacprzyk, J., Szmidt, E., Zadrożny, S., Atanassov, K.T., Krawczak, M. (eds.) IWIFSGN/EUSFLAT-2017. AISC, vol. 643, pp. 25–37. Springer, Cham (2018). https://doi.org/10.1007/978-3-319-66827-7_3
18. Murinová, P., Burda, M., Pavliska, V.: Undefined values in fuzzy logic. In: Kacprzyk, J., Szmidt, E., Zadrożny, S., Atanassov, K.T., Krawczak, M. (eds.) IWIFSGN/EUSFLAT -2017. AISC, vol. 642, pp. 604–610. Springer, Cham (2018). https://doi.org/10.1007/978-3-319-66824-6_53
19. Murinová, P., Pavliska, V., Burda, M.: Fuzzy association rules on data with undefined values. In: Medina, J., Ojeda-Aciego, M., Verdegay, J.L., Perfilieva, I., Bouchon-Meunier, B., Yager, R.R. (eds.) IPMU 2018. CCIS, vol. 855, pp. 165–174. Springer, Cham (2018). https://doi.org/10.1007/978-3-319-91479-4_14
20. Chen, Q., Kawase, S.: An approach towards consistency degrees of fuzzy theories. Fuzzy Sets Syst. **113**, 237–251 (2000)
21. Cornelis, C., Deschrijver, G., Kerre, E.E.: Implication in intuitionistic fuzzy and interval-valued fuzzy set theroy: construction, classification, application. Int. J. Approx. Reason. **35**, 55–95 (2004)
22. Zadeh, L.A.: Fuzzy sets. Inf. Control **8**, 338–353 (1965)
23. Klement, E.P., Mesiar, R., Pap, E.: Triangular Norms. Kluwer, Dordrecht (2000)
24. Dubois, D., Hüllermeier, E., Prade, H.: A systematic approach to the assessment of fuzzy association rules. Data Min. Knowl. Discov. **13**(2), 167–192 (2006)
25. Kawase, S., Chen, Q., Yanagihar, N.: On interval valued fuzzy reasoning. Trans. Japan Soc. Ind. Appl. Math 6, 285–296 (1996)
26. Geng, L., Hamilton, H.J.: Interestingness measures for data mining: a survey. ACM Comput. Surv. (CSUR) **38**(3), 9 (2006)

Fuzzy Hit-or-Miss Transform Using Uninorms

Pedro Bibiloni[1,2], Manuel González-Hidalgo[1,2], Sebastia Massanet[1,2(✉)],
Arnau Mir[1,2], and Daniel Ruiz-Aguilera[1,2]

[1] Department of Mathematics and Computer Science Soft Computing,
Image Processing and Aggregation (SCOPIA) Research Group,
University of the Balearic Islands, 07122 Palma de Mallorca, Spain
{p.bibiloni,manuel.gonzalez,s.massanet,arnau.mir,daniel.ruiz}@uib.es
[2] Balearic Islands Health Research Institute (IdISBa), 07010 Palma, Spain

Abstract. The Hit-or-Miss transform (HMT) is a morphological operator which has been successfully used to identify shapes and patterns satisfying certain geometric restrictions in an image. Recently, a novel HMT operator, called the fuzzy morphological HMT, was introduced within the framework of the fuzzy mathematical morphology based on fuzzy conjunctions and fuzzy implication functions. Taking into account that the particular case of considering a t-norm as fuzzy conjunction and its residual implication as fuzzy implication functions has proved its potential in several applications, in this paper, the case when residual implications derived from uninorms and a general fuzzy conjunction, possibly a t-norm or the same uninorm, is deeply analysed. In particular, some theoretical results related to properties desirable for the applications are proved. Finally, some experimental results are presented showing the potential of this choice of operator to detect shapes and patterns in images.

Keywords: Fuzzy hit-or-miss transform
Fuzzy mathematical morphology · Uninorm
Fuzzy implication function

1 Introduction

The hit-or-miss transform (HMT) [16,30] is an operator within the framework of mathematical morphology which is capable of identifying groups of connected pixels satisfying certain geometric restrictions or forming a certain configuration. Already introduced in the mathematical morphology developed for binary images by Matheron [20] (see also [30]) in the early sixties, its translation to

This paper has been partially supported by the Spanish Grant TIN2016-75404-P, AEI/FEDER, UE. P. Bibiloni also benefited from the fellowship FPI/1645/2014 of the *Conselleria d'Educació, Cultura i Universitats* of the *Govern de les Illes Balears* under an operational program co-financed by the European Social Fund.

V. Torra et al. (Eds.): MDAI 2018, LNAI 11144, pp. 101–113, 2018.
https://doi.org/10.1007/978-3-030-00202-2_9

the mathematical morphology for grey-level images (see [16,30,32]) has been a complex task. HMT consists in searching and locating a predefined shape, called structuring element (SE), in an image. In fact, this SE is composed of two SEs which match the geometry of the objects of interest both in the foreground and background of the image. This is the reason why the translation to grey-level images is so difficult. The concept of complement of an image must be used in the definition of the HMT but there is no clear consensus about how to define it for a grey-level image.

In spite of the aforementioned complexity, many authors have proposed extensions of HMT to grey-level images (see [10,17,22,24–26,29,31,32] and the review in [13]). Even some extensions of HMT to multivariate images have been presented in [1,33,34]. Besides all these approaches, a particularly important extension in the context of the fuzzy mathematical morphology has been introduced recently [13]. The so-called Fuzzy Mathematical HMT (FMHMT for short) uses the concepts and techniques from the fuzzy sets theory [5,21] in order to allow a better treatment and a representation with greater flexibility of the uncertainty and ambiguity present in any level of an image. Indeed, FMHMT uses the concept of fuzzy negation to model the complement of the image solving straightforwardly the problems that affect to other approaches. Moreover, it does not only detect the parts of the image which are equal to the structuring element, but also it can be interpreted as a similarity degree obtained as the aggregation of how similar are, on the one hand, the foreground SE to the image and on the other hand, the background SE to the complement image.

The FMHMT operator uses in its definition, in addition to the already mentioned fuzzy negation, the fuzzy mathematical erosion generated from a fuzzy implication function [21] and a fuzzy conjunction. In [13], several desirable theoretical properties of the FMHMT operator were studied for general fuzzy implication functions and fuzzy conjunctions. However, the main efforts were devoted to study the FMHMT operator when a t-norm and its residual implication were considered. Some experimental results showed there the potential of the FMHMT which, in addition to having a solid theoretical background, performed notably well from the applicational point of view in comparison with other existing approaches.

Following this line of research, the main goal of this paper is to study the fulfilment of the desirable theoretical properties and to analyse the performance of the FMHMT operator when residual implications derived from uninorms and a general conjunction, possibly a t-norm or the same uninorm, are considered. The fuzzy mathematical morphology based on uninorms was introduced in [9] and fully developed in [11]. This theory has been applied to image processing, providing remarkable results, especially in edge detection and noise removal [12, 14,15]. The most important asset of this theory is that it generally improves the results obtained by the fuzzy mathematical morphology generated by t-norms and consequently, this paper seeks to establish the theoretical background of the FMHMT using uninorms and to show some preliminary experimental results.

This paper is organized as follows. In Sect. 2, we recall the definitions of the fuzzy mathematical operators and the underlying fuzzy logical connectives which will be used in subsequent sections to make the paper self-contained. In Sect. 3, we briefly recall the binary HMT and the FMHMT, which generalizes the binary HMT in the context of the fuzzy mathematical morphology. Then, Sect. 4 is devoted to the study of the theoretical desirable properties of the FMHMT when residual implications derived from uninorms and a general conjunction are considered. In Sect. 5, we show how this operator can be used to perform object detection and we illustrate its abilities by applying it to different situations. The paper ends with some conclusions and future work we want to develop.

2   Preliminaries

In this section we will introduce some preliminaries about fuzzy logic and fuzzy mathematical morphology operators that will be used throughout the paper.

Fuzzy mathematical morphology operators are defined from fuzzy conjunctions, fuzzy negations and and fuzzy implication functions. More details on these logical connectives can be found in [2,4]. First, we introduce fuzzy conjunctions.

Definition 1. *A non-decreasing binary operator $C : [0,1]^2 \to [0,1]$ is called a fuzzy conjunction if it satisfies $C(0,1) = C(1,0) = 0$ and $C(1,1) = 1$.*

A deeply studied kind of fuzzy conjunctions is the class of t-norms [18].

Definition 2. *A fuzzy conjunction T on $[0,1]$ is called a t-norm when it is commutative, associative and it satisfies $T(1,x) = x$ for all $x \in [0,1]$.*

Well-known t-norms are the minimum t-norm $T_{\mathbf{M}}(x,y) = \min(x,y)$, the product t-norm $T_{\mathbf{P}}(x,y) = x \cdot y$, the Łukasiewicz t-norm $T_{\mathbf{LK}}(x,y) = \max(x+y-1,0)$, and the nilpotent minimum t-norm

$$T_{\mathbf{nM}}(x,y) = \begin{cases} 0, & \text{if } x+y \le 1, \\ \min(x,y), & \text{otherwise.} \end{cases}$$

Another kind of operators are uninorms, introduced by Yager and Rybalov in [35].

Definition 3. *A non-decreasing binary operator $U : [0,1]^2 \to [0,1]$ is called a uninorm if it is associative, commutative and there exists $e \in [0,1]$ (called neutral element) such that $U(x,e) = U(e,x) = x$ for all $x \in [0,1]$.*

All uninorms satisfy $U(0,1) \in \{0,1\}$, being a fuzzy conjunction when $U(0,1) = 0$. Different classes of uninorms have been studied and characterized (see [19]). In this paper we will use two of them: idempotent and representable ones.

Definition 4. *A uninorm U is* idempotent *if $U(x,x) = x$ for all $x \in [0,1]$.*

Idempotent uninorms have been characterized in [28] by means of a decreasing function g satisfying certain properties. Any idempotent uninorm U with neutral element e and associated function g will be denoted by $U \equiv \langle g, e \rangle_{\text{ide}}$.

Definition 5. *A conjunctive uninorm U with neutral element $e \in]0,1[$ is* representable *if there exists a continuous and strictly increasing function $h : [0,1] \to [-\infty, +\infty]$ (called* additive generator *of U), with $h(0) = -\infty$, $h(e) = 0$ and $h(1) = +\infty$ such that U is given by*

$$U_h(x,y) = h^{-1}(h(x) + h(y))$$

for all $(x,y) \in [0,1]^2 \setminus \{(0,1),(1,0)\}$ and $U(0,1) = U(1,0) = 0$.

Any representable uninorm U with neutral element e and additive generator h will be denoted by $U \equiv \langle h, e \rangle_{\text{rep}}$.

Now we introduce fuzzy negations and fuzzy implication functions.

Definition 6. *A non-increasing function $N : [0,1] \to [0,1]$ is called a* strong fuzzy negation *if it is an involution, i.e., if $N(N(x)) = x$ for all $x \in [0,1]$.*

Definition 7. *A binary operator $I : [0,1]^2 \to [0,1]$ is a* fuzzy implication function *if it is non-increasing in the first variable, non-decreasing in the second one and it satisfies $I(0,0) = I(1,1) = 1$ and $I(1,0) = 0$.*

A well-known way to obtain fuzzy implication functions from fuzzy conjunctions is the residuation method. Given a fuzzy conjunction C such that $C(1,x) > 0$ for all $x > 0$, the binary operator

$$I_C(x,y) = \sup\{z \in [0,1] \mid C(x,z) \leq y\}$$

is a fuzzy implication function called the *residual implication* or *R-implication* of C (see [23]). When the considered conjunction is an idempotent uninorm $U \equiv \langle g, e \rangle_{\text{ide}}$ with $g(0) = 1$, then its residual implication has the following expression [27]:

$$I_U(x,y) = \begin{cases} \min(g(x), y), & \text{if } x < y, \\ \max(g(x), y), & \text{if } x \geq y. \end{cases}$$

On the other hand, if the considered conjunction is a representable uninorm $U \equiv \langle h, e \rangle_{\text{rep}}$, its residual implication is given by [7]:

$$I_U(x,y) = \begin{cases} h^{-1}(h(y) - h(x)) & \text{if } (x,y) \in [0,1]^2 \setminus \{(0,0),(1,1)\}, \\ 1 & \text{otherwise.} \end{cases}$$

Using the previous operators, the basic fuzzy morphological operators such as dilation and erosion can be defined. We use the framework introduced by De Baets in [6] that is based on the duality under negation. We will use the following notation: C denotes a fuzzy conjunction; I, a fuzzy implication function; A, a grey-level image; and B, a grey-level structuring element (see [8] for formal definitions). In addition, d_A denotes the set of points where A is defined and $T_v(A)$ is the translation of a fuzzy set A by $v \in \mathbb{R}^n$ defined by $T_v(A)(x) = A(x - v)$.

Definition 8 ([21]). *The fuzzy dilation $D_C(A, B)$ and the fuzzy erosion $E_I(A, B)$ of A by B are the grey-level images defined by*

$$D_C(A, B)(y) = \sup_{x \in d_A \cap T_y(d_B)} C(B(x - y), A(x)),$$
$$E_I(A, B)(y) = \inf_{x \in d_A \cap T_y(d_B)} I(B(x - y), A(x)).$$

Some of the algebraic properties satisfied by the fuzzy erosion and the fuzzy dilation are studied in [6,8].

3   From Binary to Fuzzy Hit-or-Miss Transform

In this section, we will recall for the sake of completeness the binary hit-or-miss transform and how this operator is generalized to the fuzzy hit-or-miss transform. First, the hit-or-miss transform of a binary image is a classical morphological operator [30,32], that uses two structuring elements B_{FG} (or foreground structuring element) and B_{BG} (or background structuring element). The basic idea is to extract all those pixels of a binary image that are surrounded by areas on the image where both foreground and background structuring elements match predefined patterns. By definition, B_{FG} and B_{BG} share the same origin and $B_{FG} \cap B_{BG} = \emptyset$. We use $B = (B_{FG}, B_{BG})$ to denote the composite structuring element (SE).

Formally, the HMT of a binary image A by the composite SE B is the set of points x such that when the origin of B coincides with x, B_{FG} fits A while B_{BG} fits A^c (the complement of A):

$$A \circledast B = \{x \ : \ (B_{FG})_x \subseteq A, \ (B_{BG})_x \subseteq A^c\} = (A \ominus B_{FG}) \cap (A^c \ominus B_{BG}),$$

where $(\cdot)_x$ denotes the translation by x and $\ominus$ is the binary erosion operator $A \ominus B = \{x \ : \ B_x \subseteq A\}$.

While this operator for binary images is well-defined and easily interpretable from a geometric point of view, its extension to grey-level images is tough because it is not an increasing operator and there is no universally accepted notion of complement for grey-level images. Thus, in most extensions to this type of images [3,17,22,24,26,29,31,32], the use of A^c is systematically avoided and it is only implemented using a fuzzy negation in some of the fuzzy approaches [10,25]. For an exhaustive review on the state of the art of the hit-or-miss transform, the different approaches presented in the literature and an experimental comparison, we refer the reader to [13].

Focusing already on the fuzzy hit-or-miss transform introduced in [13], or FMHMT for short, this approach is able to directly generalize the binary HMT to the fuzzy mathematical morphology paradigm by modeling the complement operation by a fuzzy negation and the intersection of sets by a fuzzy conjunction. Moreover, since B_{FG} and B_{BG} can be considered now as fuzzy sets, there is no need for a condition such as $B_{FG} \cap B_{BG} = \emptyset$, which was mandatory in the binary case, since in the fuzzy case, when this condition does not hold, the fuzzy hit-or-miss transform still may be a non-empty set.

In the following, for any operator $C : [0,1]^2 \rightarrow [0,1]$, the expression $C(A, B)$ represents the function that at any point x is such that $C(A, B)(x) = C(A(x), B(x))$ and for any strong fuzzy negation N, the N-dual of a fuzzy set A, denoted by $N(A)$, is defined by $N(A)(x) = N(A(x))$. Thus, the formal definition of the FMHMT is given as follows.

Definition 9 ([13, **Definition 4.1**]). *Let C be a fuzzy conjunction, I be a fuzzy implication function and N be a strong fuzzy negation. The fuzzy morphological hit-or-miss transform (FMHMT) of the grey-level image A with respect to the grey-level structuring element $B = (B_1, B_2)$ is defined, for any $y \in d_A$, by*

$$FMHMT_{C,I,N}(A, B)(y) = C\left(E_I(A, B_1)(y), E_I(N(A), B_2)(y)\right), \qquad (1)$$

where $N(A)(x) = N(A(x))$ for all $x \in d_A$.

In [13] several general properties of the FMHMT are presented and studied. As we have already commented, the first requirement of any extension is accomplished. Indeed, we retrieve the binary hit-or-miss transform when we apply the FMHMT to a binary image A and a binary structuring element B. Moreover, it is proved that the FMHMT is invariant under translations and whenever C is a fuzzy conjunction with right neutral element 1 (that is, $C(x, 1) = x$ for all $x \in [0,1]$) the erosion and the dilation are particular cases of the FMHMT.

4   Fuzzy Hit or Miss Transform Using Uninorms: Definition and Results

In this section, we perform an in-depth study of the theoretical desirable properties of the FMHMT when residual implications derived from uninorms are considered. Thus, in the next results we consider the particular case of the FMHMT given in general by Eq. (1), when $I = I_U$ where U is a uninorm, that is,

$$\text{FMHMT}_{C,I_U,N}(A, B) = C(E_{I_U}(A, B_1), E_{I_U}(N(A), B_2)).$$

Note that we consider a general fuzzy conjunction C. However, in some of the results, the same uninorm U which has already been used to construct the residual implication I_U is considered as the fuzzy conjunction or even a t-norm T.

First of all, we have to check that the FMHMT detects the structuring element at a point y when it is a part of the image at that point and the considered uninorm is used as a fuzzy conjunction.

Definition 10. *Let B_1 be a grey-level image (a structuring element) and let A be a grey-level image. We say that B_1 is a part of A if there exists a point $y \in d_A$ such that if we translate B_1 to y, we have $B_1(x - y) = A(x)$ for all $x \in d_{T_y(B_1)} \cap d_A$. In this case we say that B_1 is a part of A at the point y.*

In fact, as the following result proves, the value of the FMHMT of the image at that point is the neutral element of the uninorm.

Theorem 1. *Let N be a strong negation, $B = (B_1, B_2)$ be a grey-scale structuring element where $B_2 = N(B_1)$ and B_1 is a part of A at the point y. Let U be a conjunctive uninorm with neutral element $e \in [0, 1]$. If one of the following two cases hold:*

- *U is a representable uninorm, or*
- *U is a left-continuous idempotent uninorm $U \equiv \langle N, e \rangle_{ide}$ and there is a point t such that $B(t) = e$;*

then $FMHMT_{U, I_U, N}(A, B)(y) = e$.

Proof. As B_1 is a part of A at the point y, $B_1(x - y) = A(x)$ for all x, and we have

$$E_{I_U}(A, B_1)(y) = \inf_x I_U(B_1(x - y), A(x)) = \inf_x I_U(A(x), A(x)),$$

and, on the other hand, $E_{I_U}(N(A), B_2)(y) = \inf_x I_U(N(B_1)(x - y), N(A(x))) = \inf_x I_U(N(A(x)), N(A(x)))$.

Now, if U is a representable uninorm, we have that I_U satisfies $I_U(x, x) = e$, and then

$$E_{I_U}(A, B_1)(y) = \inf_x I_U(A(x), A(x)) = \inf_x e = e.$$

Similarly, $E_{I_U}(N(A), B_2)(y) = e$, and the value of the Fuzzy Hit-or-Miss transform at the point y is:

$$FMHMT_{U, I_U, N}(A, B)(y) = U(E_{I_U}(A, B_1)(y), E_{I_U}(N(A), B_2)(y)) = U(e, e) = e.$$

Now, if U is a left-continuous idempotent uninorm associated to the strong negation N, its residual implicator I_U satisfies $I_U(x, x) = \max(x, N(x))$, and then

$$E_{I_U}(A, B_1)(y) = \inf_x I_U(A(x), A(x)) = \inf_x \max(N(A(x)), A(x)) = e,$$
$$E_{I_U}(N(A), B_2)(y) = \inf_x I_U(N(A(x)), N(A(x))) = \inf_x \max(N(A(x)), A(x)) = e.$$

Thus, $FMHMT_{U, I_U, N}(A, B)(y) = U(E_{I_U}(A, B_1)(y), E_{I_U}(N(A), B_2)(y)) = U(e, e) = e.$ □

Next, we study the general case. What happens when the structuring element is not a part of the image? To obtain a behaviour desirable for the application point of view, we have to consider a t-norm as the fuzzy conjunction. More concretely, in the following theorems, we see that if the structuring element is not a part of the image, the neutral element of the uninorm acts as an upper bound of the FMHMT.

We will distinguish the case when the uninorm is representable or idempotent because in the latter, the proof is more technical and we need an additional proposition.

Theorem 2. *Let T be a t-norm, N a strong negation and U be a conjunctive representable uninorm. If A is a grey level image and B_1 is a structuring element that is not a part of A. Then for all point $y \in d_A$ we have that*

$$FMHMT_{T,I_U,N}(A,B)(y) = T(E_{I_U}(A,B_1)(y), E_{I_U}(N(A),N(B_1))(y)) \le e.$$

Proof. Suppose by contradiction that there exists a point y such that

$$FMHMT_{T,I_U,N}(A,B)(y) = T(E_{I_U}(A,B_1)(y), E_{I_U}(N(A),N(B_1))(y)) > e.$$

As T is a t-norm and $T(a,b) \le \min(a,b)$ for all $a,b \in [0,1]$, we have that

$$E_{I_U}(A,B_1)(y) > e \implies \inf_x I_U(B_1(x-y), A(x)) > e, \text{ and}$$

$$E_{I_U}(N(A),N(B_1))(y) > e \implies \inf_x I_U(N(B_1(x-y)), N(A(x))) > e.$$

Then $I_U(B_1(x-y), A(x)) > e$ and $I_U(N(B_1(x-y)), N(A(x))) > e$ for all x. As U is a representable uninorm and I_U its residual implicator, it satisfies that $I_U(a,a) = e$ for all $a \in [0,1]$, then we have for all x,

$$I_U(B_1(x-y), A(x)) > e = I_U(A(x), A(x)) \implies A(x) \le B_1(x-y), \text{ and}$$

$$I_U(N(B_1(x-y)), N(A(x))) > e \implies N(A(x)) \le N(B_1(x-y)).$$

So, we have that $B_1(x-y) = A(x)$ for all x. In other words, B_1 is a part of A at the point y, which contradicts the hypothesis of the theorem. □

Proposition 1. *Let N be a strong negation and U be a conjunctive left-continuous idempotent uninorm with neutral element e. Suppose that there are $x, y \in [0,1]$ such that $I_U(x,y) > e$ and $I_U(N(x),N(y)) > e$, then $x = y$.*

Proof. If $x > y$, as U is a idempotent uninorm and I_U its residual implicator we have

$$I_U(x,y) = \min(N(x), y) > e, \text{ then } N(x) > e, \ y > e \text{ so } x < e, N(y) < e. \quad (2)$$

But, also we have that $I_U(N(x), N(y)) > e$ then $\max(x, N(y)) > e$ but this is impossible by (2). So, the case $x > y$ can not be given.

If $x < y$, we have $I_U(x,y) = \min(x, N(y)) > e$, then $x > e$, $N(y) > e$ so $y < e, N(x) < e \implies \max(y, N(x)) < e$.

But using that $I_U(x,y) > e$ we obtain $\max(y, N(x)) > e$, a contradiction. □

Theorem 3. *Let T be a t-norm, N a strong negation and U be a conjunctive left-continuous idempotent uninorm. If A is a grey level image, B_1 is a structuring element and y a point where*

$$FMHMT_{T,I_U,N}(A,B)(y) = T(E_{I_U}(A,B_1)(y), E_{I_U}(N(A),N(B_1))(y)) > e$$

then B_1 is a part of A in the point y.

Proof. If $FMHMT_{T,I_U,N}(A,B)(y) > e$, we have that $I_U(B_1(x-y), A(x)) \geq e$ for all x. In the same way, $I_U(N(B_1(x-y)), N(A(x))) \geq e$ for all x. Then, using Proposition 1 we obtain that $B_1(x-y) = A(x)$ for all x. So B_1 is a part of A at the point y. □

To have an idea of the behaviour of the FMHMT in the case of representable uninorms, we can compute the value of the FMHMT when the structuring element and the image are constants in all their domains. While Proposition 2 gives the general expression for a general fuzzy conjunction, Corollary 1 particularizes the result when the corresponding representable uninorm or a t-norm of the ones collected in Sect. 2 is considered.

Proposition 2. *Let C be a fuzzy conjunction, $U = \langle h,e \rangle_{rep}$ a conjunctive representable uninorm, I_U its R-implication, N a strong negation, A a grey-level image, $B = (B_1, N(B_1))$ a grey-level structuring element such that $B_1(x) = m$ for all $x \in d_{T_y(B_1)}$, and $y \in d_A$. Suppose that $A(x) = k$ for all $x \in d_{T_y(B_1)}$. Then we have that*

$$FMHMT_{C,I_U,N}(A,B)(y) = C(h^{-1}(h(k)-h(m)), h^{-1}(h(N(k))-h(N(m)))).$$

Proof. Simply apply the definition of FMHMT and take into account that $B_1(x) = m$ for all $x \in d_{T_y(B_1)}$ and $y \in d_A$ and $A(x) = k$ for all $x \in d_{T_y(B_1)}$.

Corollary 1. *Let T be a t-norm, $U = \langle \ln \frac{x}{1-x}, e \rangle_{rep}$, I_U its R-implication, $N(x) = 1 - x$, A a grey-level image, $B = (B_1, N(B_1))$ a grey-level structuring element such that $B_1(x) = m$ for all $x \in d_{T_y(B_1)}$, and $y \in d_A$. Suppose that $A(x) = k$ for all $x \in d_{T_y(B_1)}$ with $m < k$. Then we have that*

$$FMHMT_{T,I_U,N}(A,B)(y) = \begin{cases} \min\left(\frac{k-km}{k+m-2km}, \frac{m-km}{k+m-2km}\right), & \text{if } T = T_{\mathbf{M}}, \\ \frac{(1-k)(1-m)km}{(k+m-2km)^2}, & \text{if } T = T_{\mathbf{P}}, \\ 0, & \text{if } T = T_{\mathbf{LK}}, \\ 0, & \text{if } T = T_{\mathbf{nM}}. \end{cases}$$

and $FMHMT_{U,I_U,N}(A,B)(y) = h^{-1}(h(k)-h(m)+h(N(k))-h(N(m)))$.

5   Experimental Results

In this section, we check the functionality of the Hit-or-Miss transform using two synthetic images with geometric shapes. In these experiments, we consider as fuzzy conjunction the minimum t-norm and the residual implication derived from the representable uninorm given by $U = \langle \ln \left(\frac{x}{1-x} \right), 0.5 \rangle_{rep}$ and $N(x) = 1 - x$.

The geometric shapes of the first image (Fig. 1(a)) consist of some squares, a rhombus and an ellipse. The squares given in Fig. 1(b) are considered as the structuring element. The aim of this experiment is to detect all the squares with a size greater or equal that the size of the structuring element. Figure 1(c) and (d) show the results of the FMHMT operator. As it can be shown, all the squares with the previous characteristics are detected. The grey level is 0.5 (maximum) in the pixels pertaining to the square that exactly matches the structuring element (in shape and grey level), as predicted by Theorem 1.

The geometric shapes of the second image (Fig. 2(a)) are blobs and the structuring element is a blob (Fig. 2(b)). The aim of this second experiment is to detect all the blobs with the same size than the structuring element. Figure 2(c) and (d) show the results where we can see that all the blobs we wanted have been detected.

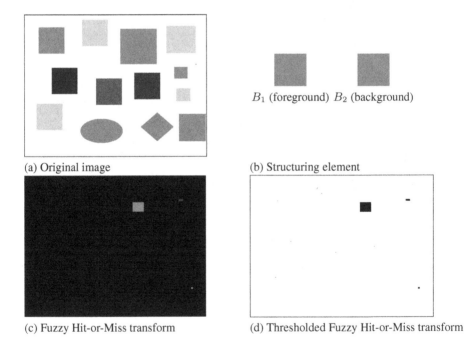

B_1 (foreground) B_2 (background)

(a) Original image

(b) Structuring element

(c) Fuzzy Hit-or-Miss transform

(d) Thresholded Fuzzy Hit-or-Miss transform

Fig. 1. FMHMT, displayed in (c) and a thresholded version in (d), of the original image given in (a) using $C = T_M$, $U = \langle \ln \left(\frac{x}{1-x} \right), 0.5 \rangle_{rep}$ and $N(x) = 1 - x$ by the structuring element $B = (B_1, B_2)$ shown in (b). We can see that the squares with a size greater than the size of the structuring element have been detected.

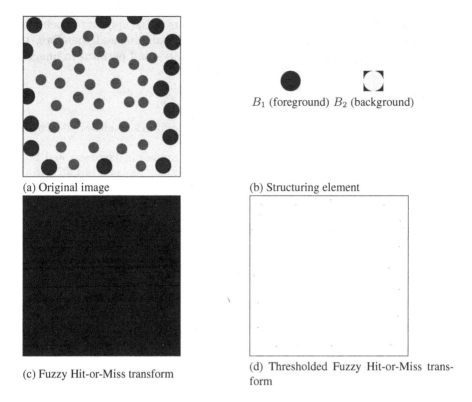

B_1 (foreground) B_2 (background)

(a) Original image

(b) Structuring element

(c) Fuzzy Hit-or-Miss transform

(d) Thresholded Fuzzy Hit-or-Miss transform

Fig. 2. FMHMT, displayed in (c) and a thresholded version in (d), of the original image given in (a) using $C = T_M$, $U = \langle \ln\left(\frac{x}{1-x}\right), 0.5 \rangle_{rep}$ and $N(x) = 1 - x$ by the structuring element $B = (B_1, B_2)$ shown in (b). We can see that the blobs with the same size than the structuring element have been detected.

6   Conclusions and Future Work

In this paper, the Fuzzy Morphological Hit-or-Miss transform, useful to identify shapes and patterns in images, has been analysed when a residual implication derived from uninorms is considered. We have studied the properties of this operator for a general fuzzy conjunction and for the particular cases of a uninorm or a t-norm. In the first case, we proved that it detects the corresponding shape when the structuring element is a part of the image and in the second case, we have seen that the neutral element of the uninorm acts as an upper bound of the value of the operator.

In the experimental section, we have considered some synthetic images with geometric figures to show how the FMHMT detects some figures that have a similar shape than the structuring element. The FMHMT detects those desired geometric figures whose size is greater or equal than the structuring element.

As a future work, we plan to study the behaviour of the operator depending on the choice of the representative or idempotent uninorm and therefore, to determine which uninorms are better to identify shapes and patterns in images.

References

1. Aptoula, E., Lefèvre, S., Ronse, C.: A hit-or-miss transform for multivariate images. Pattern Recognit. Lett. **30**(8), 760–764 (2009)
2. Baczyński, M., Jayaram, B.: Fuzzy Implications. Studies in Fuzziness and Soft Computing, vol. 231. Springer, Heidelberg (2008). https://doi.org/10.1007/978-3-540-69082-5
3. Barat, C., Ducottet, C., Jourlin, M.: Pattern matching using morphological probing. In: Proceedings of the International Conference on Image Processing, ICIP 2003, vol. 1, pp. 369–372 (2003)
4. Beliakov, G., Pradera, A., Calvo, T.: Aggregation Functions: A Guide for Practitioners. Studies in Fuzziness and Soft Computing, vol. 221. Springer, Heidelberg (2007). https://doi.org/10.1007/978-3-540-73721-6
5. Bloch, I., Maître, H.: Fuzzy mathematical morphologies: a comparative study. Pattern Recognit. **28**, 1341–1387 (1995)
6. De Baets, B.: Fuzzy morphology: a logical approach. In: Ayyub, B.M., Gupta, M.M. (eds.) Uncertainty Analysis in Engineering and Science: Fuzzy Logic. Statistics, and Neural Network Approach, pp. 53–68. Kluwer Academic Publishers, Norwell (1997)
7. De Baets, B., Fodor, J.: Residual operators of uninorms. Soft Comput. **3**, 89–100 (1999)
8. De Baets, B., Kerre, E., Gupta, M.: The fundamentals of fuzzy mathematical morphologies part I: basics concepts. Int. J. Gen. Syst. **23**, 155–171 (1995)
9. De Baets, B., Kwasnikowska, N., Kerre, E.: Fuzzy morphology based on uninorms. In: Proceedings of the Seventh IFSA World Congress, Prague, pp. 215–220 (1997)
10. Deng, T.-Q.: Fuzzy logic and mathematical morphology. Technical report. Centrum voor Wiskunde en Informatica, Amsterdam, The Netherlands, October 2000
11. González, M., Ruiz-Aguilera, D., Torrens, J.: Algebraic properties of fuzzy morphological operators based on uninorms. In: Artificial Intelligence Research and Development. Frontiers in Artificial Intelligence and Applications, vol. 100, pp. 27–38. IOS Press (2003)
12. González-Hidalgo, M., Massanet, S., Mir, A., Ruiz-Aguilera, D.: On the choice of the pair conjunction-implication into the fuzzy morphological edge detector. IEEE Trans. Fuzzy Syst. **23**(4), 872–884 (2015)
13. González-Hidalgo, M., Massanet, S., Mir, A., Ruiz-Aguilera, D.: A fuzzy morphological hit-or-miss transform for grey-level images. Fuzzy Sets Syst. **286**(C), 30–65 (2016)
14. González-Hidalgo, M., Mir-Torres, A.: Noise reduction using alternate filters generated by fuzzy mathematical operators using uninorms (ϕMM-U morphology). In: Burillo, P. (ed.) EUROFUSE Workshop 2009. Preference Modelling and Decision Analysis, pp. 233–238. Public University of Navarra, Pamplona (2009)
15. González-Hidalgo, M., Mir-Torres, A., Ruiz-Aguilera, D., Torrens, J.: Image analysis applications of morphological operators based on uninorms. In: Carvalho, P., et al. (eds.) Proceedings of the IFSA-EUSFLAT 2009 Conference, Lisbon, Portugal, pp. 630–635 (2009)

16. Heijmaans, H.: Morphological Image Operators. Academic Press, Boston (1994)
17. Khosravi, M., Schafer, R.W.: Template matching based on a grayscale hit-or-miss transform. IEEE Trans. Image Process. **5**(6), 1060–1066 (1996)
18. Klement, E., Mesiar, R., Pap, E.: Triangular Norms. Kluwer Academic Publishers, London (2000)
19. Mas, M., Massanet, S., Ruiz-Aguilera, D., Torrens, J.: A survey on the existing classes of uninorms. J. Intell. Fuzzy Syst. **29**, 1021–1037 (2015)
20. Matheron, G.: Random Sets and Integral Geometry. Wiley, Hoboken (1975)
21. Nachtegael, M., Kerre, E.: Classical and fuzzy approaches towards mathematical morphology, Chap. 1. In: Kerre, E.E., Nachtegael, M. (eds.) Fuzzy Techniques in Image Processing. Studies in Fuzziness and Soft Computing, vol. 52, pp. 3–57. Physica-Verlag, New York (2000)
22. Naegel, B., Passat, N., Ronse, C.: Grey-level hit-or-miss transforms-part I: unified theory. Pattern Recognit. **40**, 635–647 (2007)
23. Ouyang, Y.: On fuzzy implications determined by aggregation operators. Inf. Sci. **193**, 153–162 (2012)
24. Perret, B., Lefèvre, S., Collet, C.: A robust hit-or-miss transform for template matching applied to very noisy astronomical images. Pattern Recognit. **42**, 2470–24890 (2009)
25. Popov, A.T.: General approach for fuzzy mathematical morphology. In: Proceedings of 8th International Symposium on Mathematical Morphology, ISMM, pp. 39–48 (2007)
26. Raducanu, B., Grana, M.: A grayscale hit-or-miss transform based on level sets. In: Proceedings of the IEEE International Conference on Image Processing, Vancouver, BC, Canada, pp. 931–933 (2000)
27. Ruiz, D., Torrens, J.: Residual implications and co-implications from idempotent uninorms. Kybernetika **40**, 21–38 (2004)
28. Ruiz-Aguilera, D., Torrens, J., De Baets, B., Fodor, J.: Some remarks on the characterization of idempotent uninorms. In: Hüllermeier, E., Kruse, R., Hoffmann, F. (eds.) IPMU 2010. LNCS, vol. 6178, pp. 425–434. Springer, Heidelberg (2010). https://doi.org/10.1007/978-3-642-14049-5_44
29. Schaefer, R., Casasent, D.: Nonlinear optical hit–miss transform for detection. Appl. Opt. **34**(20), 3869–3882 (1995)
30. Serra, J.: Image Analysis and Mathematical Morphology, vol. 1. Academic Press, London (1982)
31. Sinha, D., Sinha, P., Douherty, E., Batman, S.: Design and analysis of fuzzy morphological algorithms for image processing. IEEE Trans. Fuzzy Syst. **5**(4), 570–584 (1997)
32. Soille, P.: Morphological Image Analysis, 2nd edn. Springer, Heidelberg (2004). https://doi.org/10.1007/978-3-662-05088-0
33. Velasco-Forero, S., Angulo, J.: Hit-or-miss transform in multivariate images. In: Blanc-Talon, J., Bone, D., Philips, W., Popescu, D., Scheunders, P. (eds.) ACIVS 2010. LNCS, vol. 6474, pp. 452–463. Springer, Heidelberg (2010). https://doi.org/10.1007/978-3-642-17688-3_42
34. Weber, J., Lefèvre, S.: Spatial and spectral morphological template matching. Image Vis. Comput. **30**, 934–945 (2012)
35. Yager, R., Rybalov, A.: Uninorm aggregation operators. Fuzzy Sets Syst. **80**, 111–120 (1996)

Learning Fuzzy Measures for Aggregation in Fuzzy Rule-Based Models

Emran Saleh[1(✉)], Aida Valls[1], Antonio Moreno[1], Pedro Romero-Aroca[2],
Vicenç Torra[3], and Humbert Bustince[4]

[1] Departament d'Enginyeria Informàtica i Matemàtiques,
Universitat Rovira i Virgili, Tarragona, Spain
{emran.saleh,aida.valls,antonio.moreno}@urv.cat
[2] Ophthalmic Service, University Hospital Sant Joan de Reus,
Institut d'Investigació Sanitària Pere Virgili (IISPV),
Universitat Rovira i Virgili, Reus, Spain
pedro.romero@urv.cat
[3] School of Informatics, University of Skövde, Skövde, Sweden
vtorra@his.se
[4] Departamento de Automàtica y Computación, Universidad Pública de Navarra,
Institute of Smart Cities, Pamplona, Spain
bustince@unavarra.es

Abstract. Fuzzy measures are used to express background knowledge of the information sources. In fuzzy rule-based models, the rule confidence gives an important information about the final classes and their relevance. This work proposes to use fuzzy measures and integrals to combine rules confidences when making a decision. A Sugeno λ-measure and a distorted probability have been used in this process. A clinical decision support system (CDSS) has been built by applying this approach to a medical dataset. Then we use our system to estimate the risk of developing diabetic retinopathy. We show performance results comparing our system with others in the literature.

Keywords: Fuzzy measures · Aggregation functions
Choquet integral · Sugeno integral · Fuzzy rule-based systems
Diabetic retinopathy

1 Introduction

Aggregation operators are mathematical functions to merge a set of numerical arguments into a single one that summarizes them. They are widely used in many knowledge fields, such as sensor data fusion and decision making [16].

Among the vast number of aggregation operators found in the literature, fuzzy integrals are one of the most general onces. Due to their parametrisation, fuzzy integrals as Choquet and Sugeno show a great flexibility in aggregating the inputs. Choquet integral generalizes both the weighted mean and the OWA

© Springer Nature Switzerland AG 2018
V. Torra et al. (Eds.): MDAI 2018, LNAI 11144, pp. 114–127, 2018.
https://doi.org/10.1007/978-3-030-00202-2_10

operator [2] and Sugeno integral generalizes weighted maximum, weighted minimum and the median operators [4,16].

Fuzzy integrals rely on a fuzzy measure (or capacity), which is a set function that indicates the importance of the information sources (i.e., of each of possible groups of input sources). Defining a proper fuzzy measure for each problem is a crucial point in order to make a suitable aggregation of the inputs and obtain the correct corresponding outputs.

In this paper we will focus on solving a classification problem in the medical field. We have been working on the definition and construction of a clinical decision support system for improving the diagnosis of diabetic retinopathy (DR). This disease is one of the major complications of diabetes and one of the most important causes of loss vision in young diabetic people all over the world. The effects of this disease can be controlled if it is detected at an early stage. With the collaboration of experts from difference medical centers in Catalonia we have collected a dataset of information of 3,000 diabetic patients. This data has been used to train and test a binary DR classification model using Fuzzy Random Forests (FRF) [12].

One of the characteristics of FRF is that a large number of classification rules are generated using different samples of the data. In our case, we have 100 trees with about 100 rules each one. When a new patient has to be classified, his data is introduced into the system and all rules are fired at different levels of satisfaction. Merging the outcome of all these rules is usually done with the Winner strategy, which consists on taking as answer the output of the rule with maximum activation [11]. However, the information provided by the rest of rules is lost.

In this paper we propose to use other aggregation methods in order to merge the contribution of the different rules that are activated by a certain patient's data. In particular, we study the use of fuzzy integrals and a new way of constructing the fuzzy measure is proposed, based on the confidence score of each of the contributing rules.

The rest of the paper is organized as follows. Section 2 presents the main concepts used in this work. In Sect. 3, we introduce the induction algorithms of fuzzy decision trees (FDT) and fuzzy random forest (FRF) models, the proposed fuzzy measures and aggregation process. In Sect. 4, we describe the dataset and discuss the experimental results. Finally, Sect. 5 shows the conclusion and future work.

2 Preliminaries

In this section, we define the basic concepts that are used in this work. We follow [16] for the definitions.

Definition 1. *A function $agg : [0,1]^n \rightarrow [0,1]$ is an aggregation function if and only if it fulfills the following properties:*

- $agg(x, ..., x) = x$ *(Identity)*

– $agg(0, ..., 0) = 0$ and $agg(1, ..., 1) = 1$ (Boundary conditions)
– If $(x_1, ..., x_\eta) \leq (y_1, ..., y_\eta)$ then $agg(x_1, ..., x_\eta) \leq agg(y_1, ..., y_\eta)$ (Non decreasing).

Note that some authors use identity only in 0 and 1 as eg. [1,6].

Definition 2. *A function $T : [0, 1]^2 \to [0, 1]$ is a t-norm function if and only if it fulfills the following properties:*

– $T(x, y) = T(y, x)$ (Commutativity)
– $T(x, y) \leq T(u, v)$ if $(x \leq u)$ and $(y \leq v)$ (Increasing monotonicity)
– $T(x, T(y, z)) = T(T(x, y), z)$ (Associativity)
– $T(x, 1) = x$ (Neutral element).

Examples of T-norms include minimum and product.

Definition 3. *A fuzzy measure (also known as non-additive measure) m on a set X with cardinality η is a set function $m : 2^X \to [0, 1]$ fulfilling the following properties:*

– $m(\emptyset) = 0, m(X) = 1$, (Boundary condition)
– $A \subseteq B$ implies $m(A) \leqslant m(B)$, for all $A, B \subset X$ (Monotonicity).

Fuzzy measures are a way to represent background knowledge about the importance of the sources of some values. In that way, they are used to weight the arguments in some aggregation operators like the Choquet and Sugeno integrals. The fuzzy measure can be defined manually or it can be obtained from some domain data.

In [6], it is proposed a fuzzy measure obtained as the power mean of the cardinality of the set of values aggregated. This fuzzy measure $m_{PM} : 2^X \to [0, 1]$ is defined as follows:

$$m_{PM}(A) = \left(\frac{|A|}{\eta}\right)^q \quad \text{with } q > 0 \tag{1}$$

For classification problems, the value of q can be optimized for each of the classes considered [1].

In this work, we use Choquet and Sugeno integrals in order to aggregate the input data. The discrete Choquet integral is defined as:

Definition 4. *Let X be a reference set with cardinality η and let m be a fuzzy measure on X; then, the Choquet integral of a function $f : X \to \mathbb{R}^+$ with respect to the fuzzy measure m is defined by*

$$Choquet(f) = \sum_{i=1}^{\eta} [f(x_{s(i)}) - f(x_{s(i-1)})] \cdot m(A_{s(i)}), \tag{2}$$

where $f(x_{s(i)})$ indicates that the indices have been permuted so that $0 \leq f(x_{s(1)}) \leq \cdots \leq f(x_{s(\eta)}) \leq 1$, and where $f(x_{s(0)}) = 0$ and $A_{s(i)} = \{x_{s(i)}, \ldots, x_{s(\eta)}\}$.

For the sake of simplicity, we will use $Choquet(x_1, \ldots, x_\eta)$.

In [6] the Choquet-like Copula-based fuzzy integral (CC-integral) is defined. It uses a copula as main operator $\bullet$ instead of the product $\cdot$ as usual in the Choquet integral. I. e., when $\bullet = \cdot$ the CC-integral is Choquet integral The properties of this extended fuzzy integral have been studied in [5,7].

Definition 5. *Let X be a reference set with cardinality η and let m be a fuzzy measure on X; then, the* CC-integral *of a function $f : X \to \mathbb{R}^+$ with respect to the fuzzy measure m is defined by*

$$CC\text{-}integral(f) = \sum_{i=1}^{\eta}[f(x_{s(i)}) \bullet m(A_{s(i)}) - f(x_{s(i-1)}) \bullet m(A_{s(i)})], \qquad (3)$$

where $f(x_{s(i)})$ indicates that the indices have been permuted so that $0 \leq f(x_{s(1)}) \leq \cdots \leq f(x_{s(\eta)}) \leq 1$, and where $f(x_{s(0)}) = 0$ and $A_{s(i)} = \{x_{s(i)}, \ldots, x_{s(\eta)}\}$.

For the sake of simplicity, we will use $CC\text{-}integral(x_1, \ldots, x_\eta)$.

Definition 6. *Let m be a fuzzy measure on X with cardinality η; then, the* Sugeno integral *of a function $f : X \to [0,1]$ with respect to m is defined by*

$$Sugeno(f) = \max_{i=1,\eta} \min(f(x_{s(i)}), m(A_{s(i)})), \qquad (4)$$

where $f(x_{s(i)})$ indicates that the indices have been permuted so that $0 \leq f(x_{s(1)}) \leq \ldots \leq f(x_{s(\eta)}) \leq 1$ and $A_{s(i)} = \{x_{s(i)}, \ldots, x_{s(\eta)}\}$.

For the sake of simplicity, we will use $Sugeno(x_1, \ldots, x_\eta)$.

3   Methodology

In this section, we explain how to build a fuzzy decision tree (FDT) and fuzzy random forest (FRF) and then we propose how to use fuzzy integrals to merge the conclusions of the rules when the FDT or FRF is used to classify a new instance.

3.1   Fuzzy Rule-Based Models Construction

There are many techniques to build fuzzy rule-based models. In this section, we describe the main steps of constructing them based on Yuan and Shaw [17]. That is, an induction method is used to build a fuzzy decision tree (FDT), and a bag of FDTs is used to build a fuzzy random forest (FRF). The following is the notation which is used in the induction procedure.

Let us consider a set of labeled examples $U = \{u_1, u_2, \ldots, u_m\}$. Each u_i is an example described by attributes $A = \{a_1, \ldots, a_n\}$.

Each attribute $a \in A$ takes values on a linguistic fuzzy partition [3] $T_a = \{t_1, ..., t_s\}$ with membership functions μ_{t_i}. The membership values on the universe can be understood as a possibility distribution.

The *U-uncertainty* (or non-specificity measure) of a possibility distribution π on any set with cardinality d is defined in [17] as:

$$g(\pi) = \sum_{i=1}^{d} (\pi_i^* - \pi_{i+1}^*) \ln i \tag{5}$$

where $\pi^* = \{\pi_1^*, \pi_2^*, ..., \pi_d^*\}$ is a permutation of $\pi = \{\pi(1), \pi(2), ..., \pi(d)\}$ such that $\pi_i^* \geq \pi_{i+1}^*$, for $i = 1, ..., d$, and $\pi_{d+1}^* = 0$.

Fuzzy Decision Tree Induction. The induction algorithm proposed in [17] is an extension of the classic ID3 method for crisp data. It incorporates two parameters to manage the uncertainty:

- The *significance level* (α) is used to ignore insignificant evidences. If the fuzzy evidence membership value is lower than α then turns it to 0.
- The *truth level threshold* (β) controls the growth of the tree. Very high β may lead to overfitting and very low β may lead to low classification accuracy.

The main steps of the fuzzy decision tree induction process are the following ones:

1. Choose the attribute with the **smallest ambiguity** (see the expression below) for the root node.
2. For each value of the attribute, create a branch if it has examples with support higher than α.
3. For each branch, calculate the truth level of classification to each class.
4. If **the truth level of classification** (see the expression below) is higher than β then end the branch with the class label which has the highest truth level of classification.
5. If no then check if an additional attribute will reduce the classification ambiguity.
6. If so, choose the attribute with **smallest classification ambiguity with the accumulated evidence** (see the expression below) for the new node, and repeat from step 2 to 6 until no more tree growth is possible.
7. If no, end the branch with the label of the class that has the highest truth level of classification.

The ambiguity of an attribute is calculated as an average of the uncertainty of this attribute for an example using the following equation:

$$Ambiguity(a) = \frac{1}{m} \sum_{j=1}^{m} g(\pi_j),$$

$$where \tag{6}$$

$$\pi_j = \{\mu'_{t_1}(u_j), ..., \mu'_{t_s}(u_j)\}$$

and $\mu'_{t_i}(u_j)$ is the normalized possibility distribution of $\mu_{t_i}(u_j)$:

$$\mu'_{t_i}(u_j) = \mu_{t_i}(u_j)/max_{1 \le k \le s}\{\mu_{t_k}(u_j)\} \tag{7}$$

The truth level of classification defines the possibility of classifying an object u_i into a class $C_k \in C$ where $C = \{C_1, ..., C_p\}$ given the fuzzy evidence E.

$$Truth(C_k|E) = S(E, C_k)/max_{1 \le j \le p}\{S(E, C_j)\} \tag{8}$$

where S is the subsethood of the fuzzy set X on the fuzzy set Y

$$S(X, Y) = \frac{M(X \cap Y)}{M(X)} = \frac{\sum_{i=1}^{m} min(\mu_X(u_i), \mu_Y(u_i))}{\sum_{i=1}^{m} \mu_X(u_i)}. \tag{9}$$

and $M(X)$ is the cardinality or sigma count of the fuzzy set X. The truth level of classification can be understood as the possibility distribution on the set U. $\pi(C|E)$ is the normalisation of the truth level. It has been defined above to be used in the calculation of *Classification ambiguity*.

Classification ambiguity: Suppose we have a fuzzy partition $P = \{E_1, ..., E_k\}$ on a fuzzy evidence F, the following equation is used to calculate the classification ambiguity of a fuzzy partition on a fuzzy evidence denoted by $G(P|F)$.

$$G(P|F) = \sum_{i=1}^{k} W(E_i|F)g(\pi(C|E_i \cap F)) \tag{10}$$

where $W(E_i|F)$ is the weight. The weight is calculated using the following equation: $W(E_i|F) = M(E_i \cap F)/\sum_{i=1}^{k} M(E_i \cap F))$.

Fuzzy Random Forests Construction: The main steps to build a fuzzy random forest are as follows:

1. Randomly, select a subset of the training examples (bootstrap) for training. It has to have a balanced distribution of each class. It is recommended that the size of each dataset (bootstrap) has to be 2/3 of the total training dataset size. The repetition of examples is acceptable. Use each bootstrap to construct a fuzzy decision tree (see Sect. 3.1).
2. While constructing the FDT, a random subset of the remaining attributes with size γ will be used when deciding for next tree node.
3. Repeat steps 1 and 2 until the number of the FDTs n is reached.

3.2   Fuzzy Measure Based on the Rule Confidence

Fuzzy measures are used to give background knowledge in relation to the elements which are going be aggregated. In our context, we aggregate data from a set of rules and we have a degree of support for each rule (*rule confidence*). These degrees define a possibility distribution of the data. These values give an important information about the system. Taking them into account while we are giving the decision of the model is valuable. In this paper we propose the use of two fuzzy measures that will be built from these *rule confidence* values. The first measure is a distorted probability. The second measure is a Sugeno λ-measure.

Let us define the notation used in the following equations: $R=$ the total set of all rules, $RC_i=$ *Rule Confidence* of the ith rule and $n =$ the total number of rules.

Distorted Probability Based Fuzzy Measures: The proposed distorted probability is defined using the following equation:

$$m_{DP}(A) = \left(\frac{\sum_{RC_j \in A} RC_j}{\sum_{RC_i \in R} RC_i} \right)^q , with \quad q > 0 \tag{11}$$

where the value q needs to be optimised. Different methods can be used to optimize q like evolutionary algorithms [1,6]. We use here a gradient descent and wide search. Note that this fuzzy measure is a distorted probability because $m = f \odot P$ with

$$P_j = \frac{RC_j}{\sum_{RC_i \in R} RC_i}, \quad and \quad f(x) = x^q \tag{12}$$

Sugeno λ-measures Based Fuzzy measures: Another way of using domain knowledge to construct a fuzzy measure is by means of the defintion of a Sugeno λ-measure as proposed in [16].

Definition 7. *Let $v : X \rightarrow [0, 1]$ and $\lambda > -1$ be such that*

$$- (1/\lambda)(\prod_{x_i \in X}[1 + \lambda v(x_i)] - 1) = 1 \qquad if \ \lambda \neq 0$$

$$- \sum_{x_i \in X} v(x_i) = 1 \qquad\qquad if \ \lambda = 0$$

then, the fuzzy measure defined by

$$m_{SL}(A) = \begin{cases} v(x_i) & if \ A = \{x_i\} \\ (1/\lambda)(\prod_{x_i \in A}[1 + \lambda v(x_i)] - 1) & if \ |A| \neq 1 \quad and \quad \lambda \neq 0 \\ \sum_{x_i \in A} v(x_i) & if \ |A| \neq 1 \quad and \quad \lambda = 0 \end{cases} \tag{13}$$

is a Sugeno λ-measure. In our proposal, the weights $v(x_i) = RC_i$ are the *rule confidence* values. Therefore, first the *rule confidence* values are used to build the Sugeno λ-measures finding an appropriate λ and then this fuzzy measure is used in the aggregation process.

3.3   Classification Using the Fuzzy Rules

A binary classification is done using the Mamdani inference procedure. Class 0 represents that patients do not suffer from DR and class 1 that they suffer. All rules are applied and the rule membership degree to the conclusion class *(RMCC)* values of the same class are aggregated to obtain the final decision. The proposed procedure is the following:

1. Use a t-norm function to calculate the satisfaction degree of each rule $\mu_R(u)$.
2. Use the product between the satisfaction degree of each rule $\mu_R(u)$ and the degree of support of the rule (*rule confidence*) to obtain the membership degree to the conclusion class *(RMCC)*.
3. Calculate a fuzzy measure (distorted probability $m_{DP}(A)$ or Sugeno λ-measure $m_{SL}(A)$) using the degree of support of the rule (rule confidence).
4. Aggregate the final value of each class using a fuzzy integral (CC-integral (Eq. 3) or Sugeno integral (Eq. 4)). In the aggregation process, the obtained *RMCCs* from the same class are weighted using fuzzy measures as explained above.
5. Compare the aggregation values, the final decision is the class label which has the maximum aggregation value.

4   Experimental Results

In this section, we describe the data used to train and validate the proposed models. The results achieved by these models are discussed in Sect. 4.2.

4.1   The Diabetic Retinopathy Problem and Dataset

An early diagnosis of DR is crucial to improve the quality of life of these patients. At the moment, the detection of DR is done by screening of the eye fundus with a non-midriatic camera. This technique requires a lot of resources from the medical centers both in terms of cost, specialized personnel and time [9]. Due to the large amount of diabetic people it is not possible to perform this test early as recommended by the medical guidelines. Therefore, tests are done every two or three years. For some patients, the detection arrives too late.

The clinical decision support system that we are developing may significantly decrease these costs because it will be used by the family physicians during the regular visits that diabetic people have. The incidence of DR is scarce, which means that most of the people do not need an eye fundus screening. Therefore, the resources could be used to the patients that really need it, facilitating the detection of DR in its first signs.

Sant Joan de Reus University Hospital (SJRUH) in Catalonia (Spain) has been systematically collecting the data of the diabetic patients of many years. These data include demographic, metabolic and analytical information which is stored in the Electronic Health Records (EHR) of the people who has diabetes. The dataset used in the work has the information of 3346 diabetic patients and it is labeled regarding to diabetic retinopathy presence. This dataset has been split into a training dataset with 2243 diabetic patients (1605 not suffering from RD and 638 who suffer from DR) and a testing dataset with 1103 examples (863 not suffering from RD and 240 who suffer from DR). The datasets are imbalanced because the patients with DR are less than healthy people. This imbalance distribution hampers the performance of some machine learning techniques. To solve this problem with FDT model, random over sampling technique has been done to the minor class until both classes have the same number of examples. FRF model internally does random under sampling technique which is a technique to deal with imbalanced datasets.

A statistical analysis on the data was done by the ophthalmologists in SJRUH [8]. Out of that study, nine attributes were identified as the important ones to detect the risk of RD development. Most of the attributes are numerical but there are some categorical ones too. With the collaboration of the experts, the numerical attributes have been fuzzified into linguistic variables according to the medical knowledge.

4.2   Tests, Results and Discussion

In this section, we study the results achieved by using the proposed aggregators with FRF and FDT models on the testing dataset. A comparison between the different aggregation proposals and the traditional methods is done as well. The aim is to improve the performance of the models and achieve a good performance that is acceptable in the medical treatments.

To evaluate the performance of the models on such kind of problems, we use specificity and sensitivity (*recall*). They are usually used in the medical field. To make it easier to the reader to follow the performance results, the harmonic mean (HM) of specificity and sensitivity is calculated as well (Eq. 14).

$$Sensitivity = \frac{TP}{TP + FN}, \qquad Specificity = \frac{TN}{TN + FP},$$
$$HM = 2 * \frac{Sensitivity * Specificity}{Sensitivity + Specificity} \tag{14}$$

The parameters of FDT and FRF were studied in previous works [11,14] and the best values have been used in this paper.

In FRF two ways of aggregating the outcome of the trees have been tested. On the one hand, the direct aggregation of all the rules of all the trees into a unique result (one-step). On the other hand, the aggregation first of the rules of each tree and in a second step the aggregation of the outcome of each tree (two-steps). To avoid effect of randomness, all one-step and two-steps FRF models are experimented with the same parameters 10 times then we take as result the ones of the model with the median HM performance. The median is more robust to outliers than the arithmetic mean.

The different methods tested are shown in Table 1. The basic winner rule (WR) for making decisions is well-known in rule-based models, it uses max

Table 1. Notation of the aggregation methods used in this work

Short name	Aggregation method name	Aggregator T-norm	Rules T-norm
WR	Max	-	Min
CCPM	CC-integral with power mean	Min	Min
ICMM	RC based CC-integral, distorted probability	Min	Min
ICMMS	RC based CC-integral, Sugeno λ-measures	Min	Min
ICPM	RC based CC-integral, distorted probability	Product	Min
ICPMS	RC based CC-integral, Sugeno λ-measures	Product	Min
ISM	RC based Sugeno, distorted probability	Min	Min
ISMS	RC based Sugeno, Sugeno λ-measures	Min	Min

Table 2. Classification results of fuzzy decision tree with $\beta = 0.70$ and $\alpha = 0.30$

q	HM	Sensit.	Specif.	Accuracy	TP	FN	FP	TN	Method
2	76.51	76.67	76.36	76.34	184	56	204	659	WR
1	77.15	76.67	77.64	77.43	184	56	193	670	CCPM
2	76.85	75.42	78.33	77.70	181	59	187	676	ICMM
2	76.89	76.67	76.31	76.38	187	53	208	655	ICMMS
3	77.65	80.00	75.43	76.43	192	48	212	651	ICPM
2	71.63	86.67	63.41	68.48	210	48	311	552	ICPMS
2	**77.98**	**78.33**	**77.64**	**77.79**	**188**	**52**	**193**	**670**	**ISM**
2	76.37	76.67	76.07	76.20	185	55	210	653	ISMS

t-conorm to aggregate the outputs of the rules. To verify the quality of the aggregation methods proposed in this work in comparison with the state of art, first the aggregation method based on Choquet-like Copula-based integral is used with the power mean as fuzzy measure (Eq. 1) as proposed in [6]. The rest of methods correspond to the different versions of Choquet and Sugeno integrals using the Rule Confidence (RC) for the fuzzy measure construction. Table 1 indicates the t-norm operator used in the fuzzy integral and the t-norm used in to calculate the degree of activation of each rule.

With each fuzzy measure, several q values were tested to find the optimal value. Notice that if $q = 1$, then the aggregation (Choquet or Sugeno integral) corresponds to the weighted mean. When q value increases, the performance of the models decreases. Low q values always showed better results, being the best ones $q = 2$, $q = 3$ for most of FRF and FDT models. The models with the best performance are highlighted in Tables 2, 3 and 4.

Observing the basic method WR (winning rule), in FRFs this aggregation method shows high specificity (around 81%) but it shows low sensitivity as well (between 71% and 73%) and HM value is around 76%. The FDT model

124   E. Saleh et al.

Table 3. Classification results of two-steps fuzzy random forest with $\alpha = 0.40$, $\beta = 0.80$

q	HM	Sensit.	Specif.	Accuracy	TP	FN	FP	TN	Method
2	76.05	71.67	81.00	78.97	172	68	164	699	WR
2	78.31	78.75	77.87	78.06	189	51	191	672	CCPM
3	**79.77**	**80.42**	**79.14**	**79.42**	**193**	**47**	**180**	**683**	**ICMM**
2	76.89	77.92	75.90	76.34	187	53	208	655	ICMMS
3	78.61	76.67	80.65	79.78	184	56	167	696	ICPM
2	71.63	81.40	63.96	67.98	210	48	311	552	ICPMS
2	79.74	78.75	80.76	80.33	189	51	166	697	ISM
2	76.37	77.08	75.67	75.97	185	55	210	653	ISMS

Table 4. Classification results of one-step fuzzy random forest with $\alpha = 0.40$, $\beta = 0.80$

q	HM	Sensit.	Specif.	Accuracy	TP	FN	FP	TN	Method
2	77.11	72.92	81.81	79.87	175	65	157	706	WR
2	76.96	75.00	79.03	78.15	180	60	181	682	CCPM
2	79.31	78.33	80.30	79.87	188	52	170	693	ICMM
2	76.59	76.25	76.94	76.79	183	57	199	664	ICMMS
2	78.10	77.08	79.14	78.69	185	55	180	683	ICPM
3	73.27	83.75	65.12	69.17	201	39	301	562	ICPMS
3	**79.73**	**79.17**	**80.30**	**80.05**	**190**	**50**	**170**	**693**	**ISM**
2	78.00	79.58	76.48	77.15	191	49	203	660	ISMS

with WR achieved specificity=76.36%, sensitivity=76.67% and HM=76.51%. By checking the models' performance in Tables 2, 3 and 4, the first conclusion is that the models based on Choquet and Sugeno integrals offer better results than WR. Method ICPMS is an exception (low HM) because it achieves a very good sensitivity but then specificity is too low to be acceptable for medical diagnosis.

Two different methods of calculating fuzzy measures have been proposed in this work. The first one is distorted probability and the second one is a Sugeno λ-measure. The results of FDTs models are presented in Table 2. ISM has the best performance with a very good HM, near 78%, and has the highest sensitivity value (78.33%). We can see that models based on distorted probability improve the ones based on λ-Sugeno measures.

In Table 3 when using FRF with two-steps aggregation, CCPM method obtained HM = 77.38%, sensitivity = 76.67% and specificity = 78.10%. Whereas, all the distorted probability measures based on the *rule confidence* values (ICMM, ICPM and ISM) obtain HM greater than 79%. ICMM achieved the highest performance (HM around 80%). In Table 4, the results of one-step FRF are presented. ICMM, ICPM achieved HM higher than 78% while and ISM has HM around 80%. Methods based on the λ-Sugeno fuzzy measure obtain quite good performance in sensitivity (see ICPMS) but the specificity decrases too much. These results clearly show that the aggregator that uses a fuzzy measure with distorted probability based on the *rule confidence* values outperforms the one using the number of the rules in the fuzzy measure (CCPM). We see a difference in the best method when using one or two steps in the aggregation of the rules of the set of trees. However, both ICMM and ISM achive quite similar HM with values of sensitivity and specificity close to 80%.

By looking to the results presented in Tables 2, 3 and 4, FRF models usually offer better performance than FDTs with every aggregation method. In FDT model with (ICPM) aggregation method, the model obtains specificity = 80%, which is higher than FRFs models results. The same model shows sensitivity = 75.43% which is lower than the sensitivity obtained by FRF models. In general, FRF models show more balance in sensitivity and specificity values.

5   Conclusion and Future Work

The use of fuzzy measures in aggregation operators shows good performance. *Rule confidence* values showed that it can play an important role in the aggregation process.

In this work, a Sugeno λ-measure and a distorted probability are used with Choquet and Sugeno integrals. These new aggregation approaches are used within fuzzy random forests (FRF) and Fuzzy decision trees (FDT). The models with these new aggregation approaches outperforms the same models with max t-conorm aggregation operator.

In comparison with the models that use the same Choquet and Sugeno integrals with fuzzy measures based only on the number of rules, the new approach obtains better performance results as well. Experiments also showed that one-step and two-steps FRFs have better performance results than FDTs. Moreover, two-steps FRF is recommended because it offers better results than one-step FRF, with 80% of sensitivity and 79% of specificity on the testing dataset. With these results, we conclude that the new aggregation operators based on the proposed fuzzy measures improve the performance of our previous works [10,12–14].

This work is oriented to build a clinical decision support system (CDSS). The CDSS will be used in the medical centers by family physicians who are not expert ophthalmologist. The goal is to help the physicians to estimate the risk of developing DR with the new patients. The proposed methods can be easily integrated into the CDSS in order to merge the rule's predictions made with the data of each patient.

Future work includes studying how to improve the performance of current fuzzy measures. Sugeno λ-measures and distorted probabilities do not permit to structure the information sources. As the data to be aggregated is highly dimensional, other families of measures as the hierarchically decomposable ones can be useful [15]. We plan to work in this line. The current approach is going to be validated using other datasets in order to see if the same conclusions about the performance different proposals depend on the problem data or not.

Acknowledgements. This work is supported by the URV grant 2017PFR-URV-B2-60, and by the Spanish research projects no: PI12/01535 and PI15/01150 for (Instituto de Salud Carlos III and FEDER funds). Mr. Saleh has a Pre-doctoral grant (FI 2017) provided by the Catalan government and an Erasmus+ travel grant by URV. Prof. Bustince acknowledges the support of Spanish project TIN2016-77356-P.

References

1. Barrenechea, E., Bustince, H., Fernandez, J., Paternain, D., Sanz, J.A.: Using the choquet integral in the fuzzy reasoning method of fuzzy rule-based classification systems. Axioms **2**(2), 208–223 (2013)
2. Beliakov, G., Pradera, A., Calvo, T.: Aggregation Functions: A Guide for Practitioners, vol. 221. Springer, Heidelberg (2007). https://doi.org/10.1007/978-3-540-73721-6
3. Bodjanova, S.: Fuzzy sets and fuzzy partitions. In: Opitz, O., Lausen, B., Klar, R. (eds.) Information and Classification. STUDIES CLASS, pp. 55–60. Springer, Heidelberg (1993). https://doi.org/10.1007/978-3-642-50974-2_6
4. Grabisch, M., Labreuche, C.: A decade of application of the Choquet and sugeno integrals in multi-criteria decision aid. Ann. Oper. Res. **175**(1), 247–286 (2010)
5. Lucca, G., Sanz, J.A., Dimuro, G.P., Bedregal, B., Bustince, H., Mesiar, R.: CF-integrals: a new family of pre-aggregation functions with application to fuzzy rule-based classification systems. Inf. Sci. **435**, 94–110 (2018)
6. Lucca, G.: CC-integrals: Choquet-like copula-based aggregation functions and its application in fuzzy rule-based classification systems. Knowl. Based Syst. **119**, 32–43 (2017)
7. Mesiar, R., Stupňanová, A.: A note on CC-integral. Fuzzy Sets Syst. (2018). (In press)
8. Romero-Aroca, P., de la Riva-Fernandez, S., Valls-Mateu, A., Sagarra-Alamo, R., Moreno-Ribas, A., Soler, N.: Changes observed in diabetic retinopathy: eight-year follow-up of a Spanish population. Br. J. Ophthalmol. **100**(10), 1366–1371 (2016)
9. Romero-Aroca, P., et al.: Cost of diabetic retinopathy and macular oedema in a population, an eight year follow up. BMC ophthalmol. **16**(1), 136 (2016)
10. Romero-Aroca, P., et al.: A clinical decision support system for diabetic retinopathy screening: creating a clinical support application. Telemed. e-Health (2018). (In press)
11. Saleh, E., et al.: Learning ensemble classifiers for diabetic retinopathy assessment. Artif. Intell. Med. **85**, 50–63 (2018)
12. Saleh, E., Moreno, A., Valls, A., Romero-Aroca, P., de la Riva-Fernandez, S.: A fuzzy random forest approach for the detection of diabetic retinopathy on electronic health record data. In: Artificial Intelligence Research and Development, vol. 288, p. 169. IOS Press (2016)
13. Saleh, E., Valls, A., Moreno, A., Romero-Aroca, P.: Integration of different fuzzy rule-induction methods to improve the classification of patients with diabetic retinopathy. In: Recent Advances in Artificial Intelligence Research and Development, pp. 6–15 (2017)
14. Saleh, E., Valls, A., Moreno, A., Romero-Aroca, P., de la Riva-Fernandez, S., Sagarra-Alamo, R.: Diabetic retinopathy risk estimation using fuzzy rules on electronic health record data. In: Torra, V., Narukawa, Y., Navarro-Arribas, G., Yañez, C. (eds.) MDAI 2016. LNCS (LNAI), vol. 9880, pp. 263–274. Springer, Cham (2016). https://doi.org/10.1007/978-3-319-45656-0_22

15. Torra, V.: On hierarchically S-decomposable fuzzy measures. Int. J. Intell. Syst. **14**(9), 923–934 (1999)
16. Torra, V., Narukawa, Y.: Modeling Decisions: Information Fusion and Aggregation Operators. Springer, Heidelberg (2007). https://doi.org/10.1007/978-3-540-68791-7
17. Yuan, Y., Shaw, M.J.: Induction of fuzzy decision trees. Fuzzy Sets Syst. **69**(2), 125–139 (1995)

Decision Making

Extraction of Patterns to Support Dairy Culling Management

M. López-Suárez[1]([✉]), E. Armengol[2], S. Calsamiglia[1], and L. Castillejos[1]

[1] Animal Nutrition and Welfare Service, Department of Animal and Food Sciences,
Universitat Autònoma de Barcelona, Bellaterra, Barcelona, Spain
{Montserrat.Lopez.Suarez,Sergio.Calsamiglia,Lorena.Castillejos}@uab.cat
[2] Artificial Intelligence Research Institute, (IIIA-CSIC), Campus UAB,
Camí de Can Planes, s/n, Bellaterra, 08193 Barcelona, Spain
eva@iiia.csic.es

Abstract. The management of a dairy farm involves taking decisions such as culling a subset of cows to improve the dairy production. Culling is the departure of cows from the herd due to sale, slaughter or death. Commonly the culling process is based on the farmer experience but there is not a general procedure to carry it out. In the present paper we use both, a method based on indistinguishability relations and the anti-unification concept, to extract patterns that characterise the cows according to their average milk production of the first lactation. Our goal is to identify as soon as possible poorly productive cows during her first lactation, which may be candidates to be culled.

Keywords: Veterinary · Dairy farms · Milk production
Voluntary culling · Artificial intelligence · Machine learning
Indistinguishability relations · Anti-unification concept

1 Introduction

The management of a dairy farm involves taking difficult technical and economic decisions such as the replacement of some cows to either maintain or increase the productivity of the dairy. The process of *culling* is defined as the departure of cows from the herd due to sale, slaughter or death. Culling reasons have been classified as voluntary (or also economic [6]), or involuntary (or also biological [6]). Biological culls are those cows for which no possible productive future exists due to disease, injury or infertility. Thus, this class of culls are mainly involuntary as most of the times are "forced" decisions. Economic culls mean that a cow is removed because a replacement is expected to produce greater profit. In this case, farmer has freedom of choice over which cows are removed from the herd, although they are healthy [2,6]. Hence, the farmer can do a voluntary selection of cows to cull based in the herd size and herd production level. We propose to analyse first lactation production data to identify those animals in a herd which are candidates to be culled following milk yield improvement criteria.

© Springer Nature Switzerland AG 2018
V. Torra et al. (Eds.): MDAI 2018, LNAI 11144, pp. 131–142, 2018.
https://doi.org/10.1007/978-3-030-00202-2_11

We are interested in constructing a model able to characterise cows according to their milk production level. For this purpose we already have used artificial intelligence techniques such as Decision Trees (DT) [9] to obtain a model that can classify a cow as *Good* or *Bad* milk producer, supporting in this way the culling process. Most of the work focused on modelling the culling task or aspects related to it, try to construct statistical models based on the analysis of past cases of a dairy [4]. The use of artificial intelligence techniques is still not widely used for managing the culling although they have been used for other purposes. Among other works, Cavero et al. [5] developed a fuzzy logic model for mastitis detection; Kamphuis et al. [8] used decision trees and [12,14] used neural networks for the detection of clinical mastitis; Shainfar et al. [11] used fuzzy neural networks to predict breeding values for dairy cattle; Grzesiak et al. [7] also used neural networks to predict milk production.

In the present paper we propose to use the method called JADE [3] to assess the relevance of the attributes describing a cow and then the anti-unification concept [1] to, based on these relevances, construct descriptions of the classes of cows according to their milk production. We experimented with a data base with information about the first lactation of around 98000 cows.

Section 2 explains the JADE method used and Sect. 3 explains the *Anti-Unification* concept. Section 4 describes the data base we used in the experiments we performed to evaluate the feasibility of the approach. Section 5 describes how we have modelled the culling task, and the data base we used in our experiments. Section 6 explains the experiments we carried out to construct a model supporting the culling task and also analyses the results focusing on the class of lowest milk producer cows. Finally, Sect. 7 is devoted to conclusions.

2   The JADE Method

In this section we introduce JADE [3], a method useful for feature selection and classification. Conceptually, the idea is to minimise the distance between two indistinguishability relations [13]: the one that gives the correct classification of the known examples and the other one that is a linear combination of the indistinguishable operators generated by the attributes describing the examples. Such distance is calculated using the Euclidean distance, so the function to be minimised is a quadratic one. Thus, the problem of assessing the weights of the attributes has been reformulated as an optimisation problem like the methods in [10,15].

Let X be a set of labeled domain objects described by a set of attributes $A = \{a_1, a_2, \ldots, a_n\}$, where the attributes a_i are considered fuzzy subsets of X; and E_{a_i} the L-indistinguishability operator generated by a_i. Each $x_i \in X$ belongs to one solution class $\{C_1, C_2, \ldots, C_k\}$, i.e., there are k classes where an unseen domain object could be classified. On the set X we can induce two kinds of partitions:

– The *correct partition* that is the one that separates the objects in X according to the solution classes $C = \{C_1 \ldots C_k\}$. This partition can be represented by a $m \times m$ matrix (m being the number of objects in X) R where each element r_{hl} is equal to 1 if the objects x_h and x_l belong to the same solution class and 0 otherwise.

- The *partitions induced by each attribute* in A. Given an attribute $a_j \in A$, the objects in X can be separated according to the value that they hold in the attribute a_j. Each one of these partitions can be represented by a $m \times m$ matrix E_{a_j} where each element e_{hl} is the similarity that the objects x_h and x_l with respect to the attribute a_j.

The global similarity between objects will be assessed by the relation

$$E = p_1 \cdot E_{a_1} + p_2 \cdot E_{a_2} + ... + p_n \cdot E_{a_n} \tag{1}$$

Based on this, JADE considers the objective function as a distance function that measures how different (or similar) are the relations R and E. The goal of JADE is to assess the weights p_i to minimise the Euclidean distance d between both relations. In other words, our goal is

$$\text{minimise} \quad d(E, R) = \sqrt{\sum_{i,j=1..k} (E(x_i, x_j) - R(x_i, x_j))^2}$$

$$\text{subject to} \quad p_1, p_2, ..., p_n \geq 0$$

$$\sum_{i=1}^{n} p_i = 1.$$

The attributes generating similarity relations with higher weights help more to the resemblance of E to R. The weights of the attributes give an idea of which of them are the more relevant to describe a class (notice that the weight of some attributes could be zero). This is similar to the statistical regression that indicates how each attribute contributes to the explanation of the free variable. See a complete description of JADE and their mathematical foundations in [3].

The result of JADE is a set of weights indicating the relevance of each attribute. Differently than in regression, the weights of JADE allow the classification of unseen objects (see in [3] how to classify) however, they do not give a model characterising the classes. It is important to obtain patterns supporting the decision of why a cow should be culled. For this reason, we propose the use of the *Anti-Unification* (AU) concept to obtain useful patterns.

3 The Anti-unification Concept

The *Anti-Unification* (AU) of a set of objects $x_1, ..., x_k$ described by a set of pairs attribute-value (where *value* is categorical) is the most specific generalisation of all the generalisations satisfied by the objects, i.e. the least general generalisation. The $AU(x_1, ..., x_n)$ is a description that includes all the attribute-value pairs shared by all the objects $x_1, ..., x_n$, i.e., it describes *all* aspects in which two or more objects are similar. Formally, for each attribute $a_i \in \{a_1, a_2, \ldots, a_n\}$

$$\text{if } \forall x_j \in X, \quad x_j.a_i = v \text{ then include } (a_i, v) \text{ in } AU(x_1, ..., x_k),$$
$$\text{otherwise reject } a_i$$

Notice that according to the definition above, the AU is formed by the set of attribute-value pairs such as each attribute has the same value in all the objects $x_1, ..., x_k$ and attributes holding different values are not considered.

We use the same concept to construct patterns from the weights provided by JADE however, we perform some modifications in the original definition of the AU. Let us suppose that an attribute a_h holds the value v_1 in all the objects except in one of them that has value v_2. In such situation, a_h will not be included in AU, but it could be interesting for us to know this especial case. Even, we could be interested in constructing a pattern including a_h. For this reason, we modify the definition of the AU and propose to construct the description taking into account all the attributes describing the objects and each attribute having as value the set of all the values taken in all the objects. Attributes that take all the possible values will not be included in the AU.

4   The Data Base

We used a data base containing 97987 objects. These objects are descriptions of Holstein-Frisian cows which lived from 2006 to 2016, belonging to dairy farms within the CONAFE register system[1]. Because our goal is to detect poorly productive cows as soon as possible, we decided to use only information relative to the first lactation. The attributes we considered for every cow were the following ones:

- BirthMonth. Month (season) in which the cow was born.
- Month1Calving: Month (season) of the first calving of a cow.
- KI: Milk production genetic index.
- ICO: Official cattle breeding index in Spain.
- Morpho: Morphologic qualification of a cow.
- KgMilkPeak: Average test-day milk yield (kg/day) of the second and third control of the first lactation (lactation peak).
- Fat: Fat average percentage from the second and third controls of the first lactation.
- Protein: Protein average percentage from the second and third control of the first lactation.
- SCC: Somatic cell count in the milk. It is an indicator of the quality of milk as it expresses the likeliness to contain harmful bacteria.
- OpenDays: Days from calving to conception.
- Calving1stAI: Interval of days between the first calving and the first insemination after it.
- AI: Number of artificial insemination attempts to conceive after the first calving.

---

[1] The *Confederación Nacional de la Raza Frisona* (CONAFE) is a Spanish entity whose goal is to develop programs oriented to the improvement and selection of the Holstein-Frisian herd.

Table 1. Intervals corresponding to each quartile of the attributes KgMilkPeak and Production/DIM.

Attribute	VL	L	H	VH
KgMilkPeak	(5, 28]	(28, 32]	(32, 36]	(36, 65]
Production/DIM	(6, 25]	(25, 29]	(29, 32]	(32, 60]

– Production/DIM: Average daily milk production of the first lactation (kg/day) calculated dividing total amount of milk produced by a cow during the whole lactation by the total days in milk (DIM).

All the attributes above have numerical values and we have discretised them. For the attributes BirthMonth and Month1Calving we divided the months according to seasons. For the remaining attributes, we calculated the quartiles and divided the whole interval of values in four parts according to these quartiles. We associated to each of the 4 quartile interval the labels: *VeryLow (VL), Low (L), High (H),* and *VeryHigh (VH)*. Table 1 shows the quartiles of the attributes KgMilkPeak and Production/DIM.

We considered Production/DIM as the solution class, i.e., we want to model and predict the first lactation milk production performance of a cow (kg/day).

5   Modelling the Culling Task

In this paper we propose to construct a model composed of patterns extracted by combining the weights calculated by JADE with the AU. Particularly, we consider four classes of cows according to the values of Production/DIM (*very low, low, high,* and *very high*) and use JADE on the objects of each one of the classes. The main goal in using the model is to clearly identify the worst milk producers, i.e., the cows belonging to the class *very low* avoiding as much as possible *false negatives*, that in this case means that a cow with Production/DIM = *VL* is not classified as *high* or *very high*. The procedure we propose is the following one:

1. Given the data base, we divide it in subsets S_i according to the values of the solution class (Production/DIM).
2. For each S_i,
   (a) Use JADE to compute the weights of each attribute. Let $p_{i1}, \ldots, p_{in}$ the weights associated to the attributes with respect to the objects of the subset S_i.
   (b) Reject the attributes having a weight under a given threshold t_a (see below).
   (c) Form the AU of all the objects in S_j taking into account only the remaining attributes.

Table 2. Weights of the attributes with respect to the class Production/DIM $= VL$.

Attribute	Weight
KgMilkPeak	0.597
ICO	0.240
KI	0.163
AI	4.979 e–06
BirthMonth	4.914 e–06
OpenDays	4.676 e–06
Calving1stAI	4.218 e–06
Month1Calving	0.0
Morpho	0.0
Fat	0.0
Protein	0.0
SCC	0.0

To use JADE is not necessary to separate the objects by classes as we proved in [3]. Nevertheless, when it is used on a set of objects belonging to several classes, the weights are only useful for classification and do not say anything about the relevance of the attributes with respect to the classes. When JADE is used on a set of objects belonging all them to the same class C_j, the weights represent the relevance of each attribute with respect to the class C_j. Therefore, in this case we have a ranking of attributes similarly to the one that can be obtained using statistical regression. By analysing the weights, we see that many attributes have a relevance near to zero, i.e., they are not relevant. For this reason we propose to fix a threshold t_a under which an attribute can be rejected. The idea is that each class can have different relevant attributes and the rejection of the ones with weight almost zero will produce more general and useful patterns.

Table 2 shows (in descendent order) the weights of the attributes of the subset of cows with Production/DIM $= VL$. Notice that, in fact, only the three first attributes (KgMilkPeak, ICO and KI) have some relevant weight whereas the others are irrelevant in practice since the weight is 0 or near to 0. Therefore, only the mentioned three attributes will be considered for AU.

The next step is to determine the values of these attributes. As we already explained, we modified the AU by assessing as value for each attribute a_j the union of the values that a_j takes in the objects to be anti-unified. For instance, for the class Production/DIM $= VL$, all the objects have KgMilkPeak $= VL$, whereas the attributes ICO and KI can take two values: VL or L. Therefore, the pattern obtained from the AU of the objects belonging to the class Production/DIM $= VL$ is

$$[\text{KgMilkPeak} = VL, \text{ICO} = \{VL, L\}, \text{KI} = \{VL, L\}] \rightarrow \text{Production/DIM} = VL$$

This pattern can be interpreted as follows: a cow with very low production in the lactation peak, and having low or very low ICO coefficient and low or very low milk production genetic index, will have a very low average daily milk production in the first lactation.

We performed preliminar experiments using this procedure. The analysis of the patterns show us that most of times they are too specific and this produces a high number of unclassified objects. For instance, let us suppose that a cow with Production/DIM $= VL$ that has [KgMilkPeak $= VL$, ICO $= \{VL, L\}$, KI $= H$]. Such a cow does not satisfy the pattern above since its value in the attribute KI (H) is not one of the values included in the pattern for this attribute (either VL or L) therefore it will not be classified as a bad producer. Nevertheless, the attribute KI has a low weight compared with the one of the attribute KgMilkPeak that by its own could classify the cow correctly as VL. For this reason, we introduced a *pattern threshold* (t_p) that controls the attributes that will form the final pattern. Thus, on one hand, only the attributes with a weight higher than a threshold t_a will be taken into account; but on another hand, we take only the necessary attributes to form a pattern with a global weight higher than t_p. For instance, if we take $t_p = 0.5$, the pattern will be composed only of KgMilkPeak $= VL$ (the weight of KgMilkPeak is 0.597), whereas if $t_p = 0.6$ the pattern will be [KgMilkPeak $= VL$, ICO $= \{VL, L\}$] (the sum of the weights of KgMilkPeak and ICO is $0.597 + 0.240 = 0.837$). This assures that the pattern is both general enough avoiding unclassified objects and accurate enough to correctly identify objects of the class.

6   Experiments

The goal of the experiments is twofold. On the one hand we want to get a model for the classification of cows according to their milk production level (Production/DIM). On the other hand, we are especially interested in identifying cows with Production/DIM $= VL$ and to avoid to classify them as good producers.

6.1   The Model

We followed the procedure explained in Sect. 5 on the whole database and experimented with several values of both t_a and t_p. In particular, when $t_p = 0$ means that all the attributes with a weight above t_a will be included in the pattern. For instance, Table 3 shows the weights of all the attributes for the classes L and H. If $t_a = 0.1$ the patterns are formed by the following attributes:

- Production/DIM $= L$: KgMilkPeak, ICO, AI, and Calving1stAI, with global weight 0.856.
- Production/DIM $= H$: KgMilkPeak, AI, Calving1stAI, and KL, with global weight 0.663.

Table 3. Weights of the attributes with respect to the classes Production/DIM $= H$ and Production/DIM $= L$.

Attribute	Production/DIM $= L$	Production/DIM $= H$
KgMilkPeak	0.369	0.337
ICO	0.191	0.039
AI	0.171	0.172
Calving1stAI	0.125	0.154
KI	0.045	0.051
Protein	0.030	2.99 e–08
Month1Calving	0.028	2.78 e–09
BirthMonth	0.018	0.087
Fat	0.013	0.049
SCC	0.006	0.063
Morpho	0.004	0.041
OpenDays	0.000	0.006

If $t_a = 0.2$ the patterns of both classes only will include the attribute KgMilk-Peak. The global weight for the pattern of the class Production/DIM $= L$ is 0.369, and for the class Production/DIM $= H$ is 0.337.

Therefore, for low values of t_a, being $t_p = 0$, the result is a model composed of patterns that are too specific (i.e., they have many attributes) and as a consequence, many times the model cannot classify unseen objects.

The global weight will be used when a object satisfies more than one pattern of different classes. In this situation the object will be classified in the class of the pattern having the highest weight. For instance, if we take $t_a = 0.20$ then the patterns for L and H are the following:

- [KgMilkPeak $= L$] $\rightarrow$ Production/DIM $= L$, with weight 0.369.
- [KgMilkPeak $= \{H, L\}$] $\rightarrow$ Production/DIM $= H$, with weight 0.337.

Therefore, if an object satisfies both patterns it will be classified as belonging to the class L because its associated weight is higher than the one associated to the pattern of the class H.

When $t_a = 0$, that means that all the attributes eventually could be included in the pattern. In that case, the pattern is composed of the minimum number of attributes necessary to reach t_p. For instance, following with the weights of Table 3, if $t_p = 0.7$ the patterns are formed by the following attributes:

- Production/DIM $= L$: KgMilkPeak, ICO, and AI, with global weight 0.731.
- Production/DIM $= H$: KgMilkPeak, AI, Calving1stAI, and KL, with global weight 0.714.

If $t_p = 0.5$ the patterns are formed by the following attributes:

- Production/DIM = L : KgMilkPeak and ICO, with global weight 0.560.
- Production/DIM = H : KgMilkPeak and AI, with global weight 0.509.

In this case, the patterns are more general than the ones obtained when $t_p = 0$ but we also detected some inconveniences since sometimes the classification of an unseen object could depend on the value of an attribute with a very low weight. For instance, let us take $t_p = 0.6$ and $t_a = 0$, and suppose the following situation:

- For the class Production/DIM = L the following pattern is constructed: [KgMilk Peak = L, AI = H,Protein = L], where each individual attribute has weight w(Kg MilkPeak) = 0.296, w(AI) = 0.197, w(Protein) = 0.117, and the total weight of the pattern is 0.610.
- For the class Production/DIM = H the following patten is constructed: [KgMilk Peak = $\{H, L\}$, AI = $\{H, L\}$, Protein = $\{H, L\}$], where each individual attribute has weight w(KgMilkPeak) = 0.396, w(AI) = 0.198, w(Protein) = 0.029, and the total weight of the pattern is 0.623.

A cow satisfying both patterns will be classified as belonging to the class Production/DIM = H because the associated weight of the pattern of this class is the highest. However, the difference of weights is due to the inclusion of the attribute Protein with a very low weight 0.029]. For this reason, we think that is important to establish a threshold t_a under which an attribute should not be included in the pattern.

Taking into account all the considerations above, the procedure followed in the experiments has been the following:

1. Given the data base, we divide it in subsets S_i according to the values of the solution class (Production/DIM).
2. For each S_i,
   (a) Use JADE to compute the weights of each attribute. Let $p_{i1}, \ldots, p_{in}$ be the weights associated to the attributes with respect to the objects of the subset S_i.
   (b) Reject the attributes having a weight under a given threshold t_a.
   (c) Form the AU of all the objects in S_j taking into account only the remaining attributes.
   (d) Form the pattern P with the attribute in AU having the highest weight.
   (e) While the global weight of the attributes in P is lower than t_p, add to P the next attribute.
   (f) Repeat the previous step until the weight of P is higher than t_p or until all the attributes in AU have been added to P.

In the previous example, if $t_a = 0.1$, the attribute Protein will not be included in the pattern for the class Production/DIM = H. Notice than without that attribute, the global weight of the pattern will be 0.594 that is under t_p. Nevertheless, we prefer to have patterns with lower confidence than having patterns with high weight but composed of many attributes with low weight.

We carried out experiments with different values of both t_p and t_a. Figure 1 shows the model we obtained taking $t_p = 0.60$ and $t_a = 0.30$. Notice that, in fact, only the pattern corresponding to the class Production/DIM = VL has a weight higher than the threshold t_p due to the attributes have weight lower than t_a.

The patterns for Production/DIM = VL and Production/DIM = VH coincide with the ones we obtained using both decision trees and also an statistical model (see [9]). In both models, the attribute KgMilkPeak is the most relevant to determine the classification of a cow with respect to the class Production/DIM. However, using decision trees we have the advantage over the statistic model that the patterns explicit in which way the values of KgMilkPeak = VH are related with those of Production/DIM VH. Thus, in [9] we seen that KgMilkPeak = VH correspond to Production/DIM VH; the reverse is also true: KgMilkPeak = VL corresponds to Production/DIM VL. Using the weights of JADE we obtained the same result. Nevertheless, differently than with decision trees, now we have also patterns for Production/DIM = L and Production/DIM = H.

- [KgMilkPeak = VL] → Production/DIM = VL with weight 0.744
- [KgMilkPeak = L] →Production/DIM = L with weight 0.422
- [KgMilkPeak = $\{L,H\}$] → Production/DIM = H with weight 0.321
- [KgMilkPeak = VH] → Production/DIM = VH with weight 0.315.

Fig. 1. Model formed by four patterns to predict the classification of a cow according to its milk production level (Production/DIM).

6.2   The Class Production/DIM = VL

The class Production/DIM = VL corresponds to cows having a very low milk production level. This means that the cows in this class are the main candidates to be culled. For this reason, it is very important to assure the maximum accuracy in predicting the membership to this class. In other words, our goal is to avoid that a cow with very low production level (Production/DIM = VL) is classified as a good producer (i.e., as belonging to the classes Production/DIM = VH or Production/DIM = H). To measure this error, we defined *false negatives* and *false positives* as follows:

- *Positives.* Cows belonging to the class Production/DIM = VL.
- *False Negatives (FN).* Cows belonging to the class Production/DIM = VL that the model has classified as Production/DIM = VH or Production/DIM = H.
- *False Positives (FP).* Cows belonging to the class Production/DIM = VH that the model has classified as Production/DIM = VL or Production/DIM = L.

Notice that in the definitions above, we do not consider a false negative a cow with Production/DIM $= VL$ classified as Production/DIM $= L$ since, eventually, cows with Production/DIM $= L$ could also be culled if necessary. After one trial of 10-fold cross-validation there is an accuracy of 74.69% in classifying positive cows. The percentage of FN is 2.34% and the percentage of FP is 1.60%.

Also, we have detected a percentage of 6.04% of cows with Production/DIM $= VL$ that have not been classified because they do not satisfy any of the patterns of the model. We have to analyze why these cows do not satisfy the patterns since the pattern for the class Production/DIM $= VL$ has a high weight and we consider it as having·a high confidence. We also think that no classifications could be due to some errors in the data base and the patterns could be useful to clean it.

7   Conclusions

In the present work we used a method based on indistinguishable relations called JADE to obtain a model supporting the culling decision process of a dairy farm. JADE assesses weights to the attributes describing the domain objects, and by fixing two minimum thresholds (one for the attributes and another for the patterns) it is possible to construct patterns to form a model for the classifications of cows according to their milk production level (Production/DIM). Although having a global model is very interesting, it is also important to focus the model on the identification of very low milk producers. In this sense, we obtained a pattern that characterises cows of the class Production/DIM $= VL$ with an accuracy of around 75% with a 2.34% of false negatives and 1.60% of false positives.

Around a 6% of cows belonging to the class Production/DIM $= VL$ are not classified because they do not satisfy the pattern of that class. A future works, we want to accurately analyse these kind of cows kind of cows in order to reduce this percentage.

Acknowledgments. The authors acknowledge data support from CONAFE (Confederación Nacional de la Raza Frisona). This research is partially funded by the projects (Project AGL2015-67409-C2-01-R) from the Spanish Ministry of Economy and Competitiveness; RPREF (CSIC Intramural 201650E044); and the grant 2014-SGR-118 from the Generalitat de Catalunya. Authors also thank to Àngel García-Cerdaña his helpful comments.

References

1. Symbolic explanation of similarities in case-based reasoning: Comput. Inf. **25**(2–3), 153–171 (2006)
2. Ansari-Lari, M., Mohebbi-Fani, M., Rowshan-Ghasrodashti, A.: Causes of culling in dairy cows and its relation to age at culling and interval from calving in Shiraz, Southern Iran. Vet. Res. Forum **3**, 233–237 (2012)
3. Armengol, E., Boixader, D., García-Cerdaña, A., Recasens, J.: t-generable indistinguishability operators and their use for feature selection and classification. Fuzzy Sets Syst. (2018)

4. Calsamiglia, S., Castillejos, L., Astiz, S., Lopez-DeToro, C., Baucells, J.: A dairy farm simulation model as a tool to explore the technical and economical consequences of management decisions. In: Proceedings of the World Buiatrics Congress 2016 World Association for Buiatrics, p. 406 (2016)
5. Cavero, D., Tölle, K.H., Buxadé, C., Krieter, J.: Mastitis detection in dairy cows by application of fuzzy logic. Livest. Sci. **105**, 207–213 (2006)
6. Fetrow, J., Nordlund, K.V., Norman, H.D.: Culling: nomenclature, definitions, and recommendations. J. Dairy Sci. **89**, 1896–1905 (2006)
7. Grzesiak, W., Blaszczyk, P., Lacroix, R.: Methods of predicting milk yield in dairy cows: predictive capabilities of wood's lactation curve and artificial neural networks (ANNs). Comput. Electron. Agric. **54**(2), 69–83 (2006)
8. Kamphuis, C., Mollenhorst, H., Heesterbeek, J.A., Hogeveen, H.: Detection of clinical mastitis with sensor data from automatic milking systems is improved by using decision-tree induction. J. Dairy Sci. **93**(8), 3616–27 (2010)
9. Lopez-Suarez, M., Armengol, E., Calsamiglia, S., Castillejos, L.: Using decision trees to extract patterns for dairy culling management. In: Iliadis, L., Maglogiannis, I., Plagianakos, V. (eds.) AIAI 2018. IAICT, vol. 519, pp. 231–239. Springer, Cham (2018). https://doi.org/10.1007/978-3-319-92007-8_20
10. Rodriguez-Lujan, I., Huerta, R., Elkan, C., Cruz, C.S.: Quadratic programming feature selection. J. Mach. Learn. Res. **11**, 1491–1516 (2010)
11. Shahinfar, S., Mehrabani-Yeganeh, H., Lucas, C., Kalhor, A., Kazemian, M., Weigel, K.A.: Prediction of breeding values for dairy cattle using artificial neural networks and neuro-fuzzy systems. Comput. Math. Methods Med. Artical ID 127130 (2012)
12. Sun, Z., Samarasinghe, S., Jago, J.: Detection of mastitis and its stage of progression by automatic milking systems using artificial neural networks. J. Dairy Res. **77**, 168–175 (2009)
13. Valverde, L.: On the structure of F-indistinguishability operators. Fuzzy Sets Syst. **17**, 313–328 (1985)
14. Wang, E., Samarasinghe, S.: On-line detection of mastitis in dairy herds using artificial neural networks. In: Proceedings of the International Congress on Modelling and Simulation (MODSIM 2005), Melbourne, Australia (2005)
15. Zhang, L., Coenen, F., Leng, P.H.: An attribute weight setting method for k-NN based binary classification using quadratic programming. In: van Harmelen, F. (ed.), ECAI, pp. 325–329. IOS Press (2002)

An Axiomatisation of the Banzhaf Value and Interaction Index for Multichoice Games

Mustapha Ridaoui[1(✉)], Michel Grabisch[1], and Christophe Labreuche[2]

[1] Paris School of Economics, Université Paris I - Panthéon-Sorbonne, Paris, France
{mustapha.ridaoui,michel.grabisch}@univ-paris1.fr
[2] Thales Research and Technology, Palaiseau, France
christophe.labreuche@thalesgroup.com

Abstract. We provide an axiomatisation of the Banzhaf value (or power index) and the Banzhaf interaction index for multichoice games, which are a generalisation of cooperative games with several levels of participation. Multichoice games can model any aggregation model in multicriteria decision making, provided the attributes take a finite number of values. Our axiomatisation uses standard axioms of the Banzhaf value for classical games (linearity, null axiom, symmetry), an invariance axiom specific to the multichoice context, and a generalisation of the 2-efficiency axiom, characteristic of the Banzhaf value.

Keywords: Banzhaf value · Multicriteria decision aid
Multichoice games · Interaction

1 Introduction

In cooperative game theory, a central problem is to define a *value*, that is, a payoff to be given to each player, taking into account his contribution into the game. Among the many values proposed in the literature, two of them have deserved a lot of attention, namely the Shapley value [24] and the Banzhaf value [1]. Both of them satisfy basic properties as linearity, symmetry, which means that the payoff given does not depend on the way the players are numbered, and the null player property, saying that a player who does not bring any contribution in coalitions he joins should receive a zero payoff. A value satisfying these three properties has necessarily the form of a weighted average of the marginal contribution of a given player into coalitions. The Shapley and Banzhaf values differ on the weights used when computing the average. In the Shapley value, the marginal contributions are weighted according to the size of the coalition, in order to satisfy efficiency, that is, the total payoff given to the players is equal to the total worth of the game. In other words, the "cake" is divided among the players with no waste. For the Banzhaf value, the weights are simply equal, and so do not depend on the size of the coalition. As a consequence, the Banzhaf value is not efficient in general.

© Springer Nature Switzerland AG 2018
V. Torra et al. (Eds.): MDAI 2018, LNAI 11144, pp. 143–155, 2018.
https://doi.org/10.1007/978-3-030-00202-2_12

Lack of efficiency could be perceived, in the context of cooperative game theory, as an undesirable feature. This explains why in this domain, the Shapley value is much more popular. However, there are contexts where efficiency is not a relevant issue or even does not make sense. This is the case for voting games and in multicriteria decision aid (MCDA). A voting game is a cooperative game which is 0-1-valued, the value 1 indicating that the coalition wins the election. In this context, the relevant notion is the *power index*, and the Banzhaf value is used as such. A power index indicates how central a player is for making a coalition winning (this is called a swing). Banzhaf [1,5] has shown that for counting swings, no weight should be applied, and this directly leads to the Banzhaf value (called in this context Banzhaf power index or Banzhaf index). In MCDA, criteria can be interpreted as voters in a voting game, and here a power index becomes an *importance index*, quantifying how important in the final decision a criterion is. In both, domains, efficiency simply does not make sense, so that the Banzhaf value/index should be considered perhaps more relevant than the Shapley value.

There are other reasons to consider the Banzhaf value as a natural concept. In order to establish this, we need to generalize the notion of value or power index to the notion of interaction index, especially meaningful in a MCDA context [7,12,13,20]. The interaction index for a set S of criteria quantifies the way the criteria in S interact, that is, how the scores on criteria in S contribute to the overall score. It can be considered that the interaction index when S is a singleton amounts to the importance index, which leads to two types of interaction indices, one based on the Shapley value and the other based on the Banzhaf index. This being said, aggregation models in MCDA which are based on capacities (monotone cooperative games) can be of the Choquet integral type, multilinear type, or other integrals like Pan-integral, concave integral, decomposition integral, etc. (beside other types such as the Sugeno integral, suitable in an ordinal context). It has been proved by Grabisch et al. [11] that if the Choquet integral is used, the relevant interaction index is the Shapley interaction index, while in the case of the multilinear model, the Banzhaf index should be used. In addition, in computer sciences, the notion of Fourier Transform is defined and widely used, e.g., in cryptography (see, e.g., [4]). It turns out that the Banzhaf interaction index and the Fourier transform differ only by some coefficient (see details in [8, Ch. 2.16.2]. Other connections exist, e.g., with the Sobol indices in statistics (see [10]).

The aim of the paper is to establish the Banzhaf index and Banzhaf interaction index for multichoice games, which are a generalisation of cooperative games. Multichoice games allow each player to choose a certain level of participation, among k possible levels. Their counterpart in MCDA are very interesting since they encode any aggregation model with discrete attributes [21,22]. To our knowledge, there is no definition of an interaction index for multichoice games. Nevertheless, Lange and Grabisch [17] have provided a general form of interaction index for games on lattices. This does not fit our analysis, that focuses on interaction index defined for groups of criteria. Our approach is to build these indices in an axiomatic way, using an approach similar to Weber [26].

2   Preliminary Definitions

We consider throughout a finite set of elements $N = \{1, \ldots, n\}$, which could be players, agents in cooperative game theory, criteria, attributes in multi-criteria decision analysis, voters or political parties in voting theory. We often denote cardinality of sets $S, T, \ldots$ by corresponding small letters $s, t, \ldots$, otherwise by the standard notation $|S|, |T|, \ldots$. Moreover, we will often omit braces for singletons, e.g., writing $N \setminus i$ instead of $N \setminus \{i\}$.

Let $L_i := \{0, 1, \ldots, k_i\}, (k_i \in \mathbb{N}, k_i \geq 1)$ and define $L = \times_{i \in N} L_i$. The set L is endowed with the usual partial order $\leq$: for any $x, y \in L$, $x \leq y$ if and only if $x_i \leq y_i$ for every $i \in N$. For each $x \in L$, we define the support of x by $\Sigma(x) = \{i \in N | x_i > 0\}$ and the kernel of x by $\mathrm{K}(x) = \{i \in N | x_i = k_i\}$. Their cardinalities are respectively denoted by $\sigma(x)$ and $\kappa(x)$. For any $x \in L$ and $S \subseteq N$, x_S denotes the restriction of x to the set S, while x_{-S} denotes the restriction of x to the set $N \setminus S$. For all alternatives $x, y \in L$ and $S \subseteq N$, the notation (x_S, y_{-S}) denotes the compound alternative z such that $z_i = x_i$ if $i \in S$ and y_i otherwise. The same meaning is intended for L_S and L_{-S}.

In cooperative game theory, the set L_i is interpreted as the set of activity levels of player $i \in N$, and any $x \in L$ is called an *activity profile*. In an MCDA context, L_i is the set of all possible values taken by (discrete) attribute $i \in N$, while $x \in L$ is called an *alternative*. Throughout the paper, we adopt without limitation the terminology of game theory.

For convenience, we assume that all players have the same number of levels, i.e., $k_i = k$ for every $i \in N$, $(k \in \mathbb{N})$.

A *(cooperative) game* on N is a set function $v : 2^N \to \mathbb{R}$ vanishing on the empty set. A game v is said to be a *capacity* [2] or *fuzzy measure* [25] if it satisfies the monotonicity condition: $v(A) \leq v(B)$ for every $A \subseteq B \subseteq N$.

Cooperative games can be seen as pseudo-Boolean functions vanishing at 0_N. A *pseudo-Boolean function* [3,14] is any function $f : \{0, 1\}^N \to \mathbb{R}$. Noting that any subset S of N can be encoded by its characteristic function 1_S, where $1_S = (x_1, \ldots, x_n)$, with $x_i = 1$ if $i \in S$ and $x_i = 0$ otherwise, there is a one-to-one correspondence between set functions and pseudo-Boolean functions: $f(1_S) = v(S)$ for every $S \subseteq N$. Therefore, a natural generalisation of games is multichoice games. A *multichoice game* [15] on N is a function $v : L \to \mathbb{R}$ such that $v(0, \ldots, 0) = 0$. A multichoice game v is monotone if $v(x) \leq v(y)$ whenever $x \leq y, \forall x, y \in L$. A monotonic multichoice game is called a *k-ary capacity* [9]. In a MCDA context, $v(x)$ is the overall score of alternative x. For any $x \in L$, $x \neq 0_N$, the *Dirac game* δ_x is defined by $\delta_x(y) = 1$ iff $y = x$, and 0 otherwise. We denote by $\mathcal{G}(L)$ the set of all multichoice games defined on L.

The *derivative* of $v \in \mathcal{G}(L)$ w.r.t. $T \subseteq N$ at $x \in L$ such that for any $i \in T, x_i < k_i$ is given by: $\Delta_T v(x) = \sum_{S \subseteq T} (-1)^{t-s} v(x + 1_S)$.

3   Banzhaf Value and Interaction Indices

In this section we recall the concepts of value and interaction indices introduced in cooperative game theory. The notion of power index or value is one of the most important concepts in cooperative game theory. A *value* [24] on N is a function $\phi : \mathcal{G}(2^N) \rightarrow \mathbb{R}^N$ which assigns to each player $i \in N$ in a game $v \in \mathcal{G}(2^N)$ a payoff $\phi_i(v)$, which is most often a share of $v(N)$, the total worth of the game. In the context where N is the set of voters, $\phi_i(v)$ can be interpreted as the voting power of player $i \in N$ in game $v \in \mathcal{G}(2^N)$, i.e., to what extent the fact that i votes 'yes' makes the final decision to be 'yes'. In such a case, ϕ is called a power index. Obviously, power indices in voting theory are close to importance indices in MCDA. In cooperative game theory, diverse kinds of values/power indices have been proposed, among which a large part have the following form: $\phi_i(v) = \sum_{S \subseteq N \setminus i} p_S^i \big(v(S \cup i) - v(S)\big), p_S^i \in \mathbb{R}$. If the family of real constants $\{p_S^i, S \subseteq N \setminus i\}$ forms a probability distribution, the value ϕ_i is said to be a *probabilistic value* [26].

The exact form of a value/power index depends on the axioms that are imposed on it. The two best known are due to Shapley [24] and Banzhaf [1]. The Banzhaf value [5] of a player $i \in N$ in a game $v \in \mathcal{G}(2^N)$ is defined by

$$\phi_i^B(v) = \sum_{S \subseteq N \setminus i} \frac{1}{2^{n-1}} \big(v(S \cup i) - v(S)\big).$$

It is uniquely axiomatized by a set of four axioms [5,18]: linearity axiom, dummy axiom, symmetry axiom and 2-efficiency axiom. They will be recalled below.

Another interesting concept is that of interaction among criteria. An *interaction index* on N of the game $v \in \mathcal{G}(2^N)$ is a function $I^v : 2^N \rightarrow \mathbb{R}$ that represents the amount of interaction (it can be positive or negative) among any subset of players. Grabisch and Roubens [12] proposed an axiomatic characterisation of the Shapley and the Banzhaf interaction indices. For this, they introduce the following definitions:

Let v be a game on N, and T a nonempty subset of N. The restriction of v to T is a game of $\mathcal{G}(2^T)$ defined by $v^T(S) = v(S), \forall S \subseteq T$. The restriction of v to T in the presence of a set $A \subseteq N \setminus T$ is a game $\mathcal{G}(2^T)$ defined by $v_{\cup A}^T(S) = v(S \cup A) - v(A)$ for every $S \subseteq T$. The reduced game with respect to T is a game denoted $v_{[T]}$ defined on the set $(N \setminus T) \cup [T]$ where $[T]$ indicates a single hypothetical player, which is the union (or representative) of the players in T. It is defined as follows for any $S \subseteq N \setminus T$:

$$v_{[T]}(S) = v(S),$$
$$v_{[T]}(S \cup [T]) = v(S \cup T).$$

The following axioms have been considered by Grabisch and Roubens [12]:

- Linearity axiom (L): $I^v(S)$ is linear on $\mathcal{G}(2^N)$ for every $S \subseteq N$.

  $i \in N$ is said to be *dummy* for $v \in \mathcal{G}(2^N)$ if $\forall S \subseteq N \setminus i$, $v(S \cup i) = v(S) + v(i)$.
- Dummy player axiom (D): If $i \in N$ is a dummy player for $v \in \mathcal{G}(2^N)$, then
  1. $I^v(i) = v(i)$,
  2. for every $S \subseteq N \setminus i$, $S \neq \varnothing$, $I^v(S \cup i) = 0$.

- Symmetry axiom (S): for all $v \in \mathcal{G}(2^N)$, for all permutation π on N,

$$I^v(S) = I^{\pi v}(\pi S).$$

- 2-efficiency axiom (2-E): For any $v \in \mathcal{G}(2^N)$,

$$I^v(i) + I^v(j) = I^{v_{[ij]}}([ij]), \forall i, j \in N.$$

- Recursive axiom (R): For any $v \in \mathcal{G}(2^N)$,

$$I^v(S) = I^{v^{N \setminus j}}_{\cup j}(S \setminus j) - I^{v^{N \setminus j}}(S \setminus j), \forall S \subseteq N, s \geq 2, \forall j \in S.$$

Theorem 1 *(Grabisch and Roubens [12]). Under (L), (D), (S), (2-E) and (R),*

$$\forall \in \mathcal{G}(2^N), I^v(S) = \sum_{T \subseteq N \setminus S} \frac{1}{2^{n-t}} \sum_{L \subseteq S} (-1)^{s-l} v(T \cup L), \forall S \subseteq N, S \neq \varnothing.$$

In particular, for a pair $S = \{i, j\}$, we obtain $I^v(\{i, j\}) = \sum_{T \subseteq N \setminus \{i,j\}} \frac{1}{2^{n-t}} \delta_{i,j} v(S)$, where $\delta_{i,j} v(S) := v(S \cup \{i, j\}) - v(S \cup \{i\}) - v(S \cup \{j\}) + v(S)$. Moulin interprets the quantity $v(\{i, j\}) - v(\{i\}) - v(\{j\})$ as the cost/surplus of mutual externalities of players i and j [19]. More generally, $\delta_{i,j} v(S)$ can be seen as the cost/surplus of mutual externalities of players i and j, in the presence of coalition S. The interaction index $I_v(\{i, j\})$ is thus the *expected cost/surplus of mutual externalities* of players i and j.

In MCDA, recall that $v(S)$ is the overall score of an option that is perfectly satisfactory (with score 1) on criteria S and completely unacceptable (with score 0) on the remaining criteria. The interaction index $I_v(\{i, j\})$ can also be interpreted as the variation of the mean weight of criterion i when criterion j switches from the least satisfied criterion to the most satisfied criterion [16]. Positive interaction depicts situations where there is *complementarity* among criteria i and j: criteria i and j deserve to be well-satisfied together (the more criterion i is satisfied, the more it is important to satisfy as well criterion j). On the opposite side, negative interaction occurs when there is *substitutability* among criteria i and j: it is not rewarding to improve both criteria i and j together.

Now we present an axiomatization of Fujimoto et al. [6] based on the concept of partnership coalition. For this, they introduce the following axiom:

Reduced-partnership-consistency axiom (RPC): If P is a partnership in a game v then $I^v(P) = I^{v_{[P]}}([P])$.

A coalition $P \subseteq N, P \neq \varnothing$, is said to be a partnership in a game $v \in \mathcal{G}(2^N)$ if, for all $S \subseteq P, v(S \cup T) = v(T)$, for all $T \subseteq N \setminus P$.

Theorem 2 *(Fujimoto et al. [6]). Under the linear axiom, the dummy axiom, the symmetry axiom, the 2-efficiency axiom and the reduced-partnership-consistency axiom,*

$$\forall v \in \mathcal{G}(2^N), I^v(S) = \sum_{T \subseteq N \setminus S} \frac{1}{2^{n-t}} \sum_{L \subseteq S} (-1)^{s-l} v(T \cup L), \forall S \subseteq N, S \neq \varnothing.$$

4   Axiomatisation of the Banzhaf Value for Multichoice Games

In this section, we give a characterisation of Banzhaf value for multichoice games, in the spirit of what was done by Weber [26] for cooperative games. Ridaoui et al. [22] have already generalized and axiomatized the Shapley value for multichoice games. The axiomatisation given in [22] is based on five axioms, linearity, nullity, symmetry, invariance and efficiency. We present the first four axioms used in [22], as some of them will be used in our characterisation. It is worth mentioning that the use of such axioms is common in axiomatisation of values. Let ϕ be a value defined for any $v \in \mathcal{G}(L)$.

Linearity axiom (L): ϕ is linear on $\mathcal{G}(L)$, i.e., $\forall v, w \in \mathcal{G}(L), \forall \alpha \in \mathbb{R}$,

$$\phi_i(v + \alpha w) = \phi_i(v) + \alpha \phi_i(w), \forall i \in N.$$

A player $i \in N$ is said to be *null* for $v \in \mathcal{G}(L)$ if $v(x + 1_i) = v(x), \forall x \in L, x_i < k$.

Null axiom (N): If a player i is null for $v \in \mathcal{G}(L)$, then $\phi_i(v) = 0$.

Let π be a permutation on N. For all $x \in L$, we denote $\pi(x)_{\pi(i)} = x_i$. For all $v \in \mathcal{G}(L)$, the game $\pi \circ v$ is defined by $\pi \circ v(\pi(x)) = v(x)$.

Symmetry axiom (S): For any permutation π of N,

$$\phi_{\pi(i)}(\pi \circ v) = \phi_i(v), \forall i \in N.$$

Invariance axiom (I): Let us consider two games $v, w \in \mathcal{G}(L)$ such that, for some $i \in N$,

$$v(x + 1_i) - v(x) = w(x) - w(x - 1_i), \forall x \in L, x_i \notin \{0, k\}$$
$$v(x_{-i}, 1_i) - v(x_{-i}, 0_i) = w(x_{-i}, k_i) - w(x_{-i}, k_i - 1), \forall x_{-i} \in L_{-i},$$

then $\phi_i(v) = \phi_i(w)$.

The linearity axiom means that if several multichoice games are combined linearly, the value of the resulting multichoice game is a linear combination of the values of each individual multichoice game. Axiom (N) states that a player having no influence on a multichoice game is not important. Axiom (S) says that the numbering of the players plays no role in the computation of value. Axiom (I) indicates that the computation of the value does not depend on the position on the grid. More precisely, if the game w is simply a shift of v of one unit on the grid, then v and w shall have the same value (importance).

Ridaoui et al. [22] have shown the following result.

Theorem 3. *Let ϕ be a value defined for any $v \in \mathcal{G}(L)$. If ϕ fulfils (L), (N), (I) and (S) then there exists a family of real constants $\{b_{n(x_{-i})}, x_{-i} \in L_{-i}\}$ such that*

$$\phi_i(v) = \sum_{x_{-i} \in L_{-i}} b_{n(x_{-i})}\big(v(x_{-i}, k_i) - v(x_{-i}, 0_i)\big), \forall i \in N, \tag{1}$$

where $n(x_{-i}) = (n_0, \ldots, n_k)$ with n_j the number of components of x_{-i} being equal to $j \in \{0, 1, \ldots, k\}$.

We introduce two additional axioms, and first some notation. For $i, j \in N$, and $v \in \mathcal{G}(L)$, denote by $v^{[ij]}$ the multichoice game defined on the set ($N^{[ij]} = N \setminus \{i, j\} \cup [ij]$), where $[ij]$ indicates a single player, which is the merge of the distinct players i and j. The multichoice game $v^{[ij]}$ is defined as follows,

$$\forall y \in \{0, 1, \ldots, k\}^{N^{[ij]}}, v^{[ij]}(y) = v(y_{-ij}, \ell_{ij}) \quad \text{if } y_{[ij]} = \ell, \ell \in \{0, 1, \ldots, k\}.$$

2-Restricted efficiency (2-RE): For all $x \in L \setminus 0_N$,

$$\phi_i(\delta_x) + \phi_j(\delta_x) = \phi_{[ij]}(\delta_x^{[ij]}),$$

where, $\forall y \in \{0, 1, \ldots, k\}^{N^{[ij]}}$, with $y_{[ij]} = \ell, \ell \in \{0, 1, \ldots, k\}$,

$$\delta_x^{[ij]}(y) = \begin{cases} \delta_x(y_{-ij}, \ell_{ij}) & \text{if } x_i, x_j \in \{1, 2, \ldots, k-1\}, \text{ or } \{x_i, x_j\} = \{0, k\}, \\ \delta_{(x_{-ij}, k_i, k_j)}(y_{-ij}, \ell_{ij}) & \text{else if } x_i \vee x_j = k, \\ \delta_{(x_{-ij}, 0_i, 0_j)}(y_{-ij}, \ell_{ij}) & \text{otherwise (i.e., if } x_i \wedge x_j = 0). \end{cases}$$

The original 2-Efficiency [18] says that the worth alloted to a coalition of two players when they form a partnership shall be divided into the worth alloted to its members. Here this axiom is considered only for the Dirac multichoice games. In the definition of $\delta_x^{[ij]}$, we need to change x by adding some symmetry between i and j in the last two cases. The 2-Restricted efficiency axiom means that for the Dirac multichoice game, the sum of the values of two players equals to the value of the merge of these players in the corresponding reduced game. The first situation of $\delta_x^{[ij]}(y)$ is standard and generalizes the classical case. The last two cases are limit cases. If only one of the elements x_i, x_j belong to $\{0, k\}$ but not the other one, then one shall take, for symmetry reasons, the same value for i and j. We need to take, for consistency reasons, the extreme value 0 or k that is reached by x_i or x_j.

For the classical Banzhaf value, the *dummy player axiom* (stronger than the null axiom) is used as a calibration property. When there is only one player left, the player shall get its worth $v(\{i\})$. We generalize this idea by the following calibration axiom restricted to Dirac games.

Calibration axiom (C): Let $i \in N$, with $n = 1$. $\phi_i(\delta_{k_N}) = 1$.

Theorem 4. *Under axioms (L), (N), (I), (S), (2-RE) and (C), for all $v \in \mathcal{G}(L)$*

$$\phi_i(v) = \frac{1}{2^{n-1}} \sum_{x_{-i} \in L_{-i}} 2^{\sigma(x_{-i}) - \kappa(x_{-i})} \big(v(x_{-i}, k_i) - v(x_{-i}, 0_i)\big), \forall i \in N \quad (2)$$

Proof: It is easy to check that the formula (2) satisfies the axioms.

Conversely, we consider ϕ satisfying the axioms (L), (N), (I), (S), (2-RE) and (C). Let $x \in L$, we write $x = (0_{N \setminus S \cup T}, x_S, k_T)$, with $x_S \in L_S \setminus \{0, k\}^S, S = \Sigma(x) \setminus K(x)$, and $T = K(x)$. From axioms (L), (N), (I) and (S) and Theorem 3, we have

$$\phi_i(\delta_x) = b_{n(x_{-i})}\big(\delta_x(x_{-i}, k_i) - \delta_x(x_{-i}, 0_i)\big),$$

then we obtain,

$$\phi_i(\delta_{(x_{-i},k_i)}) = b_{n(x_{-i})} = -\phi_i(\delta_{(x_{-i},0_i)}), \tag{3}$$

and

$$\phi_i(\delta_{(x_{-i},x_i)}) = 0, \text{for } x_i \in L_i \setminus \{0,k\}. \tag{4}$$

From (3) and (4), we have, for any $i \in T$

$$\phi_i(\delta_x) + \phi_j(\delta_x) = b_{(n-s-t,n(x_S),t-1)}, \forall j \in S, \tag{5}$$

$$\phi_i(\delta_x) + \phi_j(\delta_x) = 2b_{(n-s-t,n(x_S),t-1)}, \forall j \in T, \tag{6}$$

and,

$$\phi_i(\delta_x) + \phi_j(\delta_x) = b_{(n-s-t,n(x_S),t-1)} - b_{(n-s-t-1,n(x_S),t)}, \forall j \in N \setminus S \cup T. \tag{7}$$

By axiom (2-RE) we have,

– from (7), $\forall s \in \{0, \ldots, n-1\}$, $\forall t \in \{1, \ldots, n\}$, with $s + t \le n - 1$,

$$b_{(n-s-t-1,n(x_S),t)} = b_{(n-s-t,n(x_S),t-1)}, \tag{8}$$

– from (6), $\forall s \in \{0, \ldots, n-1\}$, $\forall t \in \{2, \ldots, n\}$, with $s + t \le n$,

$$b_{(n-s-t,n(x_S),t-2)} = 2b_{(n-s-t,n(x_S),t-1)}, \tag{9}$$

– from (5), $\forall s \in \{1, \ldots, n-1\}$, $\forall t \in \{1, \ldots, n-1\}$, with $s + t \le n$,

$$b_{(n-s-t,n(x_S),t-1)} = b_{(n-s-t,n(x_{S\setminus j}),t-1)}, j \in S, \tag{10}$$

and from (C) and (9), we have

$$b_{0,\ldots,n-1} = \frac{1}{2^{n-1}}, \forall i \in N. \tag{11}$$

We distinguish the two following cases:

1. If $S = \varnothing$,
   – from (11) and (8), we have

$$b_{n-1,0,\ldots,0} = b_{n-2,0,\ldots,0,1} = \ldots = b_{1,0,\ldots,0,n-2} = b_{0,\ldots,0,n-1} = \frac{1}{2^{n-1}}, \tag{12}$$

   then, for every $\ell \in \{1, \ldots, n\}$,

$$b_{n-\ell,0,\ldots,0,\ell-1} = \frac{1}{2^{n-1}}, \tag{13}$$

   – by (9) and (12), we have: $b_{n-2,0,\ldots,0} = \ldots = b_{0,\ldots,0,n-2} = \frac{1}{2^{n-2}}$,
   then, for every $\ell \in \{2, \ldots, n\}$, $b_{n-\ell,0,\ldots,0,\ell-2} = \frac{1}{2^{n-1}}$,

2. If $S \neq \emptyset$, by (10) and (9), we have

$$b_{(n-s-t,n(x_S),t-1)} = 2b_{(n-1-s_1-t,n(x_{S_1}),t)}, S_1 = S \setminus j, j \in S$$

$$b_{(n-1-s_1-t,n(x_{S_1}),t)} = 2b_{(n-2-s_2-t,n(x_{S_2}),t+1)}, S_2 = S_1 \setminus j, j \in S_1$$

$$\vdots$$

$$b_{(n-s-t,n(x_j),t-s)} = 2b_{(n-s-t,0,...,0,t+s-1)},$$

hence, by (13) we have, $\forall s \in \{1, \ldots, n-1\}, \forall t \in \{1, \ldots, n-1\}$, with $s+t \leq n$,

$$b_{(n-s-t,n(x_S),t-1)} = \frac{2^s}{2^{n-1}}.$$

The result is proved.   ∎

We finally show that our value $\phi_i(v)$ can be written as the sum of Banzhaf values over games derived from the multichoice game. This is related to some additivity property. More precisely, the power index $\phi_i(v)$ takes the form of the sum over $x \in \{0, \ldots, k-1\}^N$ of a classical Banzhaf value over the restriction of function v on $\times_{i \in N}\{x_i, x_i + 1\}$.

Proposition 1. *For every* $v \in \mathcal{G}(L), \phi_i(v) = \displaystyle\sum_{x \in \{0,...,k-1\}^N} \phi_i^B(\mu_x^v), \forall i \in N,$ *with,* $\mu_x^v(S) = v(x + 1_S) - v(x), \forall S \subseteq N, \forall x \in L,$ *such that* $x_i < k, \forall i \in N.$

Proof: Let $v \in \mathcal{G}(L)$ and for any $x \in L$, such that $x_i < k, \forall i \in N$, we define the game μ_x^v for every $S \subseteq N$ by $\mu_x^v(S) = v(x + 1_S) - v(x)$. We have

$$\phi_i(v) = \frac{1}{2^{n-1}} \sum_{x_{-i} \in L_{-i}} 2^{\sigma(x_{-i}) - \kappa(x_{-i})} \left(v(x_{-i}, k_i) - v(x_{-i}, 0_i) \right)$$

$$= \sum_{\substack{y_{-i} \in L_{-i} \\ \forall j \in N \setminus i, y_j < k}} \frac{1}{2^{n-1}} \sum_{y_{-i} \leq x_{-i} \leq (y+1)_{-i}} \left(v(x_{-i}, k_i) - v(x_{-i}, 0_i) \right)$$

$$= \sum_{\substack{y \in L \\ \forall j \in N, y_j < k}} \frac{1}{2^{n-1}} \sum_{x_{-i} \in \{0,1\}^{N \setminus i}} \left(v(x_{-i} + y_{-i}, y_i + 1) - v(x_{-i} + y_{-i}, y_i) \right)$$

$$= \sum_{\substack{y \in L \\ \forall j \in N, y_j < k}} \frac{1}{2^{n-1}} \sum_{A \subseteq N \setminus i} \left(\mu_y^v(A \cup i) - \mu_y^v(A) \right).$$

∎

5   Axiomatisation of the Banzhaf Interaction Index

An interaction index of a multichoice game v is a function $I^v : 2^N \to \mathbb{R}$. The interaction of a single player i is the value related to player i. In this section, we present an axiomatisation of the interaction index based on the Banzhaf value. To this aim, we use the following generalised axioms introduced in [23]:

Linearity axiom (L): I^v is linear on $\mathcal{G}(L)$, i.e., $\forall v, w \in \mathcal{G}(L), \forall \alpha \in \mathbb{R}$,

$$I^{v+\alpha w} = I^v + \alpha I^w.$$

Null axiom (N): If a player i is null for $v \in \mathcal{G}(L)$, then for all $T \subseteq N$ such that $T \ni i$, $I^v(T) = 0$.

Invariance axiom (I): Let us consider two functions $v, w \in \mathcal{G}(L)$ such that, for all $i \in N$,

$$v(x + 1_i) - v(x) = w(x) - w(x - 1_i), \forall x \in L, x_i \notin \{0, k\}$$

$$v(x_{-i}, 1_i) - v(x_{-i}, 0_i) = w(x_{-i}, k_i) - w(x_{-i}, k_i - 1), \forall x_{-i} \in L_{-i},$$

then $I^v(T \cup i) = I^w(T \cup i), \forall T \subseteq N \setminus i$.

Symmetry axiom (S): For all $v \in \mathcal{G}(L)$, for all permutation π on N,

$$I^{\pi \circ v}(\pi(T)) = I^v(T), \forall T \subseteq N, T \neq \varnothing.$$

Let v be a multichoice game in $\mathcal{G}(L)$ and $S \subseteq N$. The restriction of v to $N \setminus S$, denoted by v^{-S}, is defined by $v^{-S}(x_{-S}) = v(x_{-S}, 0_S), \forall x_{-S} \in L_{-S}$. The restriction of v on $N \setminus i$ in the presence of i denoted by v_i^{-i} is the multichoice game on L_{-i} defined by $v_i^{-i}(x_{-i}) = v(x_{-i}, k_i) - v(0_{-i}, k_i), \forall x_{-i} \in L_{-i}$.

Recursivity axiom (R): For any $v \in \mathcal{G}(L)$,

$$I^v(T) = I^{v_i^{-i}}(T \setminus i) - I^{v^{-i}}(T \setminus i), \forall T \subseteq N, T \neq \varnothing, \forall i \in T.$$

The Recursivity axiom is the exact counterpart of the one for classical games in [12].

Ridaoui et al. [23] proved the following Lemma.

Lemma 1. *Under axioms* **(L)**, **(N)**, **(I)**, **(S)** *and* **(R)**, *for any $v \in \mathcal{G}(L)$, $\forall T \subseteq N, T \neq \varnothing$,*

$$I^v(T) = \sum_{\substack{A \subseteq T \\ A \neq \emptyset}} (-1)^{t-a} I_{[A]}^{v_{[A]}^{(-T) \cup [A]}}([A]), \tag{14}$$

where $v_{[A]}^{(-T) \cup [A]}$ is the restriction of v to T with respect to $A \subseteq T$ defined on the set $\{0, \ldots, k\}^{(N \setminus T) \cup [A]}$ as follows: $v_{[A]}^{(-T) \cup [A]}(x_{-T}, \ell_{[A]}) = v(x_{-T}, \ell_A, 0_{T \setminus A})$.

Our main result shows that there is a unique index fulfilling the previous axioms.

Theorem 5. *Under axioms* **(L)**, **(N)**, **(I)**, **(S)**, **(C)**, **(2-E)** *and* **(R)**, *for all $v \in \mathcal{G}(L)$*

$$I^v(T) = \sum_{x_{-T} \in L_{-T}} \frac{2^{\sigma(x_{-T}) - \kappa(x_{-T})}}{2^{n-t}} \sum_{A \subseteq T} (-1)^{t-a} v(0_{T \setminus A}, k_A, x_{-T}), \forall T \subseteq N, T \neq \varnothing.$$

Proof: Let $v \in \mathcal{G}(L)$, and $T \subseteq N, T \neq \varnothing$. By axioms **(L)**, **(N)**, **(I)**, **(S)**, **(C)** and **(2-E)**, we have $I^{v_{[A]}^{(-T)\cup[A]}}([A]) = \sum\limits_{\substack{x_{-T} \in L_{-T}}} b_{n(x_{-T})}\left(v_{[A]}^{(-T)\cup[A]}(x_{-T}, k_{[A]}) - v_{[A]}^{(-T)\cup[A]}(x_{-T}, 0_{[A]})\right)$, with $b_{n(x_{-T})} = \dfrac{2^{\sigma(x_{-T}) - \kappa(x_{-T})}}{2^{n-t}}$.

By Lemma (1), we have

$$I^v(T) = \sum_{\substack{A \subseteq T \\ A \neq \varnothing}} (-1)^{t-a} I^{v_{[A]}^{(-T)\cup[A]}}([A])$$

$$= \sum_{\substack{A \subseteq T \\ A \neq \varnothing}} (-1)^{t-a} \sum_{x_{-T} \in L_{-T}} b_{n(x_{-T})}\left(v(x_{-T}, k_A, 0_{T \setminus A}) - v(x_{-T}, 0_T)\right)$$

$$= \sum_{x_{-T} \in L_{-T}} b_{n(x_{-T})} \sum_{\substack{A \subseteq T \\ A \neq \varnothing}} (-1)^{t-a}\left(v(x_{-T}, k_A, 0_{T \setminus A}) - v(x_{-T}, 0_T)\right)$$

$$= \sum_{x_T \in L_{-T}} b_{n(x_{-T})} \sum_{A \subseteq T} (-1)^{t-a} v(k_A, 0_{T \setminus A}, x_{-T}).$$

∎

As for the power index ϕ_i, the interaction index $I^v(T)$ can be written as the sum of Banzhaf interaction indices over games derived from the multichoice game.

Proposition 2. *Let $v \in \mathcal{G}(L)$.* $I^v(T) = \sum\limits_{x \in \{0,\ldots,k-1\}^N} I_B^{\mu_x^v}(T), \forall T \subseteq N, T \neq \varnothing$, *with,* $\mu_x^v(S) = v(x + 1_S) - v(x), \forall S \subseteq N, \forall x \in L$, *such that $x_i < k_i, \forall i \in N$.*

Proof:

$$I_B^{\mu_x^v}(T) = \frac{1}{2^{n-1}} \sum_{x \in \{0,\ldots,k-1\}^N} \sum_{S \subseteq N \setminus T} \Delta_T \mu_x^v(S)$$

$$= \frac{1}{2^{n-1}} \sum_{x \in \{0,\ldots,k-1\}^N} \sum_{S \subseteq N \setminus T} \Delta_T v(x + 1_S)$$

$$= \frac{1}{2^{n-1}} \sum_{S \subseteq T} (-1)^{t-s} \sum_{x_{-T} < k_{-T}} \left(v(0_{T \setminus A}, k_A, x_{-T}) + v(0_{T \setminus A}, k_A, x_{-T} + 1_{-T})\right)$$

$$= \frac{1}{2^{n-1}} \sum_{S \subseteq T} (-1)^{t-s} \sum_{x_{-T} \leq k_{-T}} 2^{\sigma(x_{-T}) - \kappa(x_{-T})} v(0_{T \setminus A}, k_A, x_{-T}).$$

∎

References

1. Banzhaf, J.: Weighted voting doesn't work: a mathematical analysis. Rutgers Law Rev. **19**, 317–343 (1965)
2. Choquet, G.: Theory of capacities. Ann. L'Institut Fourier **5**, 131–295 (1953)
3. Crama, Y., Hammer, P.: Boolean Functions. Number 142 in Encyclopedia of Mathematics and Its Applications. Cambridge University Press, Cambridge (2011)
4. de Wolf, R.: A brief introduction to Fourier analysis on the Boolean cube. Theory Comput. Libr. Grad. Surv. **1**, 1–20 (2008)
5. Dubey, P., Shapley, L.S.: Mathematical properties of the Banzhaf power index. Math. Oper. Res. **4**(2), 99–131 (1979)
6. Fujimoto, K., Kojadinovic, I., Marichal, J.-L.: Axiomatic characterizations of probabilistic and cardinal-probabilistic interaction indices. Games Econ. Behav. **55**(1), 72–99 (2006)
7. Grabisch, M.: k-order additive discrete fuzzy measures and their representation. Fuzzy Sets Syst. **92**(2), 167–189 (1997)
8. Grabisch, M.: Set Functions, Games and Capacities in Decision Making. Springer, Heidelberg (2016). https://doi.org/10.1007/978-3-319-30690-2
9. Grabisch, M., Labreuche, C.: Capacities on lattices and k-ary capacities. In: International Conference Of the Euro Society for Fuzzy Logic and Technology (EUSFLAT), Zittau, Germany, 10–12 September 2003
10. Grabisch, M., Labreuche, C.: A note on the Sobol' indices and interactive criteria. Fuzzy Sets Syst. **315**, 99–108 (2017)
11. Grabisch, M., Marichal, J.-L., Roubens, M.: Equivalent representations of set functions. Math. Oper. Res. **25**(2), 157–178 (2000)
12. Grabisch, M., Roubens, M.: An axiomatic approach to the concept of interaction among players in cooperative games. Int. J. Game Theory **28**(4), 547–565 (1999)
13. Grabisch, M., Roubens, M.: Application of the Choquet integral in multicriteria decision making. Fuzzy Meas. Integr.-Theory Appl. 348–374 (2000)
14. Hammer, P., Rudeanu, S.: Boolean Methods in Operations Research and Related Areas Econometrics and Operations Research, 7th edn. Springer, Heidelberg (1986). https://doi.org/10.1007/978-3-642-85823-9
15. Hsiao, C.R., Raghavan, T.E.S.: Shapley value for multi-choice cooperative games I. Discussion Paper of the University of Illinois at Chicago, Chicago (1990)
16. Kojadinovic, I.: A weight-based approach to the measurement of the interaction among criteria in the framework of aggregation by the bipolar Choquet integral. Eur. J. Oper. Res. **179**, 498–517 (2007)
17. Lange, F., Grabisch, M.: The interaction transform for functions on lattices. Discret. Math. **309**(12), 4037–4048 (2009)
18. Lehrer, E.: An axiomatization of the Banzhaf value. Int. J. Game Theory **17**(2), 89–99 (1988)
19. Moulin, H.: Fair Division and Collective Welfare. MIT Press, Cambridge (2003)
20. Murofushi, T., Soneda, S.: Techniques for reading fuzzy measures (III): interaction index. In: 9th Fuzzy System Symposium, Sapporo, Japan, pp. 693–696 (1993)
21. Ridaoui, M., Grabisch, M., Labreuche, C.: An alternative view of importance indices for multichoice games. In: Rothe, J. (ed.) ADT 2017. LNCS (LNAI), vol. 10576, pp. 81–92. Springer, Cham (2017). https://doi.org/10.1007/978-3-319-67504-6_6

22. Ridaoui, M., Grabisch, M., Labreuche, C.: Axiomatization of an importance index for generalized additive independence models. In: Antonucci, A., Cholvy, L., Papini, O. (eds.) ECSQARU 2017. LNCS (LNAI), vol. 10369, pp. 340–350. Springer, Cham (2017). https://doi.org/10.1007/978-3-319-61581-3_31

23. Ridaoui, M., Grabisch, M., Labreuche, C.: An interaction index for multichoice games. arXiv:1803.07541 (2018)

24. Shapley, L.S.: A value for n-person games. In: Kuhn, H.W., Tucker, A.W. (eds.), Contributions to the Theory of Games. Number 28 in Annals of Mathematics Studies, vol. II, pp. 307–317. Princeton University Press (1953)

25. Sugeno, M.: Theory of fuzzy integrals and its applications. Ph.D thesis. Tokyo Institute of Technology (1974)

26. Weber, R.J.: Probabilistic values for games. In: Roth, A.E. (ed.), The Shapley Value: Essays in Honor of Lloyd S. Shapley, pp. 101–120. Cambridge University Press (1988)

Fuzzy Positive Primitive Formulas

Pilar Dellunde[1,2,3]($\boxtimes$)

[1] Universitat Autònoma de Barcelona, Barcelona, Spain
pilar.dellunde@uab.cat
[2] Barcelona Graduate School of Mathematics, Barcelona, Spain
[3] Artificial Intelligence Research Institute IIIA-CSIC, Barcelona, Spain

Abstract. Can non-classical logic contribute to the analysis of complexity in computer science? In this paper, we give a step towards the solution of this open problem, taking a logical model-theoretic approach to the analysis of complexity in fuzzy constraint satisfaction. We study fuzzy positive-primitive sentences, and we present an algebraic characterization of classes axiomatized by this kind of sentences in terms of homomorphisms and finite direct products. The ultimate goal is to study the expressiveness and reasoning mechanisms of non-classical languages, with respect to constraint satisfaction problems and, in general, in modelling decision scenarios.

Keywords: Fuzzy constraint satisfaction · Preference modeling
Fuzzy logics · Model theory

1 Introduction

Can non-classical logic contribute to the analysis of complexity in computer science? The motivation to answer this question comes, in the first place, from the reading of [22], where some open problems were proposed by the authors about the relationship between fuzzy logic and valued constraint satisfaction. In our opinion, a research oriented to find a non-classical logical approach to complexity, should address, at least, the following three issues:

1. Show that there is a good trade-off between algebra and logic in the relevant fragments.
2. Identify which problems in complexity theory are naturally expressed as questions about the expressive power of the non-classical logic.
3. Prove that these complexity problems are not better addressed in other known logical formalisms.

Of course, all these issues are interrelated. To evaluate the trade-off between algebra and logic, it is important to identify which are the relevant fragments of the non-classical logic where the complexity problems have to be expressed; and to prove the relevancy of the fragments, a comparative study of different logical formalisms with respect to their expressive power has to be performed.

V. Torra et al. (Eds.): MDAI 2018, LNAI 11144, pp. 156–168, 2018.
https://doi.org/10.1007/978-3-030-00202-2_13

Revisiting the role of non-classical logics in computer science, has to be done both, in general terms, trying to find a uniform approach, but also focusing on particular classes of problems naturally addressed for some non-classical logics, as it is the case of this paper, where we contribute to the model-theoretic analysis of fuzzy constraint satisfaction using predicate fuzzy logics.

Constraint-based modeling has become a central research area in computational social choice, and in particular in preference modeling, where preferences can be seen as soft constraints [18]. Different soft constraint formalisms can be found in the literature, some prominent examples are fuzzy constraint satisfaction [10,25], possibilistic [19], probabilistic [12], and weighted [26]. More recently, the semiring-based and the valued constraint general framework have been introduced ([3] and [26], respectively), and previous formalisms can conveniently be regarded as instances of semiring-based or valued soft constraints. For a general reference to the different soft-constraint formalisms in preference modeling see [21] and [18].

The classical constraint satisfaction problem (CSP) has been proved to have strong connections with various problems in database theory and classical finite-model theory [15], where CSP can be rephrased as a homomorphism problem, a conjunctive-query evaluation problem, or a join-evaluation problem. Some problems in complexity theory are naturally expressed as questions about the expressive power of certain classical logics. With the plurality of valued structures involved in soft contraint problems, it is a natural question to ask, for the relationship between valued CSP and non-classical logical formalisms. In particular, as pointed out in [22], with mathematical fuzzy logic (MFL). Only in recent times, model theory of predicate fuzzy logics has been developed as a subarea of MFL (see for instance [5] or [9]), leaving the important area of fuzzy finite-model theory yet unexplored.

Considering a general semantics for MFL, a plethora of left continuous t-norms can be defined, going far beyond of the minimum t-norm in the interval $[0, 1]$ of the reals, most commonly used in fuzzy CSP (FCSP). Nevertheless, as pointed already in some earlier works (see for instance [25] or [27]) the minimum is the only total order semiring operator that is idempotent (see also [26]), and its *drowning effect* limits the application of FCSP to specific contexts (for a recent example of the application of fuzzy constraints in compact preference representation see [20]). t-norms in general are not good as aggregation operators, but our research do not want to focus only in aggregation, we would like rather to explore the logical properties of fuzzy languages, their expressiveness, and reasoning mechanisms with respect to constraint satisfaction problems and, in general, in modeling decision scenarios [2].

Positive-primitive formulas are one of the key elements in the logical study of classical CSP (see for instance [15]). The original contribution of the article is the mathematical proof of an axiomatization theorem for primitive-positive theories. The proof uses specific techniques of model theory and algebra in the fuzzy context, and it is included in Sect. 4. Some preliminaries on FCSP, and predicate fuzzy logics needed for the theorem are introduced in Sects. 2 and 3. A discussion section at the end of the paper presents some ideas for future work.

2   Preliminaries

Fuzzy CSP. The valued structure most commonly used in the literature of fuzzy constraint satisfaction is the standard Gödel algebra, that has as domain the $[0, 1]$ interval of the real numbers, and as t-norm the minimum. In this paper, we will work with MTL-algebras, which constitute the set of truth-values where sentences of predicate fuzzy logic are evaluated. We focus on finite MTL-algebras, but the results can be extended to the case where the valued structure is, for instance, the infinite standard Gödel or Łukasiewicz algebra. The domains of the finite MTL-algebras we consider are not necessarily totally ordered, allowing to represent some types of non-linear preferences.

MTL-algebras are defined as bounded integral commutative residuated lattices $(A, \sqcap, \sqcup, *, \Rightarrow, 0, 1)$, where $\sqcap$ and $\sqcup$ are respectively the lattice meet and join operations, $*$ is a left-continuous t-norm, and $(\Rightarrow, *)$ is a residuated pair (for an exhaustive exposition of MTL-algebras we refer to [11]).

Definition 1. *Let* **A** *be a* MTL-*algebra, D a set, and k a natural number. It is said that R is a k-ary* fuzzy relation *on D, if $R : D^k \to A$ is a function evaluated in A.*

Definition 2. *An* instance $\mathcal{I}$ *of fuzzy constraint satisfaction is a triple (V, D, C), where*

- *V is a set of variables;*
- *D is a set of values, referred to as the* domain*;*
- *C is a collection of constraints $C_1, \ldots, C_q$, where each constraint C_i is a pair $(\overline{x}, R^{\mathcal{I}})$, where $R^{\mathcal{I}}$ is a k-ary fuzzy relation on D, for some natural number $k \geq 1$, and $\overline{x}$ is a k-tuple over V, referred to as the* scope *of the constraint.*

Given an instance $\mathcal{I}$ of fuzzy constraint satisfaction with set of constraints $C = \{(\overline{x}_1, R_1^{\mathcal{I}}), \ldots, (\overline{x}_n, R_n^{\mathcal{I}})\}$, and a k-tuple $\overline{d} \in D$, we say that $R_i^{\mathcal{I}}(\overline{d})$ is the *degree of satisfaction of $\overline{d} \in D$ of constraint* $(\overline{x}_i, R_i^{\mathcal{I}})$, and that $R_1^{\mathcal{I}}(\overline{d}) * \cdots * R_n^{\mathcal{I}}(\overline{d})$ is the *degree of joint satisfaction of the constraints*, where $*$ is the t-norm of the algebra **A**. For the sake of clarity, we have restricted the definition to the case where the degree of joint satisfaction is calculated only by means of the t-norm $*$ in the standard way, but other functions could have been introduced using as base both $*$ and the min.

The *Fuzzy Constraint Satisfaction Problem* is to find an optimal solution, in the sense of maximazing the degree of joint satisfaction of the constraints. Related to this central problem there is a variety of other problems that it is possible to formulate using the graded nature of fuzzy constraints, for instance, we can ask if there is a k-tuple $\overline{d} \in D$ such that the degree of joint satisfaction is greater or lower than a given threshold.

Predicate Fuzzy Logics. Given an instance $\mathcal{I}$ of fuzzy constraint satisfaction with set of constraints $C = \{(\overline{x}_1, R_1^{\mathcal{I}}), \ldots, (\overline{x}_n, R_n^{\mathcal{I}})\}$, we can associate to $\mathcal{I}$ a fuzzy relational **A**-structure $\mathcal{I} = (D, R_1^{\mathcal{I}}, \ldots, R_n^{\mathcal{I}})$, and study its properties using model theory of predicate fuzzy logics. Now we present the syntax and semantics of the minimal predicate fuzzy logic MTL$\forall^m$, the predicate extension of the left-continuous t-norm based logic MTL introduced in [11], and we refer to [6, Chap. 1] for a complete and extensive presentation of MTL$\forall^m$.

Definition 3 (Syntax of Predicate Languages). *A predicate language* $\mathcal{P}$ *is a triple* $\langle Pred_\mathcal{P}, Func_\mathcal{P}, Ar_\mathcal{P} \rangle$, *where* $Pred_\mathcal{P}$ *is a nonempty set of predicate symbols,* $Func_\mathcal{P}$ *is a set of* function symbols *(disjoint from* $Pred_\mathcal{P}$*), and* $Ar_\mathcal{P}$ *represents the* arity function, *which assigns a natural number to each predicate symbol or function symbol. We call this natural number the* arity *of the symbol. The predicate symbols with arity zero are called* truth constants, *while the function symbols whose arity is zero are named* individual constants.

The set of $\mathcal{P}$-terms, $\mathcal{P}$-formulas and the notions of free occurrence of a variable, open formula, substitutability and sentence are defined as in classical predicate logic. We asume that the equality symbol $\approx$ of the language is interpreted in every structure as the crisp identity. Notice that, in the language we have introduced there are also function symbols. The results we present in this paper hold also for arbritrary languages, and for this reason we have presented a general proof, that could be used in further applications of pp-definability in non-relational structures, not necessarily related to FCSP.

Definition 4. *We introduce an axiomatic system for the predicate logic* MTL$\forall^m$:

(P) *Instances of the axioms of the propositional logic* MTL.
($\forall$1) $(\forall x)\varphi(x) \to \varphi(t)$, *where the term* t *is substitutable for* x *in* φ.
($\exists$1) $\varphi(t) \to (\exists x)\varphi(x)$, *where the term* t *is substitutable for* x *in* φ.
($\forall$2) $(\forall x)(\xi \to \varphi) \to (\xi \to (\forall x)\varphi(x))$, *where* x *is not free in* ξ.
($\exists$2) $(\forall x)(\varphi \to \xi) \to ((\exists x)\varphi \to \xi)$, *where* x *is not free in* ξ.

The deduction rules of MTL$\forall^m$ *are those of* MTL *and the rule of generalization: from* φ *infer* $(\forall x)\varphi$. *The definitions of proof and provability are analogous to the classical ones. A set of formulas* Φ *is* consistent, *if* $\Phi \not\vdash \overline{0}$.

From now on we fix a finite MTL-algebra **A** and consider only structures over this algebra.

Definition 5 (Semantics of Predicate Fuzzy Logics). *Consider a predicate language* $\mathcal{P} = \langle Pred_\mathcal{P}, Func_\mathcal{P}, Ar_\mathcal{P} \rangle$. *We define an* **M**-*structure* **M** *for* $\mathcal{P}$ *as a triple* $\langle M, (P_\mathbf{M})_{P \in Pred}, (F_\mathbf{M})_{F \in Func} \rangle$, *where* M *is a nonempty domain,* $P_\mathbf{M}$ *is an* n-*ary fuzzy relation for each* n-*ary predicate symbol, identified with an element of* $\mathbf{A}$, *if* $n = 0$; *and* $F_\mathbf{M}$ *is a function from* M^n *to* M, *identified with an element of* M, *if* $n = 0$.

As usual, if **M** *is an* **A**-*structure for* $\mathcal{P}$, *an* **M**-*evaluation of the object variables is a mapping* v *assigning to each object variable an element of* M. *The set of all object variables is denoted by* Var. *If* v *is an* **M**-*evaluation,* $x \in Var$ *and* $a \in M$, *we denote by* $v[x \mapsto a]$ *the* **M**-*evaluation so that* $v[x \mapsto a](x) = a$

and $v[x \mapsto a](y) = v(y)$ *for* y *an object variable such that* $y \neq x$. *If* **M** *is an* **M**-*structure and* v *is an* **M**-*evaluation, we define the* values *of terms, and the* truth values *of formulas in* M *for an evaluation* v *recursively as follows:*

$$||x||_{\mathbf{M},v} = v(x);$$
$$||F(t_1,\ldots,t_n)||_{\mathbf{M},v} = F_{\mathbf{M}}(||t_1||_{\mathbf{M},v},\ldots,||t_n||_{\mathbf{M},v}), \text{ for } F \in Func;$$
$$||P(t_1,\ldots,t_n)||_{\mathbf{M},v} = P_{\mathbf{M}}(||t_1||_{\mathbf{M},v},\ldots,||t_n||_{\mathbf{M},v}), \text{ for } P \in Pred;$$
$$||\lambda(\varphi_1,\ldots,\varphi_n)||_{\mathbf{M},v} = \lambda_{\mathbf{A}}(||\varphi_1||_{\mathbf{M},v},\ldots,||\varphi_n||_{\mathbf{M},v}), \text{ for every connective } \lambda;$$
$$||(\forall x)\varphi||_{\mathbf{M},v} = inf\{||\varphi||_{\mathbf{M},v[x \to a]} \mid a \in M\};$$
$$||(\exists x)\varphi||_{\mathbf{M},v} = sup\{||\varphi||_{\mathbf{M},v[x \to a]} \mid a \in M\}.$$

We assume that the language has an equality symbol $\approx$, interpreted as a crisp identity. We denote by $||\varphi||_{\mathbf{M}} = 1$ the fact that $||\varphi||_{\mathbf{M},v} = 1$ for all **M**-evaluation v; and given a set of sentences Φ, we say that **M** is a *model of* Φ, if for every $\varphi \in \Phi$, $||\varphi||_{\mathbf{M}} = 1$. We denote by Mod($\Phi$) the set of models of Φ, and by Th(**M**), the theory of **M**, that is, the set of sentences evaluated 1 in **M**. We say that two models are *elementary equivalent*, if they have the same theory.

Structures over a Fixed Finite MTL-Algebra. Since we work with structures over a fixed finite MTL-algebra, the infimum and the supremum in Definition 5 always exist, and they coincide with the minimum and maximum. There are two important properties that all the structures over a finite MTL-algebra have, and that we will use throughout this article. The first one is that they are existentially witnessed: given a **A**-structure **M**, we say that **M** is ∃-*witnessed* if it satisfies the following property: for every formula of the form $(\exists \overline{x})\psi(\overline{x})$, there are $\overline{d} \in M$ such that $||(\exists \overline{x})\psi(\overline{x})||_{\mathbf{M}} = ||\psi(\overline{d})||_{\mathbf{M}}$.

The second property is compactness, both for satisfiabilty and consequence (the proof can be found in [8, Theorem 4.4]). Remark that, in fuzzy logic it is not always the case, for instance the product predicate logic is neither satisfiability nor consequence compact with respect to its standard algebra. Given a set of sentences Σ, and a sentence ϕ, we denote by $\Sigma \models_{\mathbf{A}} \phi$ the fact that every **A**-model of Σ is also an **A**-model of ϕ.

Theorem 1 (A-compactness). *For every set of sentences* Σ *and sentence* ϕ, *the following holds:*

1. [Satisfiability] *If for every finite subset* $\Sigma_0 \subseteq \Sigma$, Σ_0 *has an* **A**-*model, then* Σ *has also an* **A**-*model.*
2. [Consequence] *If* $\Sigma \models_{\mathbf{A}} \phi$, *then there is a finite subset* $\Sigma_0 \subseteq \Sigma$ *such that* $\Sigma_0 \models_{\mathbf{A}} \phi$.

From now on we will refer to **A**-structures simply as *structures*, because all the structures we consider will be over the same algebra.

3 Fuzzy Positive-Primitive Formulas

Let $\mathcal{I}$ be an instance of fuzzy constraint satisfaction and $\mathcal{I} = (D, R_1^{\mathcal{I}}, \ldots, R_n^{\mathcal{I}})$ its associated fuzzy relational structure. In logical terms, the FCSP can be formulated as the problem of finding a tuple $\overline{d}$ such that

$$||R_1(\overline{d}) \& \cdots \& R_n(\overline{d})||_{\mathcal{I}} = ||(\exists \overline{x})(R_1(\overline{x}) \& \cdots \& R_n(\overline{x}))||_{\mathcal{I}}$$

where $\&$ is the strong conjunction interpreted in $\mathcal{I}$ as the t-norm. The formulas that allow us to give a logical expression of the FSCP are called *fuzzy positive-primitive* and are the object of study of this section. In particular, we show that homomorphisms and direct products preserve fuzzy positive-primitive formulas. For a general reference of the classical positive-primitive fragment see [14].

Definition 6 (Fuzzy Positive-Primitive Formula). *Given a predicate language $\mathcal{P}$, and a $\mathcal{P}$-formula ϕ, it is said that ϕ is* fuzzy positive-primitive, *if ϕ is of the form $(\exists \overline{x})\psi$, where ψ is a quantifier-free formula built from atomic formulas by using only the connectives $\wedge$ and $\&$.*

For the sake of simplicity, from now on we will refer to fuzzy positive-primitive formulas, simply as *pp-formulas*. Remark that both conjunctions, strong and weak, can appear in pp-formulas, allowing different combinations of these connectives for expressing the degree of joint satisfaction of a set of constraints. Let us recall now the definition of homomorphism introduced in [9] as a generalization of the notion of classical homomorphism.

Definition 7 (Homomorphism). *Let $\mathcal{P}$ be a predicate language, $\mathbf{M}$ and $\mathbf{N}$ be two $\mathcal{P}$-structures and g a mapping from M to N. We say that g is a* homomorphism *from $\mathbf{M}$ into $\mathbf{N}$ if and only if*

1. For every n-ary function symbol $F \in \mathcal{P}$, and $d_1, \ldots, d_n \in M$,

$$g(F_{\mathbf{M}}(d_1, \ldots, d_n)) = F_{\mathbf{N}}(g(d_1), \ldots, g(d_n)).$$

2. For every n-ary predicate symbol $P \in \mathcal{P}$, and $d_1, \ldots, d_n \in M$,

$$\text{if } ||P(d_1, \ldots, d_n)||_{\mathbf{M}} = 1, \text{ then } ||P(g(d_1), \ldots, g(d_n))||_{\mathbf{N}} = 1.$$

Moreover, we say that g is an embedding, if g is one-to-one, and that g is an isomorphism, if g is a surjective embedding.

In the following lemma we prove that pp-formulas are preserved by homomorphisms.

Lemma 1. *Let $\mathcal{P}$ be a predicate language, $\mathbf{M}$ and $\mathbf{N}$ be two $\mathcal{P}$-structures, g a homomorphism from $\mathbf{M}$ into $\mathbf{N}$, and ϕ a positive-primitive $\mathcal{P}$-formula. Then, for every $d_1, \ldots, d_n \in M$,*

$$\text{if } ||\phi(d_1, \ldots, d_n)||_{\mathbf{M}} = 1, \text{ then } ||\phi(g(d_1), \ldots, g(d_n))||_{\mathbf{N}} = 1.$$

Proof. By induction on the complexity of ϕ.

Atomic Step. Let ϕ be an atomic formula of the form $P(t_1 \ldots, t_k)$, where $P \in \mathcal{P}$ is a predicate symbol, and $t_1 \ldots, t_k$ are $\mathcal{P}$-terms. Since g is a homomorphism, we have that, in general, for every $\mathcal{P}$-term t, and $d_1, \ldots, d_n \in M$, $g(t_\mathbf{M}(d_1, \ldots, d_n)) = t_\mathbf{N}(g(d_1), \ldots, g(d_n))$ and thus

$$||P(t_1 \ldots, t_k)(d_1, \ldots, d_n)||_\mathbf{M} = 1 \Rightarrow$$
$$||P(t_{1\mathbf{M}}(d_1, \ldots, d_n), \ldots, t_{k\mathbf{M}}(d_1, \ldots, d_n))||_\mathbf{M} = 1 \Rightarrow$$
$$||P(g(t_{1\mathbf{M}}(d_1, \ldots, d_n)), \ldots, g(t_{k\mathbf{M}}(d_1, \ldots, d_n)))||_\mathbf{N} = 1 \Rightarrow$$
$$||P(t_{1\mathbf{N}}(g(d_1), \ldots, g(d_n)), \ldots, t_{k\mathbf{N}}(g(d_1), \ldots, g(d_n)))||_\mathbf{N} = 1 \Rightarrow$$
$$||P(t_1 \ldots, t_k)(g(d_1), \ldots, g(d_n))||_\mathbf{N} = 1.$$

Quantifier-free. Assume inductively that the property holds for ψ and for χ, then we have:

$$1 = ||\psi \& \chi(d_1, \ldots, d_n)||_\mathbf{M} = ||\psi(d_1, \ldots, d_n)||_\mathbf{M} * ||\chi(d_1, \ldots, d_n)||_\mathbf{M} \Rightarrow$$
$$||\psi(d_1, \ldots, d_n)||_\mathbf{M} = 1 \text{ and } ||\chi(d_1, \ldots, d_n)||_\mathbf{M} = 1 \Rightarrow$$
$$||\psi(g(d_1), \ldots, g(d_n))||_\mathbf{N} = 1 \text{ and } ||\chi(g(d_1), \ldots, g(d_n))||_\mathbf{N} = 1 \Rightarrow$$
$$||\psi(g(d_1), \ldots, g(d_n))||_\mathbf{N} * ||\chi(g(d_1), \ldots, g(d_n))||_\mathbf{N} = 1 \Rightarrow$$
$$||\psi \& \chi(g(d_1), \ldots, g(d_n))||_\mathbf{N} = 1.$$

Observe that the same argument holds for the weak conjunction $\wedge$.

Existential Step. Assume inductively that the property holds for $\psi(x)$. Since $\mathbf{M}$ is an $\exists$-witnessed structure, we have that for some $e \in M$,

$$||(\exists x)\psi(x, d_1, \ldots, d_n)||_\mathbf{M} = ||\psi(e, d_1, \ldots, d_n)||_\mathbf{M}$$

Thus, if $||(\exists x)\psi(x, d_1, \ldots, d_n)||_\mathbf{M} = 1$, then $||\psi(e, d_1, \ldots, d_n)||_\mathbf{M} = 1$ and, by inductive hypothesis,

$$1 = ||\psi(g(e), g(d_1), \ldots, g(d_n))||_\mathbf{N} \leq ||(\exists x)\psi(x, g(d_1), \ldots, g(d_n))||_\mathbf{N}.$$

Now let us introduce the notion of direct product. Unlike other definitions introduced in the literature, for instance in [22], we work in products over the same algebra $\mathbf{A}$. Notice that the product is well-defined because the algebra is finite.

Definition 8 (A-direct product). *Let $\mathcal{P}$ be a predicate language, I a nonempty set, and for every $i \in I$, $\mathbf{M}_i$ a $\mathcal{P}$-structure. The direct product of the family $\{\mathbf{M}_i : i \in I\}$, denoted by $\prod_{i \in I} \mathbf{M}_i$, is the structure that has as domain the usual classical direct product, and the usual classical interpretation for constants and function symbols, and for every n-adic predicate symbol $P \in \mathcal{P}$, and tuples of elements $\overline{d_1}, \ldots, \overline{d_n}$ of $\prod_{i \in I} M_i$,*

$$P_{\prod_{i \in I} \mathbf{M}_i}(\overline{d_1}, \ldots, \overline{d_n}) = \min\{P_{\mathbf{M}_i}(\overline{d_1}(i), \ldots, \overline{d_n}(i)) : i \in I\}$$

Notice that, so defined, the i-projection of the direct product onto $\mathbf{M}_i$ is a homomorphism, and thus, by Lemma 1, preserves pp-formulas. We will use this fact later in the proof of the axiomatization theorem. In the following lemma we prove that pp-formulas are preserved by direct products.

Lemma 2. *Let $\mathcal{P}$ be a predicate language, I a nonempty set, and for every $i \in I$, $\mathbf{M}_i$ a $\mathcal{P}$-structure. Assume that ϕ is a positive-primitive $\mathcal{P}$-formula, and $\overline{d_1}, \ldots, \overline{d_n}$ are tuples of elements of $\prod_{i \in I} M_i$. Then the following holds: if for every $i \in I$, $\|\phi(\overline{d_1}(i), \ldots, \overline{d_n}(i))\|_{\mathbf{M}_i} = 1$, then $\|\phi(\overline{d_1}, \ldots, \overline{d_n})\|_{\prod_{i \in I} \mathbf{M}_i} = 1$.*

Proof. By induction on the complexity of ϕ. The proof of the atomic and quantifier-free step is analogous to the corresponding proof in Lemma 1, by using the fact that for every $\mathcal{P}$-term t,

$$t_{\prod_{i \in I} \mathbf{M}_i}(\overline{d_1}, \ldots, \overline{d_n}) = (t_{\mathbf{M}_i}(\overline{d_1}(i), \ldots, \overline{d_n}(i)) : i \in I)$$

For the existential step, assume inductively that the property holds for $\psi(x)$. If for every $i \in I$, $\|(\exists x)\psi(x, \overline{d_1}(i), \ldots, \overline{d_n}(i))\|_{\mathbf{M}_i} = 1$, since the structures are $\exists$-witnessed, then for every $i \in I$, there is $\overline{e}(i) \in M_i$ such that

$$\|\psi(\overline{e}(i), \overline{d_1}(i), \ldots, \overline{d_n}(i))\|_{\mathbf{M}_i} = 1$$

Then, by using the inductive hypothesis,

$$1 = \|\psi(\overline{e}, \overline{d_1}, \ldots, \overline{d_n})\|_{\prod_{i \in I} \mathbf{M}_i} \leq \|(\exists x)\psi(x, \overline{d_1}(i), \ldots, \overline{d_n}(i))\|_{\prod_{i \in I} \mathbf{M}_i}.$$

4   Fuzzy Positive-Primitive Sets of Axioms

Axiomatization theorems provide a correspondence between syntactic and semantic notions in logic. Diagrams are the building blocks that, glued with compactness, allow us to build extensions of structures, and prove these axiomatization theorems. Let us thus to introduce the method of diagrams in this fuzzy setting in order to characterize homomorphisms, and prove an equivalent condition to the preservation of pp-formulas between structures.

Definition 9. *Let $\mathcal{P}$ be a predicate language, and $\mathbf{M}$ a $\mathcal{P}$-structure. The expansion of the language $\mathcal{P}$ by adding an individual constant symbol c_m for every $m \in M$, is denoted by $\mathcal{P}^M$; and the expansion of the structure $\mathbf{M}$ to $\mathcal{P}^M$ is denoted by $\mathbf{M}^\sharp$, where for every $m \in M$, $(c_m)_{\mathbf{M}^\sharp} = m$.*

Definition 10. *Let $\mathcal{P}$ be a predicate language. For every $\mathcal{P}$-structure $\mathbf{M}$ we define $\mathrm{Diag}(\mathbf{M})$ as the set of atomic $\mathcal{P}^M$-sentences σ such that $\|\sigma\|_{\mathbf{M}^\sharp} = 1$.*

Following the same lines of the proof of [7, Proposition 32], we can obtain this characterization of homomorphisms in terms of diagrams.

Corollary 1. *Let $\mathcal{P}$ be a predicate language and $\mathbf{M}$ and $\mathbf{N}$ be two $\mathcal{P}$-structures. The following are equivalent:*

1. *There is an expansion of $\mathbf{N}$ that is a model of $\mathrm{Diag}(\mathbf{M})$.*
2. *There is a homomorphism $g : M \to N$ from $\mathbf{M}$ into $\mathbf{N}$.*

Notice that, since the $\mathrm{Diag}(\mathbf{M})$ contains equalities but not inequalities, the obtained homomorphism does not need to be an embedding. Now we present a characterization in terms of extensions, of when two structures preserve pp-formulas.

Proposition 1. *Let $\mathcal{P}$ be a predicate language, and $\mathbf{M}$ and $\mathbf{N}$ be two $\mathcal{P}$-structures. Then, every pp-sentence which is evaluated 1 in $\mathbf{M}$, is also evaluated 1 in $\mathbf{N}$ if and only if there is a $\mathcal{P}$-structure $\mathbf{L}$, elementary equivalent to $\mathbf{N}$, and a homomorphism g from $\mathbf{M}$ into $\mathbf{L}$.*

Proof. First we show that $\mathrm{Diag}(\mathbf{M}) \cup \mathrm{Th}(\mathbf{N})$ has a model. We prove that for every finite subset $\{\sigma_1 \ldots, \sigma_n\}$ of $\mathrm{Diag}(\mathbf{M})$, $\{\sigma_1 \ldots, \sigma_n\} \cup \mathrm{Th}(\mathbf{N})$ has a model. Let $c_{m_1}, \ldots, c_{m_k}$ be the object constants of the expanded language that occur in $\{\sigma_1 \ldots, \sigma_n\}$. For every $1 \le i \le n$, let σ_i' be the formula obtained from σ_i by substituting the constants $c_{m_1}, \ldots, c_{m_k}$ by new variables $\overline{y} = y_{m_1}, \ldots, y_{m_k}$.

Then we have that $\|(\exists \overline{y})(\sigma_1' \wedge \cdots \wedge \sigma_n(\overline{y}))\|_{\mathbf{M}} = 1$ and thus, by the assumption of this lemma, since $(\exists \overline{y})(\sigma_1' \wedge \cdots \wedge \sigma_n'(\overline{y}))$ is a pp-sentence, $\|(\exists \overline{y})(\sigma_1' \wedge \cdots \wedge \sigma_n(\overline{y}))\|_{\mathbf{N}} = 1$. Since $\mathbf{N}$ is an $\exists$-witnessed structure, we have a sequence of elements of N, $\overline{e} = e_{m_1}, \ldots, e_{m_k}$, such that $\|((\sigma_1' \wedge \cdots \wedge \sigma_n(\overline{e})\|_{\mathbf{N}} = 1$. Thus we can conclude that an expansion of $\mathbf{N}$ satisfies $\{\sigma_1 \ldots, \sigma_n\} \cup \mathrm{Th}(\mathbf{N})$. By $\mathbf{A}$-compactness for satisfiability, there is a $\mathcal{P}$-structure $\mathbf{L}$ that has an expansion which is a model of $\mathrm{Diag}(\mathbf{M}) \cup \mathrm{Th}(\mathbf{N})$.

By Lemma 1, there is a homomorphism g from $\mathbf{M}$ into $\mathbf{L}$. Moreover, since $\mathbf{L}$ is a model of $\mathrm{Th}(\mathbf{N})$, $\mathbf{L}$ is elementary equivalent to $\mathbf{N}$.

Now we prove an axiomatization theorem for theories closed under homomorphisms and direct products. Recall that a theory T is *closed under* a class O, if the class of its models, $\mathrm{Mod}(T)$, is closed under O. And it is say that a theory T is *axiomatized* by a set of sentences Σ, if $\mathrm{Mod}(T) = \mathrm{Mod}(\Sigma)$.

Theorem 2. *Let $\mathcal{P}$ be a predicate language and T be a consistent theory. Then, T is closed under homomorphisms and direct products if and only if T is axiomatized by a set of positive primitive sentences.*

Proof. Let $T_\vee$ be the set of finite disjunctions of pp-sentences evaluated positively in every model of T (that is, evaluated with an element of the algebra $a \in A$ such that $a > 0$), and T_{pp} be the set of pp-sentences of $T_\vee$. In the proof we distinguish two parts: 1) we show that $T_\vee$ axiomatizes T, and 2) we show that T_{pp} axiomatizes $T_\vee$.

1) $\underline{T_\vee \text{ axiomatizes } T}$. First notice that $T_\vee$ is a nonempty set, for instance $(\exists x)(x \approx x) \in T_\vee$, and $T_\vee$ is satisfiable because T is a consistent theory. Let $\mathbf{N}$ be a model of $T_\vee$, we will show that $\mathbf{N}$ is also a model of T. Let Γ be the set

of all sentences of the form $\neg \delta$, where δ is a pp-sentence and $||\delta||_\mathbf{N} < 1$. Now we prove that $\Gamma \cup T$ has a model.

If $\Gamma = \emptyset$ is clear, because T is consistent. If $\Gamma \neq \emptyset$, we have that, for every nonempty subset $\{\neg \delta_1 \ldots, \neg \delta_n\}$ of Γ, $\{\neg \delta_1 \ldots, \neg \delta_n\} \cup T$ has a model. Otherwise, in every model of T the sentence $\delta_1 \vee \cdots \vee \delta_n$ will be evaluated positively and thus, $\delta_1 \vee \cdots \vee \delta_n \in T_\vee$, contradicting the fact that $\mathbf{N}$ is a model of $T_\vee$. By $\mathbf{A}$-compactness for satisfiability, there is a model $\mathbf{M}$ of $\Gamma \cup T$. Then we have that every pp-sentence which is evaluated 1 in $\mathbf{M}$, is also evaluated 1 in $\mathbf{N}$, because $\mathbf{M}$ is a model of Γ. Then, by Proposition 1, there is a structure $\mathbf{L}$, elementarily equivalent to $\mathbf{N}$, and a homomorphism g from $\mathbf{M}$ into $\mathbf{L}$. Since T is closed under homomorphisms, we can conclude that $\mathbf{N}$ is also a model of T. Consequently, $T_\vee$ is a set of axioms for T.

1) T_{pp} axiomatizes $T_\vee$. We show that, for every $\delta_1 \vee \cdots \vee \delta_n \in T_\vee$, there is $1 \leq i \leq n$ with $\delta_i \in T_{pp}$. Assume, searching for a contradiction, that for every $1 \leq i \leq n$, there is $\mathbf{M}_i$ which is a model of $T_\vee$ but not of δ_i. Consider the direct product $\prod_{1 \leq i \leq n} \mathbf{M}_i$. Since $T_\vee$ is closed under direct products, $\prod_{1 \leq i \leq n} \mathbf{M}_i$ is also a model of $T_\vee$ and, in particular, of the sentence $\delta_1 \vee \cdots \vee \delta_n$. Then, for some $1 \leq i_0 \leq n$, $\prod_{1 \leq i \leq n} \mathbf{M}_i$ is a model of δ_{i_0}. Take the i_0-projection function $i_0 : \prod_{1 \leq i \leq n} \mathbf{M}_i \to \mathbf{M}_{i_0}$. Since i_0 is a homomorphism, i_0 preserves pp-formulas, and thus, $\mathbf{M}_{i_0}$ is also a model of δ_{i_0}, contradicting our original assumption. We can conclude that there is some $1 \leq i \leq n$ with $\delta_i \in T_{pp}$. Consequently, T_{pp} is a set of axioms for $T_\vee$.

Notice that, in the proof of Theorem 2, we have only used finite direct products. Using $\mathbf{A}$-compactness for consequence we can obtain the following corollary of Theorem 2.

Corollary 2. *Let $\mathcal{P}$ be a predicate language and ϕ be a satisfiable sentence. Then, ϕ is equivalent to a pp-sentence if and only if ϕ is preserved under homomorphisms and direct products.*

5 Discussion and Future Work

Can non-classical logic contribute to the analysis of complexity in computer science? We started the paper with the statement of this general question, and in this final section, we would like to comment on how the axiomatization theorem can be regarded as a contribution to provide an answer to this question.

In one of the books of reference in the field [4], model theory is described as algebra+logic. Working in this same framework, and in the line of recent works taking an algebraic approach to valued CSP (see for instance [16] and [17]), we have presented an algebraic characterization of the preservation of pp-formulas in terms of direct products and homomorphisms. Theorem 2 tells us that there is a good trade-off between algebra and logic in the fuzzy positive-primitive fragment. This result allows us also to characterize pp-definability in terms of polymorphisms, that can be defined using homomorphisms and finite direct products.

However, the notion of fuzzy homomorphism traditionally used in the fuzzy literature, do not encompass other notions of polymorphism such as weighted or fractional polymorphisms (see for instance [16] or [17]). Further research is needed to study stronger definitions of homomorphism (for example [22], [7] or [9]) and see which are more adequate for the purpose of rephrasing FCSP using homomorphism problems. One of the main characteristics we have to impose to homomorphisms, is that they preserve positive values. Theorem 2 also sheds light to the fact that, if we introduce stronger notions of homomorphisms, we will need to redefine pp-formulas, possibly using a language expanded with constant symbols for the elements of the valued structure, in order to maintain the correspondence between algebra and logic.

The relational structures we have studied are over finite algebras, but we have proven the results both, for finite and for infinite domains, in order to cope with applications on infinite templates. Work in progress includes the generalization of Geiger's Theorem [13] to the fuzzy context, where some important preliminary results were obtained in [22], for locally finite valuation structures. In the classical case, the pp-preservation problem restricted to finite structures was solved by B. Rossman in [23], with some previous results, for instance in [1], in the context of CSP dualities. It would be interesting to prove the corresponding version in the fuzzy context, especially taking into account the improvements recently introduced in [24], with respect to the bounds on the quantifier-rank of the sentences.

Acknowledgements. The research leading to these results has received funding from RecerCaixa. This project has also received funding from the European Union's Horizon 2020 research and innovation program under the Marie Sklodowska-Curie grant agreement No 689176 (SYSMICS project), and by the projects RASO TIN2015-71799-C2-1-P, CIMBVAL TIN2017-89758-R, and the grant 2017SGR-172 from the Generalitat de Catalunya. The author would like to thank the reviewers for their comments, and the Algorithmic Decision Theory Group of Data61 (UNSW, Sydney) for hosting me during this research.

References

1. Atserias, A., Dawar, A., Kolaitis, P.G.: On preservation under homomorphisms and unions of conjunctive queries. J. ACM **53**(2), 208–237 (2006)
2. Bellman, R.E., Zadeh, L.A.: Decision-making in a fuzzy environment. Manag. Sci. **17**, 141–164 (1970)
3. Bistarelli, S., Montanari, U., Rossi, F.: Semiring-based constraint satisfaction and optimization. J. ACM **44**(2), 201–236 (1997)
4. Chang, C.C., Keisler, H.J.: Model Theory. Elsevier Science Publishers, Amsterdam (1973)
5. Cintula, P., Hájek, P.: Triangular norm based predicate fuzzy logics. Fuzzy Sets Syst. **161**, 311–346 (2010)
6. Cintula, P., Hájek, P., Noguera, C. (eds.): Handbook of Mathematical Fuzzy Logic. Studies in Logic, Mathematical Logic and Foundations, vol. 37. College Publications, London (2011)

7. Dellunde, P.: Preserving mappings in fuzzy predicate logics. J. Log. Comput. **22**(6), 1367–1389 (2011)
8. Dellunde, P.: Applications of ultraproducts: from compactness to fuzzy elementary classes. Log. J. IGPL **22**(1), 166–180 (2014)
9. Dellunde, P., García-Cerdaña, A., Noguera, C.: Löwenheim-Skolem theorems for non-classical first-order algebraizable logics. Log. J. IGPL **24**(3), 321–345 (2016)
10. Dubois, D., Fargier, H., Prade, H.: The calculus of fuzzy restrictions as a basis for flexible constraint satisfaction. In: 2nd IEEE International Conference on Fuzzy Systems, IEEE (1993)
11. Esteva, F., Godo, L.: Monoidal t-norm based logic: towards a logic for left-continuous t-norms. Fuzzy Sets Syst. **124**, 271–288 (2001)
12. Fargier, H., Lang, J.: Uncertainty in constraint satisfaction problems: a probabilistic approach. In: Clarke, M., Kruse, R., Moral, S. (eds.) ECSQARU 1993. LNCS, vol. 747, pp. 97–104. Springer, Heidelberg (1993). https://doi.org/10.1007/BFb0028188
13. Geiger, D.: Closed systems of functions and predicates. Pac. J. Math. **27**(1), 95–100 (1968)
14. Hodges, W.: Model Theory. Cambridge University Press, Cambridge (1993)
15. Kolaitis, P.G., Vardi, M.Y.: A Logical Approach to Constraint Satisfaction. In: Creignou, N., Kolaitis, P.G., Vollmer, H. (eds.) Complexity of Constraints. LNCS, vol. 5250, pp. 125–155. Springer, Heidelberg (2008). https://doi.org/10.1007/978-3-540-92800-3_6
16. Kolmogorov, V., Krokhin, A., Rolinek, M.: The Complexity of General-Valued CSPs. In: FOCS, pp. 1246–1258 (2015)
17. Krokhin, A.A., Zivny, S.: The Complexity of Valued CSPs. The Constraint Satisfaction Problem, pp. 233–266 (2017)
18. Meseguer, P., Rossi, F., Schiex, T.: Soft constraints. In: Handbook of Constraint Programming, chap. 9, pp. 281–328 (2006)
19. Moura J., Prade, H.: Logical analysis of fuzzy constraint satisfaction problems. In: 7nd IEEE International Conference on Fuzzy Systems. IEEE (1993)
20. Pini, M.S., Rossi, F., Venable, K.B.: Compact preference representation via fuzzy constraints in stable matching problems. In: Rothe, J. (ed.) Compact Preference Representation via Fuzzy Constraints in Stable Matching Problems. LNCS (LNAI), vol. 10576, pp. 333–338. Springer, Cham (2017). https://doi.org/10.1007/978-3-319-67504-6_23
21. Rossi, F., Brent, K., Walsh, T.: A short introduction to preferences. In: Synthesis Lectures on Artificial Intelligence and Machine Learning. Morgan and Claypool Pub (2011)
22. Horcík, R., Moraschini, T., Vidal, A.: An algebraic approach to valued constraint satisfaction. In: 26th EACSL Annual Conference on Computer Science Logic, pp. 42:1–42:20 (2017)
23. Rossman, B.: Homomorphism preservation theorems. J. ACM **55**(3), 15:1–15:53 (2008)
24. Rossman, B.: An improved homomorphism preservation theorem from lower bounds in circuit complexity. ITCS **27**, 1–27 (2017)
25. Ruttkay, Z.: Fuzzy constraint satisfaction. In: 3rd IEEE International Conference on Fuzzy Systems. IEEE (1994)

26. Schiex, T., Fargier, H., Verfaillie, G.: Valued constraint satisfaction problems: Hard and easy problems. In: Proceedings of the Fourteenth International Joint Conference on Artificial Intelligence (IJCAI 95), pp. 631–639 (1995)
27. Torra, V.: On considering constraints of different importance in fuzzy constraint satisfaction problems. Int. J. Uncertain. Fuzziness Knowl. Based Syst. **6**(5), 489–502 (1998)

Basic Level Concepts as a Means to Better Interpretability of Boolean Matrix Factors and Their Application to Clustering

Petr Krajča[(✉)] and Martin Trnecka

Department of Computer Science, Palacky University Olomouc,
17. listopadu 12, 77146 Olomouc, Czech Republic
petr.krajca@upol.cz, martin.trnecka@gmail.com

Abstract. We present an initial study linking in cognitive psychology well known phenomenon of basic level concepts and a general Boolean matrix factorization method. The result of this fusion is a new algorithm producing factors that explain a large portion of the input data and that are easy to interpret. Moreover, the link with the cognitive psychology allowed us to design a new clustering algorithm that groups objects into clusters that are close to human perception. In addition we present experiments that provide insight to the relationship between basic level concepts and Boolean factors.

1 Introduction

Boolean matrix factorization (BMF)—also known as the Boolean matrix decomposition—has become one of the standard methods in data mining with applications to many fields. The general aim of the BMF method is to find new more fundamental variables hidden in the data, called factors, that can be used to explain the input data in a concise and presumably comprehensible way.

Seminal work [6] shows that this problem can be efficiently approached with the formal concept analysis (FCA) [9]. It has been shown [6] that factors can be understood as maximal rectangles full of 1's in the input matrix and these rectangles correspond to formal concepts as studied in FCA. In fact, many existing BMF algorithms are based on FCA and this correspondence (see e.g. [5]).

Quality of a discovered factor is usually evaluated as a number of 1's it covers in the input matrix, i.e. how large portion of the input data is explained via the given factor (for more details see [2]). This measure along with some greedy strategy is used by almost all BMF algorithms (an overview can found e.g. in [5]). However, algorithms considering the number of covered 1's as the only criterion for the factor selection neglect an interpretability of the results. Factors are selected merely mechanically, hence there is no guarantee that the discovered factors are interesting, meaningful, or comprehensible from the viewpoint of a human expert. Note that factor interpreting is given little or no attention at all in contemporary literature that involves the BMF.

© Springer Nature Switzerland AG 2018
V. Torra et al. (Eds.): MDAI 2018, LNAI 11144, pp. 169–181, 2018.
https://doi.org/10.1007/978-3-030-00202-2_14

We present an initial study linking basic level concepts—concepts that explain data from the human point of view—and factorization based on the coverage of input data. The result of this fusion is a new algorithm producing factors that achieve high coverage of input data and that are easy to interpret (according to the cognitive psychology). Further, we propose a new clustering algorithm that groups objects into clusters that are close to human perception.

The rest of the paper is organized as follows. In the following Sect. 2 we provide basic notions and notations. Then, in Sect. 3 we describe and evaluate a classical BMF algorithm and several its modifications we propose. In Sect. 4 we discuss a new clustering algorithm based on ideas and observations made in Sect. 3. The paper concludes with notes on related works and future research.

2   Basic Notions

In the following section we provide a basic notions and notations used through the paper. We use formal concept analysis (FCA) [9] as a basic framework which allows us to establish link between BMF and basic level concepts.

2.1   Formal Concept Analysis

The input of the FCA is a two dimensional table where rows correspond to objects (e.g. *dog, cat, parrot, ant*), columns correspond to attributes (e.g. *four legs, wings, fur*), and each cell of the table indicates if a given object has a given attribute. This table can be represented as a Boolean matrix $I \in \{0,1\}^{n \times m}$. To every Boolean matrix one may associate the pair $\langle \uparrow, \downarrow \rangle$ of operators assigning to sets $C \subseteq X = \{1, \ldots, n\}$ and $D \subseteq Y = \{1, \ldots, m\}$ the sets $C^\uparrow \subseteq Y$ and $D^\downarrow \subseteq X$ defined by

$$C^\uparrow = \{j \in Y \mid \text{for each } i \in C : I_{ij} = 1\},$$
$$D^\downarrow = \{i \in X \mid \text{for each } j \in D : I_{ij} = 1\}.$$

That is, $C^\uparrow$ is the set of all attributes (columns) common to all objects (rows) in C and $D^\downarrow$ is the set of all objects having all the attributes from D, respectively.

A *formal concept* of I, a maximal rectangle or pattern in I, is any pair $\langle C, D \rangle$ satisfying $C^\uparrow = D$ and $D^\downarrow = C$. Usually the C and D are called the *extent* and the *intent* (respectively) of a formal concept $\langle C, D \rangle$. The set

$$\mathcal{B}(I) = \{\langle C, D \rangle \mid C \subseteq X, D \subseteq Y, C^\uparrow = D, D^\downarrow = C\}$$

of all formal concepts with the partial order $\leq$, defined by $\langle C_1, D_1 \rangle \leq \langle C_2, D_2 \rangle$ iff $C_1 \subseteq C_2$ (iff $D_1 \supseteq D_2$) forms a complete lattice. Note that the ordering $\leq$ models the natural hierarchy of concepts according to which more general concepts have more inclusive extents and less inclusive intents, e.g. *mammal $\leq$ animal*.

2.2   Basic Level of Concepts

The basic level phenomenon is, in a sense, encountered in everyday life. When we see a particular dog, we say "This is a dog," rather than "This is a German Shepherd" or "This is a mammal." That is, we prefer to name the object we see by "dog" to naming it by "German Shepherd" or "mammal". Put briefly, basic level concepts are the concepts we prefer in naming objects.

Basic level concepts can be seen as concepts that lies "somewhere in the middle" of the formal concept hierarchy (formed by the $\leq$ ordering) and that carve up the world well. More precisely, the basic level concepts can be seen as a compromise between the accuracy of classification at a maximally general level and the predictive power of a maximally specific level.

The notion of the basic level was formalized (via FCA) in the pioneering works [3,4]. In our work we use the Similarity approach to the definition of the basic level, i.e. a formal concept $\langle C, D \rangle$ belongs to the basic level if it satisfies the following properties:

- $\langle C, D \rangle$ has a high cohesion;
- $\langle C, D \rangle$ has a significantly larger cohesion than its upper neighbors;
- $\langle C, D \rangle$ has a slightly smaller cohesion than its lower neighbors.

Various formalization of the above mentioned conditions were proposed in [3]. Let us note that the cohesion of a formal concept $\langle C, D \rangle$ represents a measure of the mutual similarity of rows which can be accessed via formula:

$$coh(C, D) = \frac{\sum_{\{i,j\} \subseteq C, i<j} sim(I_{i\_}, I_{j\_})}{|C| \cdot (|C| - 1)/2},$$

where the similarity of two rows is defined as the well know Jaccard index:

$$sim(I_{i\_}, I_{j\_}) = \frac{|I_{i\_} \cap I_{j\_}|}{|I_{i\_} \cup I_{j\_}|}.$$

For every formal concepts $\langle C, D \rangle$ we can compute the degree $BL(A, B) = \alpha_1 \otimes \alpha_2 \otimes \alpha_3$, to which $\langle C, D \rangle$ is a concept from the basic level. α_1, α_2 and α_3 represent the degrees of validity of the above mentioned conditions and $\otimes$ represents a truth function of many-valued conjunction [10]. Note that we are using real unit interval $[0, 1]$ as a scale of truth degrees and the product (Goguen) t-norm as a conjunction, i.e. $a \otimes b = a \cdot b$, in our experiments. For more details see [3].

2.3   Boolean Matrix Decomposition

A general aim in BMF is to find for a given Boolean matrix $I \in \{0, 1\}^{n \times m}$ (and possibly other given parameters) matrices $A \in \{0, 1\}^{n \times k}$ and $B \in \{0, 1\}^{k \times m}$ for which

$$I \text{ (approximately) equals } A \circ B, \tag{1}$$

where ∘ is the Boolean matrix product, i.e. $(A \circ B)_{ij} = \max_{l=1}^{k} \min(A_{il}, B_{lj})$. In essence, matrices A and B can be seen as a concise representation of the original matrix I. An example of such decomposition can be the following.

$$I = \begin{pmatrix} 1 & 0 & 1 & 1 & 1 \\ 0 & 1 & 1 & 0 & 1 \\ 0 & 1 & 0 & 0 & 1 \\ 1 & 0 & 1 & 1 & 0 \end{pmatrix} = \begin{pmatrix} 1 & 1 & 0 \\ 0 & 1 & 1 \\ 0 & 0 & 1 \\ 1 & 0 & 0 \end{pmatrix} \circ \begin{pmatrix} 1 & 0 & 1 & 1 & 0 \\ 0 & 0 & 1 & 0 & 1 \\ 0 & 1 & 0 & 0 & 1 \end{pmatrix} = A \circ B$$

Factor (gray rectangular area) in matrix I is represented via Boolean matrix product of the third column of matrix A and the third row of matrix B. A decomposition of I into $A \circ B$ may be interpreted as a discovery of k factors that exactly or approximately explain the data: interpreting I, A, and B as the object-attribute, object-factor, and factor-attribute matrices, the model behind (1) reads: the object i has the attribute j if and only if there exists factor l such that l applies to i and j is one of the particular manifestations of l.

Formal concepts can be utilized for the BMF problem as follows. For a given set $\mathcal{F} = \{\langle C_1, D_1 \rangle, \ldots, \langle C_k, D_k \rangle\} \subseteq \mathcal{B}(I)$ (with a fixed indexing of the formal concepts $\langle C_l, D_l \rangle$), define the $n \times k$ and $k \times m$ Boolean matrices $A_{\mathcal{F}}$ and $B_{\mathcal{F}}$ by

$$(A_{\mathcal{F}})_{il} = \begin{cases} 1 & \text{if } i \in C_l, \\ 0 & \text{if } i \notin C_l, \end{cases} \quad \text{and} \quad (B_{\mathcal{F}})_{lj} = \begin{cases} 1 & \text{if } j \in D_l, \\ 0 & \text{if } j \notin D_l, \end{cases}$$

for $l = 1, \ldots, k$. That is, the lth column and lth row of A and B are the characteristic vectors of C_l and D_l, respectively. The set $\mathcal{F}$ is also called set of factor concepts. An entry of a matrix I is covered by the factor if it is included in some factor (concept) $\langle C, D \rangle$, i.e. factor $\langle C, D \rangle$ covers all entries of the matrix I that lies in the Cartesian product $C \times D$.

3   Boolean Matrix Decomposition: Algorithms

The link between formal concepts and BMF discovered in [6] allowed to design several efficient algorithms for the BMF problem (see an overview in [5]). In essence, all these algorithms use a greedy approach to identify k first formal concepts covering all 1's in a given Boolean matrix. We illustrate this approach with the GRECON[1] algorithm, depicted in Algorithm 1, proposed in [6].

At first, the algorithm initializes a set of factors $\mathcal{F}$ to an empty set and creates a copy I' of the input matrix I. Subsequently, it finds a set $\mathcal{C}$ of all formal concepts. This can be done efficiently with several algorithms [1,11,16]. In order to find a subset of $\mathcal{C}$ corresponding to a set of factors, the algorithm proceeds as follows.

The algorithm assigns to each formal concept $\langle C, D \rangle$ in $\mathcal{C}$ a score which is given by a function COVERAGE:

$$\text{COVERAGE}(I', \langle C, D \rangle) = \frac{\sum_{i=0}^{n} \sum_{j=0}^{m} I'_{ij} \cdot (C \circ D)_{ij}}{||I'||}$$

---

[1] In [6] this algorithm is named Algorithm 1.

Algorithm 1. GRECON

Input: Boolean $n \times m$ matrix I
Output: Set of factor concepts $\mathcal{F}$

1 $\mathcal{F} \leftarrow \emptyset$
2 $I' \leftarrow I$
3 $\mathcal{C} \leftarrow$ FINDALLFORMALCONCEPTS(I)
4 **while** $||I'|| > 0$ **do**
5    $bestScore \leftarrow -\infty$
6    $\langle A, B \rangle \leftarrow \langle \emptyset, \emptyset \rangle$
7   **foreach** concept $\langle C, D \rangle$ **in** $\mathcal{C}$ **do**
8      $score \leftarrow$ COVERAGE$(I', \langle C, D \rangle)$
9   **if** $score > bestScore$ **then**
10        $\langle A, B \rangle \leftarrow \langle C, D \rangle$
11        $bestScore \leftarrow score$

12    $\mathcal{F} \leftarrow \mathcal{F} \cup \{\langle A, B \rangle\}$
13    $\mathcal{C} \leftarrow \mathcal{C} - \{\langle A, B \rangle\}$
14    $I' \leftarrow I' \ominus (A \circ B)$

15 **return** $\mathcal{F}$

In other words, the score assigned to a formal concept $\langle C, D \rangle$ is a ratio of a sum of all 1's covered by this formal concept in the matrix I' to a sum of all 1's in the matrix I'. Note that each score can be seen as a measure of *how much the formal concept explains data* in the matrix. Formal concept $\langle A, B \rangle$ with the highest score is selected and is considered as a factor concept. This factor concept is inserted into the set $\mathcal{F}$ (line 12) and removed from the set $\mathcal{C}$ (line 13). Further, all 1's covered by the concept $\langle A, B \rangle$ are removed from the matrix I' (line 14). Note that $(P \ominus Q)$ is given as $(P \ominus Q)_{ij} = max(0, P_{ij} - Q_{ij})$. This step (lines 4–14) repeats until there are no more 1's in the matrix I'. It can be easily seen that the algorithm always terminates after finitely many steps and the set $\mathcal{F}$ corresponds to the solution of the BMF problem. For more details see [6].

From the BMF viewpoint this approach is reasonable and gives valid results. The potential shortcoming of the algorithm is the way it evaluates concepts— the only criterion is the number of 1's covered in the matrix. This means, factors are discovered merely mechanically with no further information or insight into data. Hence, an interesting question arises: *Are such factors good factors?* Or, alternatively: *Do such factors explain data well (from the human point of view)?*

We decided to approach these questions from the perspective of the basic level concepts. The basic level concepts are natural way humans treat and understand large collections of objects, thus it is reasonable to ask if there is a relationship between basic level concepts and algorithmically discovered factors.

3.1 Design of New Algorithms and Experimental Evaluation

We designed several algorithms and experiments that provide an initial insight into the relationship between Boolean factors and basic level concepts. We introduce these algorithms throughout this section and demonstrate their characteristics by experiments. In our experiments we focus on two aspects. (1) Basic level degree, i.e., whether factor is close to human intuition. (2) Coverage quality—if

Table 1. Real datasets and their characteristics

Datasets	Rows	Columns	Density (%)	Concepts	Max	1st Decile	Med
Animals	20	14	41.4	52	0.307	0.139	0.048
DBLP	6,980	18	13.2	2,067	0.266	0.094	0.046
Drinks	68	23	33.5	320	0.374	0.113	0.051
Sports	20	10	38.0	39	0.450	0.191	0.045
Things	508	22	25.6	1,863	0.310	0.111	0.052
Zoo	101	21	35.8	357	0.420	0.119	0.051

the result of BMF explains data correctly. Let us note that coverage quality function of $A \in \{0,1\}^{n \times l}$ and $B \in \{0,1\}^{l \times m}$ is a measure of the quality of the first l computed factors: $c(l) = 1 - E(I, A(l) \circ B(l))/\|I\|$, where $A(l)$ and $B(l)$ denote the $n \times l$ and $l \times m$ matrices that correspond to the first l factors and E is an error function defined for two matrices $I, J \in \{0,1\}^{m \times n}$ in the following way: $E(I, J) = \sum_{i,j=1}^{m,n} |I_{ij} - J_{ij}|$.

In all presented experiments we use real datasets from various fields. Namely: Animals [4], DBLP [15], Drinks [4], Sports [4], Things [8], and Zoo [7]. Table 1 displays the basic characteristics of our datasets: The numbers of rows and columns, density (percentage of 1's in the entries of the dataset), the number of formal concepts in the dataset and maximum, first decile and median value of the basic level degree over all concepts of the dataset.

3.2 Algorithms Utilizing Coverage or Basic Level only

In our first set of experiments we focus on factors discovered with the GRECON algorithm (coverage only, see Algorithm 1) and their degrees to which they are from basic level. Because being a basic level concept is not a yes/no property, it is challenging to say what is a good basic level concept. Nonetheless, basic level concepts are those with the highest degrees, hence we assume that good basic level concepts belong at least among the top ten percent. Figure 1 shows that in the majority of cases factors do not belong among such concepts, see the column 1st decile in Table 1. This observation is further confirmed in Table 2— average basic level of factors delivered by GRECON is significantly below the mark determined by the top ten percent of basic level concepts.

These results provide strong indication that the majority of discovered factors do not coincide with the basic level concepts. This observation has two possible implications: (1) The algorithm may discover factors that would not be intuitively considered by a human expert, hence can reveal useful information hidden in the data. (2) On the downside such factors may be counterintuitive, difficult to grasp, or uninteresting. In fact, it is a very common phenomenon that Algorithm 1 often returns uninteresting factors covering only a single attribute.

To investigate the other side of the relationship, i.e., whether basic level concepts are good factors, we have to adjust the Algorithm 1 first. If we want

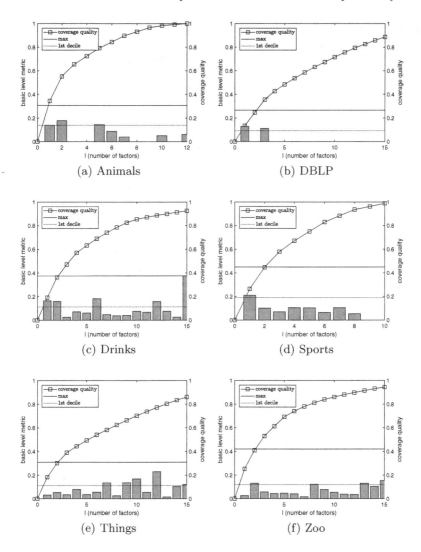

Fig. 1. Basic level degree and coverage quality of the first 15 factors on real data.

to list factors w.r.t. their basic level degree (i.e. the most basic level concepts first), we have to change the score function (line 8). In this case it is fully sufficient to replace the COVERAGE$(I', \langle C, D \rangle)$ function with a function $BL(C, D)$. This means basic level degree of a concept can serve as a score assigned to the given concept in the Algorithm 1.

Figure 2 shows that the coverage quality of factors selected via the basic level metric (line BL) is significantly worse than the coverage quality of factors delivered by the original GRECON algorithm. Table 2 (column BL) shows that in some cases average basic level degree of factors improves and in some case unexpectedly gets worse. In these cases the algorithm returns large amounts of factors (even with low basic level degrees) because it has difficulties to find concepts explaining all the data. Further, more detailed analysis revealed that the algorithm tends to return superfluous factors describing some parts of the matrix multiple times. This experiment clearly shows that the basic level degree of a formal concept is not a good criterion for selecting factors.

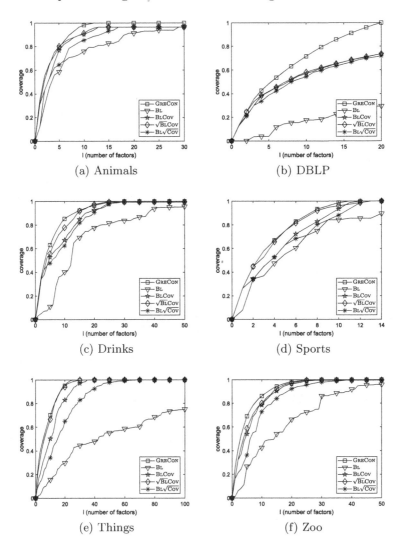

(a) Animals   (b) DBLP

(c) Drinks   (d) Sports

(e) Things   (f) Zoo

Fig. 2. Coverage quality of the first l factors on real data.

3.3   Combining Coverage and Basic Level

Both discussed experiments point out to the particular issues of the COVERAGE and BL functions as measures used to select factor concept. In the first case

Table 2. Average basic level degree of discovered factors

Dataset	GRECON	BL	BLCOV	$\sqrt{\text{BL}}$COV	BL$\sqrt{\text{COV}}$	BLESC
Animals	0.059	0.094	0.117	0.094	0.126	0.113
DBLP	0.012	0.058	0.088	0.088	0.090	0.089
Drinks	0.101	0.065	0.167	0.156	0.165	0.180
Sports	0.090	0.138	0.155	0.155	0.159	0.205
Things	0.098	0.067	0.125	0.127	0.128	0.075
Zoo	0.068	0.072	0.135	0.127	0.143	0.151

algorithm tends to return factors not coinciding with the human intuition, in the latter case it tends to return factors which can be considered redundant. On the other hand both functions have their strong sides as well. The COVERAGE function prefers factors which explain data well, the BL function prefers factors that are close to the human intuition. It opens up two interesting questions. *Is it possible to combine these two measures? Can such combination accent the strong sides of the measures and suppress their downsides?*

The score assigned with the COVERAGE function has two possible interpretations. It provides an information on how many 1's are covered with a given formal concept in a given matrix, i.e., the higher number the more 1's is covered. Furthermore, the score can be interpreted as a degree to which the given formal concept explains the data in the matrix, i.e., the 1.0 degree means that the formal concept explains the data entirely, the 0.0 degree means that the formal concept explains the data not at all. Due to this interpretation it is possible to combine COVERAGE and basic level degrees with the truth function $\otimes$, i.e.,

$$\text{BLCOV}(I', \langle C, D \rangle) = BL(C, D) \otimes \text{COVERAGE}(I', \langle C, D \rangle).$$

The score assigned with the BLCOV function can be seen as a degree to which the given formal concept is a basic level concept and also is describing the data, i.e., formal concepts with the higher degrees are basic level concept and are also well-describing the data. One may observe in Fig. 2 that the combination provides results that are "compromise" between the coverage and basic level approach.

Since the truth degrees assigned with the BL and COVERAGE functions can be disproportional, we considered two more variants of the BLCOV function:

$$\sqrt{\text{BL}}\text{COV}(I', \langle C, D \rangle) = \sqrt{BL(C, D)} \otimes \text{COVERAGE}(I', \langle C, D \rangle) \quad (2)$$

$$\text{BL}\sqrt{\text{COV}}(I', \langle C, D \rangle) = BL(C, D) \otimes \sqrt{\text{COVERAGE}(I', \langle C, D \rangle)} \quad (3)$$

Technically, the square root in (2) and (3) is used to increase numerical values of the truth degrees. However, (2) also has a natural interpretation—formal concepts with high degrees are *more or less* basic level concepts and are well-describing the data. Needless to say analogous interpretation exists for (3). One may observe (see Fig. 2) that the $\sqrt{\text{BL}}\text{Cov}$ function outperforms other functions combining coverage and basic level. Note that coverage quality of the delivered factors is almost identical with the original GRECON and these factors are from the basic level in higher degree than factors computed via GRECON, see Table 2.

4 Clustering Algorithm

Our experiments revealed appealing qualities of the $\sqrt{\text{BL}}\text{Cov}$ function—it assigns higher values to groups of object and attributes that are well-explaining the data and that are also close to human intuition (are basic level concepts). Therefore we decided to build a clustering algorithm around this function. We call it the BLESC (Basic level Elimination Strategy Clustering) algorithm.

The algorithm takes a formal context I as its input and returns a set of clusters. Each cluster is a set of objects presumably representing a basic level concept. The algorithm (see Algorithm 2 for the pseudo-code) is based on similar ideas as Algorithm 1, therefore we describe only its key parts. Basically, it finds all formal concepts and iteratively assigns them score with the $\sqrt{\text{BL}}\text{Cov}$ function. In each iteration formal concept with the highest score is picked and its extent is returned as a new cluster (line 13). Afterwards, all 1's in rows corresponding to the objects from the cluster are replaced with zeros (lines 14 and 15). The algorithm stops if there are no 1's in the auxiliary matrix I'.

The strategy that eliminates 1's from the matrix I' (lines 14 and 15) ensures that each newly discovered cluster contains at least one object not covered by the previous clusters. Furthermore, this strategy along with the $\sqrt{\text{BL}}\text{Cov}$ function forces the algorithm to focus especially on formal concepts with objects that were not considered yet and, by extension, to prefer clusters that are non-overlapping.

Algorithm 2. BLESC

Input: Boolean $n \times m$ matrix I
Output: Set of clusters $\mathcal{G}$

1 $\mathcal{G} \leftarrow \emptyset$
2 $I' \leftarrow I$
3 $\mathcal{C} \leftarrow$ FINDALLFORMALCONCEPTS(I)
4 **while** $\|I'\| > 0$ **do**
5      $bestScore \leftarrow -\infty$
6      $\langle A, B \rangle \leftarrow \langle \emptyset, \emptyset \rangle$
7   **foreach** concept $\langle C, D \rangle$ in $\mathcal{C}$ **do**
8          $score \leftarrow$ COVERAGE($I', \langle C, D \rangle) \cdot \sqrt{BL(C, D)}$
9   **if** $score > bestScore$ **then**
10             $\langle A, B \rangle \leftarrow \langle C, D \rangle$
11             $bestScore \leftarrow score$
12     $\mathcal{C} \leftarrow \mathcal{C} - \{\langle A, B \rangle\}$
13     $\mathcal{G} \leftarrow \mathcal{G} \cup \{A\}$
14   **foreach** object a in A **do**
15         $I' \leftarrow I' \ominus (\{a\} \circ \{a\}^{\uparrow})$
16 **return** $\mathcal{G}$

4.1 Algorithm Evaluation

Evaluation of the BLESC algorithm is challenging for several reasons. To the best of our knowledge there are no standard datasets or measures one could use to validate such algorithms. Furthermore, whether a concept is perceived as a basic level depends on a given context and often on a knowledge or expertise of the observer. For instance, what would be considered by a car mechanic as a *Ford Focus*, would be considered by a general population simply as *a car* (this phenomenon is already discussed in [4]). Therefore, in order to show validity of the algorithm, we decided to describe contents of the datasets we use and comment on results returned by BLESC.

The Zoo [7] dataset contains list of 101 animals which belong to 7 classes, e.g. *mammals*, *birds*, etc. The BLESC algorithm successfully identified 4 of them (*mammals*, *birds*, *insects*, *amphibians*), i.e. successfully classified 73 objects. Further, it identified cluster of fishes (13 objects), however 4 objects from other classes were incorrectly considered as fishes too. Remaining objects were classified into categories that can be described as *crustaceans*, or *sea predators*.

To confirm these results we used dataset Animals [4] with a similar content but with a different set of objects and attributes and without predefined classes of objects. In this dataset algorithm discovered 6 clusters that correspond to *mammals*, *fishes*, *terrestrial predators*, *flying creatures*, *water living egg laying animals*, *terrestrial egg laying animals*.

The Drinks [4] dataset contains description of 68 beverages and their properties. The authors of this dataset identified 11 classes of beverages, e.g. *beers*, *wines*, *mineral waters*. The algorithm discovered 12 cluster and successfully identified clusters of *beers*, *wines*, *milks*, *liquors*, and *distilled spirits*. Interestingly, the algorithm merged *mineral waters*, *fruit juices*, and some *sodas* into a single cluster that can be described as *light non-alcoholic beverages*. Further, high-calorie soft drinks (Coca Cola, energy drinks, etc.) were clustered together and Coca Cola light-like drinks were put in their own cluster. Even though these clusters do not correspond to classes defined in the dataset, we consider them as perfectly correct. That is algorithm successfully clustered 58 of 68 objects. Remaining objects were placed into not so clear clusters that can be called, for instance, *low-calorie drinks*, however, it can be seen that these cluster were created only to deal with noise.

In the Sports [4] dataset, i.e., in a dataset describing 20 sports, our algorithm found 6 clusters: *land-based collective sports with an opponent* (football, tenis, etc.), *individual timed sports* (e.g. running, triathlon), *invidual water sports*, *collective sports*, *land-based sports with a score*, and *water sports*.

For the DBLP [15] dataset our algorithm returned 103 clusters that can be described as *authors publishing in top data management conferences, AI conferences, machine learning conferences*, etc.

5   Conclusion, Related Works, and Future Research

Basic level concepts have their origin in the cognitive psychology and only little attention to their role in computer science or artificial intelligence research was given so far. One of few publications devoted to this area is [12] where basic level concepts are used to cluster rows in ranked data tables. Similar approach, based on the Minimum description length (MDL) principle is utilized in [13] or [14]. In this works, the interpretability of factors is improved via MDL. Let us note, this approach usually produces extremely small coverage of the input data.

Our experiments show that the algorithms based on basic level concepts, i.e. the GRECON algorithm with the $\sqrt{\text{BLCOV}}$ function and the BLESC algorithm, can deliver valid and interpretable results which are very close to human intuition from the cognitive psychology point of view. Therefore, we are to investigate other possible applications of this promising metric in other areas, for instance, in document classification.

References

1. Andrews, S.: Making use of empty intersections to improve the performance of CbO-type algorithms. In: Bertet, K., Borchmann, D., Cellier, P., Ferré, S. (eds.) ICFCA 2017. LNCS (LNAI), vol. 10308, pp. 56–71. Springer, Cham (2017). https://doi.org/10.1007/978-3-319-59271-8_4

2. Belohlavek, R., Outrata, J., Trnecka, M.: How to assess quality of BMF algorithms? In: Yager, R.R., Sgurev, V.S., Hadjiski, M., Jotsov, V.S. (eds.) Proceeding of International Conference on Intelligent Systems, IS 2016, pp. 227–233 (2016)

3. Belohlavek, R., Trnecka, M.: Basic level of concepts in formal concept analysis. In: Domenach, F., Ignatov, D.I., Poelmans, J. (eds.) ICFCA 2012. LNCS (LNAI), vol. 7278, pp. 28–44. Springer, Heidelberg (2012). https://doi.org/10.1007/978-3-642-29892-9_9

4. Belohlavek, R., Trnecka, M.: Basic level in formal concept analysis: interesting concepts and psychological ramifications. In: Rossi, F. (ed.) Proceedings of the 23rd International Joint Conference on Artificial Intelligence, IJCAI 2013, pp. 1233–1239 (2013)

5. Belohlavek, R., Trnecka, M.: From-below approximations in boolean matrix factorization: geometry and new algorithm. J. Comput. Syst. Sci. **81**(8), 1678–1697 (2015)

6. Belohlavek, R., Vychodil, V.: Discovery of optimal factors in binary data via a novel method of matrix decomposition. J. Comput. Syst. Sci. **76**(1), 3–20 (2010)

7. Dheeru, D., Karra Taniskidou, E.: UCI machine learning repository (2017). http://archive.ics.uci.edu/ml

8. Farhadi, A., Endres, I., Hoiem, D., Forsyth, D.A.: Describing objects by their attributes. In: Proceedings of Conference on Computer Vision and Pattern Recognition (CVPR 2009), pp. 1778–1785 (2009)

9. Ganter, B., Wille, R.: Formal Concept Analysis - Mathematical Foundations. Springer, Heidelberg (1999). https://doi.org/10.1007/978-3-642-59830-2

10. Gottwald, S.: A Treatise on Many-Valued Logics, vol. 3. Research Studies Press, Baldock (2001)

11. Krajca, P., Outrata, J., Vychodil, V.: Computing formal concepts by attribute sorting. Fundam. Inform. **115**(4), 395–417 (2012)
12. Krajča, P.: Rank-aware clustering of relational data: organizing search results. In: USB Proceedings The 13th International Conference on Modeling Decisions for Artificial Intelligence, pp. 61–72 (2016)
13. Lucchese, C., Orlando, S., Perego, R.: Mining top-k patterns from binary datasets in presence of noise. In: Proceedings of the International Conference on Data Mining, SDM 2010, pp. 165–176 (2010)
14. Lucchese, C., Orlando, S., Perego, R.: A unifying framework for mining approximate top-k binary patterns. IEEE Trans. Knowl. Data Eng. **26**(12), 2900–2913 (2014)
15. Miettinen, P.: Matrix decomposition methods for data mining: computational complexity and algorithms. Ph.D. thesis (2009)
16. Outrata, J., Vychodil, V.: Fast algorithm for computing fixpoints of galois connections induced by object-attribute relational data. Inf. Sci. **185**(1), 114–127 (2012)

Fuzzy Type Powerset Operators
and F-Transforms

Jiří Močkoř[(✉)]

Institute for Research and Applications of Fuzzy Modeling,
University of Ostrava, 30. dubna 22, 701 03 Ostrava 1, Czech Republic
Jiri.Mockor@osu.cz

Abstract. We introduce two types of aggregation operators for lattice-valued fuzzy sets, called *fuzzy type powerset operators* and *fuzzy type F-transforms*, which are derived from classical powerset operators and F-transforms, respectively. We prove that, in contrast with classical powerset operators, fuzzy type powerset operators form a subclass of fuzzy type F-transforms. Some examples of fuzzy type powerset operators are presented.

1 Introduction

Aggregation of a large set of input values into a smaller set of values is an indispensable tool not only in mathematics, but of many other sciences and engineering technics. The comprehensive state-of-art overviews on aggregation can be found in [2] (dated to 1985) and in [1] (dated to 2002) and lately, in [3]. Traditionally, by the aggregation is understood the process of combining several numerical values into a single representative, called aggregation value. The numerical function performing this process is called aggregation function, satisfying natural conditions as monotonicity and boundary conditions. Because of the simplicity of this definition of aggregation, there are practically endless numbers of specific examples of aggregation functions in both mathematics and other sciences. There is large field of applications of aggregation, including applied mathematics, computer sciences and many applied fields.

Aggregation of information is critical in any inference system, and so the study of aggregation operators is essential also for the fuzzy sets. Very soon, standard numerical aggregations were extended to aggregation of fuzzy sets. Classically, aggregation operations on fuzzy sets are operations by which several fuzzy sets are combined in a desirable way to produce a single fuzzy set and where the aggregation functions are derived from ordinary aggregation operators on the unit interval (see, e.g., [19]). Another approach how to expand aggregation operators to fuzzy sets is to reduce the size of the universe of fuzzy sets. Such aggregation operators represents mappings $[0,1]^X \to [0,1]^Y$, where

---

This research was partially supported by the project GA18-06915S provided by the Grant Agency of the Czech Republic.

V. Torra et al. (Eds.): MDAI 2018, LNAI 11144, pp. 182–192, 2018.
https://doi.org/10.1007/978-3-030-00202-2_15

$card(X) > card(Y)$. Also these aggregation operators have to satisfy the boundary and monotonicity conditions. There are two important examples of such aggregation operators in fuzzy sets, both with many theoretical and practical applications. These operators are *F-transform and powerset operators*.

The powerset theory is widely used in algebra, logic, topology and also in computer science. Recall that given a set X, there exists the set $P(X) = \{S : S \subseteq X\}$, called the powerset of X and such that every map $f : X \to Y$ can be extended to the powerset operators $f^\to : P(X) \to P(Y)$ and $f^\gets : P(Y) \to P(X)$, such that

$$f^\to(S) = f(S), \quad f^\gets(T) = f^{-1}(T).$$

Because the classical set theory can be considered to be a special part of the fuzzy set theory, introduced by [20], it is natural that powerset objects associated with fuzzy sets soon were investigated as generalizations of classical powerset objects. The first approach was done again by Zadeh [20], who defined $Z(X) = [0,1]^X$ to be a new powerset object called Zadeh's powerset functor instead of $\mathcal{P}(X)$ and introduced new powerset operators $f_Z^\to : [0,1]^X \to [0,1]^Y$ and $f_Z^\gets : [0,1]^Y \to [0,1]^X$, such that for $s \in [0,1]^X, t \in [0,1]^Y, y \in Y$,

$$f_Z^\to(s)(y) = \bigvee_{x, f(x)=y} s(x), \quad f_Z^\gets(t) = t \circ f.$$

If, instead of a general map, we consider a surjective map $f : X \twoheadrightarrow Y$, the powerset operators $f_P^\to : P(X) \to P(Y)$, and $f_Z^\to : Z(X) \to Z(Y)$ are, in fact, aggregation operators, reducing the size of underlying set X.

Zadeh's extension was for the first time intensively studied by [16], especially the relation between classical powerset extension $f^\to$ and $f_Z^\to$. The works of Rodabaugh gave very serious basis for further research of powerset objects and operators. On the other hand, very soon the powerset theory began to develop towards greater interdependence with the theory of categories, especially with the monad theory. A special example of monads in clone form was introduced by Rodabaugh [17] as a special structure describing powerset objects generated by monads. It was observed soon, that powerset objects used in the fuzzy sets theory are generated by monads. In the papers [5,7] we presented some examples of powerset theories based on fuzzy sets which are generated by monads and which are frequently used in the fuzzy set theory.

Another example of the aggregation operator in fuzzy sets theory is the F-transform operator, which was in lattice-valued form introduced by Perfilieva [14] and elaborated in many other papers (see, e.g., [11–13,15]). Aggregation F-transform is a special aggregation map that transforms general $\mathcal{L}$-valued fuzzy sets defined in the set X to $\mathcal{L}$-valued fuzzy sets defined in another (smaller) set Y, and the inverse transformations back to the original spaces then produce either the original functions or their approximations.

The aggregation powerset operators and F-transform operators have many common formal properties. Both operators enable to extend epimorphisms $f : X \twoheadrightarrow Y$ in suitable categories to the morphisms $f^\to : F(X) \to F(Y)$ between new $\bigvee$-semilattices structures associated with fuzzy sets and both operators have also similar properties with respect to the semilattice operations defined on these $\bigvee$-semilattice structures.

In the paper we show that in many concrete examples used in fuzzy sets theory, aggregation powerset theory is a special case of the aggregation F-transform theory. We introduce the notion of the *$\mathcal{L}$-fuzzy type powerset theory* and we show that many of well known and frequently used powerset theories in fuzzy set theory are of that type. As the main result we show that the powerset operators of the $\mathcal{L}$-fuzzy type powerset theory are identical to the F-transforms with respect to special fuzzy partitions. This enables us to extend tools and methods used in all these theories.

2   Preliminary Notions

In this section we present some preliminary notions and definitions which could be helpful for better understanding of results concerning sets with similarity relations and categorical tools. A principal structure used in the paper is a *complete residuated lattice* (see e.g. [10]), i.e. a structure $\mathcal{L} = (L, \wedge, \vee, \otimes, \rightarrow, 0, 1)$ such that $(L, \wedge, \vee)$ is a complete lattice, $(L, \otimes, 1)$ is a commutative monoid with operation $\otimes$ isotone in both arguments and $\rightarrow$ is a binary operation which is residuated with respect to $\otimes$, i.e.

$$\alpha \otimes \beta \leq \gamma \quad \text{iff} \quad \alpha \leq \beta \rightarrow \gamma.$$

By $\mathcal{L}$-fuzzy set in a set X we understand a map $X \rightarrow \mathcal{L}$. Recall that a set with similarity relation (or $\mathcal{L}$-set) is a couple (X, δ), where $\delta : X \times X \rightarrow \mathcal{L}$ is a map such that

(a) $(\forall x \in X)\quad \delta(x, x) = 1$,
(b) $(\forall x, y \in X)\quad \delta(x, y) = \delta(y, x)$,
(c) $(\forall x, y, z \in X)\quad \delta(x, y) \otimes \delta(y, z) \leq \delta(x, z)$ (generalized transitivity).

In the paper we use the category $\mathbf{Set}(\mathcal{L})$ of $\mathcal{L}$-sets as objects and with maps $f : X \rightarrow Y$ such that $\gamma(f(x), f(y)) = \delta(x, y)$ for all $x, y \in X$ as morphisms $f : (X, \delta) \rightarrow (Y, \gamma)$. A $\mathcal{L}$-fuzzy set $f : X \rightarrow L$ is called *an extensional map* with respect to δ, if $f(x) \otimes \delta(x, y) \leq f(y)$, for arbitrary $x, y \in X$. The set of all extensional $\mathcal{L}$-fuzzy sets with respect to (X, δ) is denoted by $F(X, \delta)$.

Recall that a *cut* in a set X is a system $(C_\alpha)_{\alpha \in L}$ of subsets of X such that $C_\alpha \subseteq C_\beta$ if $\alpha \geq \beta$ and the set $\{\alpha \in L : a \in C_\alpha\}$ has the greatest element for any $a \in X$. By $D(X)$ we denote the set of all cuts in X.

Analogously as in classical sets, in $\mathcal{L}$-sets we can define the so called f-cuts.

Definition 2.1 ([6]). *Let (X, δ) be an $\mathcal{L}$-set. Then a system* $\mathbf{C} = (C_\alpha)_\alpha$ *of subsets in A is called an* **f-cut** *in (X, δ) in the category $\mathbf{Set}(\mathcal{L})$ if*

1. $\forall a, b \in X, \quad a \in C_\alpha \Rightarrow b \in C_{\alpha \otimes \delta(a,b)}$,
2. $\forall a \in X, \forall \alpha \in \mathcal{L}, \quad \bigvee_{\{\beta : a \in C_\beta\}} \beta \geq \alpha \Rightarrow a \in C_\alpha$.

The set of all f-cuts in (X, δ) is denoted by $C(X, \delta)$.

Any system of subsets $(C_\alpha)_\alpha$ in a set X can be extended to the f-cut $(\overline{C_\alpha})_\alpha$, defined by

$$\overline{C_\alpha} = \{a \in X : \bigvee_{\{(x,\beta):x \in C_\beta\}} \beta \otimes \delta(a, x) \geq \alpha\}.$$

The set $C(X, \delta)$ of all f-cuts in a $\mathcal{L}$-set (X, δ) can be ordered by $(C_\alpha)_\alpha \leq (D_\alpha)_\alpha$ iff $C_\alpha \subseteq D_\alpha$, for each $\alpha \in \mathcal{L}$. Then $C(X, \delta)$ is a complete $\vee$-semilattice, such that $\bigvee_{i \in I}(C_\alpha^i)_\alpha = (\overline{\bigcup_{i \in I} C_\alpha^i})_\alpha$.

We recall some basic facts about F-transform method. The *core* of a $\mathcal{L}$-fuzzy set $f : X \to L$ is defined by $core(f) = \{x \in X : f(x) = 1\}$. A normal $\mathcal{L}$-fuzzy set f in a set X is such that $core(f) \neq \emptyset$.

An F-transform in a form introduced by Perfilieva [15] is based on the so called fuzzy partitions on the crisp set.

Definition 2.2 ([15]). *Let X be a set. A system $\mathcal{A} = \{A_\lambda : \lambda \in \Lambda\}$ of normal $\mathcal{L}$-fuzzy sets in X is a fuzzy partition of X, if $\{core(A_\lambda) : \lambda \in \Lambda\}$ is a partition of X. A pair $(X, \mathcal{A})$ is called a space with a fuzzy partition. The index set of $\mathcal{A}$ will be denoted by $|\mathcal{A}|$.*

In the paper [9] we introduced the category **SpaceFP** of spaces with fuzzy partitions.

Definition 2.3. *The category **SpaceFP** is defined by*

1. *Fuzzy partitions $(X, \mathcal{A})$, as objects,*
2. *Morphisms $(g, \sigma) : (X, \{A_\lambda : \lambda \in \Lambda\}) \to (Y, \{B_\omega : \omega \in \Omega\})$, such that*
   (a) $g : X \to Y$ and is $\sigma : \Lambda \to \Omega$ are mappings,
   (b) $\forall \lambda \in \Lambda$, $A_\lambda(x) \leq B_{\sigma(\lambda)}(g(x))$, for each $x \in X$.
3. *The composition of morphisms in **SpaceFP** is defined by $(h, \tau) \circ (g, \sigma) = (h \circ g, \tau \circ \sigma)$.*

Objects of the category **SpaceFP** represent ground structures for a fuzzy transform, firstly proposed by Perfilieva [14] and, in the case where it is applied to $\mathcal{L}$-valued functions with $\mathcal{L}$-valued partitions, in [15].

Definition 2.4 ([15]). *Let $(X, \mathcal{A})$ be a space with a fuzzy partition $\mathcal{A} = \{A_\lambda : \lambda \in |\mathcal{A}|\}$. An upper F-transform with respect to the space $(X, \mathcal{A})$ is a function $F_{X,\mathcal{A}}^\uparrow : \mathcal{L}^X \to \mathcal{L}^{|\mathcal{A}|}$, defined by*

$$f \in \mathcal{L}^X, \lambda \in |\mathcal{A}|, \quad F_{X,\mathcal{A}}^\uparrow(f)(\lambda) = \bigvee_{x \in X} (f(x) \otimes A_\lambda(x)).$$

In the residuated lattice $\mathcal{L}$ we can define a fuzzy partition $\mathbf{L} = \{L_\alpha : \alpha \in L\}$, such that $L_\alpha(\beta) = \alpha \leftrightarrow \beta$, $\beta \in L$.

The following notion of an extensional fuzzy set in the category **SpaceFP** extends the notion of an extensional mapping in the category $\mathbf{Set}(\mathcal{L})$.

Definition 2.5. *A mapping* $f : X \to L$ *is called an extensional* $\mathcal{L}$-*fuzzy set in a space with a fuzzy partition* $(X, \mathcal{A})$ *in the category* **SpaceFP**, *if there exists a map* $\sigma : |\mathcal{A}| \to L$, *such that* (f, σ) *is a morphism* $(X, \mathcal{A}) \to (L, \mathbf{L})$ *in the category* **SpaceFP**. *By* $F(X, \mathcal{A})$ *we denote the set of all extensional fuzzy sets in* $(X, \mathcal{A})$.

In [8] we proved that for any space with a fuzzy partition $(X, \mathcal{A})$ it is possible to construct a $\mathcal{L}$-set $(X, \delta_{X,\mathcal{A}})$ with the similarity relation called *characteristic similarity relations* of $(X, \mathcal{A})$. The similarity relation $\delta_{X,\mathcal{A}}$ is the minimal similarity relation defined in X, such that for arbitrary map $f : X \to L$, f is extensional in $(X, \mathcal{A})$ iff f is extensional with respect to $\delta_{X,\mathcal{A}}$. Hence, we have

$$F(X, \mathcal{A}) = E(X, \delta_{X,\mathcal{A}}).$$

3 $\mathcal{L}$-Fuzzy Type Powerset Theories

We repeat the basic definition of the powerset theory and we introduce a special type of a powerset theory, called $\mathcal{L}$-*fuzzy type powerset theory*.

In what follows, by $CSLAT(\vee)$ we denote the category of complete $\vee$-semilattices as objects and with $\vee$-preserving maps as morphisms. By **Set** we denote the classical category of sets with mappings.

Definition 3.1 (Rodabaugh [17]). *Let* $\mathbf{K}$ *be a ground category. Then* $\mathbf{T} = (T, \to, V, \eta)$ *is called* $CSLAT(\vee)$-**powerset theory in** $\mathbf{K}$, *if*

1. $T : |\mathbf{K}| \to |CSLAT(\vee)|$ *is an object-mapping,*
2. *for each* $f : A \to B$ *in* $\mathbf{K}$, *there exists* $f_T^{\to} : T(A) \to T(B)$ *in* $CSLAT(\vee)$,
3. *There exists a concrete functor* $V : \mathbf{K} \to \mathbf{Set}$ *(i.e., injective on morphisms) such that* η *determines in* **Set** *for each* $A \in \mathbf{K}$ *a mapping* $\eta_A : V(A) \to T(A)$,
4. *For each* $f : A \to B$ *in* $\mathbf{K}$, $f_T^{\to} \circ \eta_A = \eta_B \circ V(f)$.

For a powerset theory $\mathbf{T}$ in a category $\mathcal{K}$, for arbitrary morphism $f : X \to Y$ there exists also the inverse powerset operator $f_T^{\leftarrow} : T(Y) \to T(X)$, defined by

$$y \in Y, \quad f_T^{\leftarrow}(y) = \bigvee_{x \in T(X), f_T^{\to}(x) \leq y} x. \tag{1}$$

In the paper we will deal with several examples of $CSLAT(\vee)$-powerset theories. Some of these examples were derived by previous authors, e.g., Rodabaugh [17], Höhle [4], Solovyov [18], other examples were presented in Mockor [5,7]. It should be observed that in all these examples the object function $T : |\mathcal{K}| \to |CSLAT(\vee)|$ is, in fact, the object function of a functor $T : \mathcal{K} \to CSLAT(\vee)$, with $T(f) = f_T^{\to}$, for any morphism f in $\mathcal{K}$. These examples will be investigated in Sect. 4.

As we can see in the Sect. 4, all the examples of powerset theories have, in addition to axioms related to this theory, some other common features. These properties can be characterized as additional properties of the object function T of a powerset theory $(T, \to, V, \eta)$, defining the internal structure of objects $T(X)$. In many of concrete examples of powerset operators associated with fuzzy set theory, the supports of objects $T(X)$ are connected with the structure $\mathcal{L}^{V(X)}$, i.e., with the classical fuzzy sets in the support $V(X)$ of these objects. It is clear that this property is not a general property of powerset theories, on the other hand, this property is typical for powerset theories with object functions which are in relationships with some variants of $\mathcal{L}$-fuzzy sets and their generalization. Therefore, it is natural such object functions T to be called $\mathcal{L}$-fuzzy type functions. In the following definition we introduce the notion of a functor of $\mathcal{L}$-fuzzy type.

Definition 3.2. *Let $\mathcal{K}$ be a concrete category with a concrete functor $V : \mathcal{K} \to$ Set. We say that a functor $T : \mathcal{K} \to CSLAT(\vee)$ is of a $\mathcal{L}$-**fuzzy type**, if*

1. *There exists a natural transformation i*

$$T \xrightarrow{i} Z.V, \qquad T^{op} \xrightarrow{i} Z.V^{op},$$

   with $\bigvee$-preserving components $i_X, X \in \mathcal{K}$, where Z is the Zadeh's powerset functor,
2. *For each $X \in \mathcal{K}, \alpha \in \mathcal{L}, f \in T(X)$, there exists $\alpha \star f \in T(X)$, such that*

$$i_X(\alpha \star f) = \alpha \otimes i_X(f).$$

For an object $X \in \mathcal{K}$, a semilattice $T(X) \in CSLAT$ is called an $\mathcal{L}$-*fuzzy type object.*

As we have already stated in the Introduction, F-transform is a special aggregation map that reduces general $\mathcal{L}$-valued fuzzy sets defined in the set X to $\mathcal{L}$-valued fuzzy sets defined in a smaller set Y. In many practical examples, however, the aggregation F-transform need not to be defined for $\mathcal{L}$-valued fuzzy sets, but it can be also defined for general objects associated with fuzzy sets. In that way, instead of the aggregation F-transform $F : \mathcal{L}^X \to \mathcal{L}^Y$ we can consider the general F-transform $T(X) \to T(Y)$, where $u : X \twoheadrightarrow Y$ is an epimorphism in a category and T is a $\mathcal{L}$-fuzzy type functor. The definition of the $\mathcal{L}$-fuzzy type F-transform is the following.

Definition 3.3. *Let $\mathcal{K}$ be a category with a concrete functor $V : \mathbf{K} \to$ Set and let $T : \mathbf{K} \to CSLAT(\vee)$ be an $\mathcal{L}$-fuzzy type functor. Then T is called the $\mathcal{L}$-**fuzzy type F-transform**, if for each epimorphism $u : X \twoheadrightarrow Y$ in the category $\mathbf{K}$ there exists a subset $\mathcal{A} = \{g_y : y \in V(Y)\} \subseteq T(X)$, such that the following hold:*

1. *$\{i_X(A_y) : y \in V(Y)\}$ is a fuzzy partition of $V(X)$, such that for each $x \in V(X)$, $i_X(A_y)(x) = 1_{\mathcal{L}}$ iff $u(x) = y$,*
2. *The following diagram commutes:*

$$
\begin{array}{ccc}
T(X) & \xrightarrow{\;T(u)\;} & T(Y) \\
{\scriptstyle i_X}\big\downarrow & & \big\downarrow{\scriptstyle i_Y} \\
\mathcal{L}^{V(X)} & \xrightarrow{\;F^{\uparrow}_{V(X),i_X(\mathcal{A})}\;} & \mathcal{L}^{V(Y)}.
\end{array}
$$

The subset $\mathcal{A} \subseteq T(X)$ is called $\mathcal{L}$-*fuzzy type partition of* $T(X)$.

As we have introduced a special type of the F-transform, we introduce the special type of the powerset operator.

Definition 3.4. *Let* $\mathcal{K}$ *be a category and let* $\mathbf{T} = (T, \rightarrow, V, \eta)$ *be a* $CSLAT(\vee)$-*powerset theory in* $\mathcal{K}$. *Then* $\mathbf{T}$ *is called to be a* $\mathcal{L}$-**fuzzy type** $CSLAT(\vee)$-**powerset theory in** $\mathcal{K}$, *if the following hold.*

1. *The functor* $T : \mathcal{K} \rightarrow CSLAT(\vee)$ *is of the* $\mathcal{L}$-*fuzzy type,*
2. *For each* $X \in |\mathcal{K}|$, $\eta_X : V(X) \rightarrow T(X)$ *generates* $T(X)$, *i.e.,*
   (a) *For each* $f \in T(X)$, $\quad f = \bigvee_{x \in V(X)} i_X(f)(x)_X \star \eta_X(x)$,
   (b) *For each* $x, z \in V(X)$, $\quad i_X(\eta_X(x))(z) = i_X(\eta_X(z))(x)$,
   (c) *For each* $X \in |\mathcal{K}|, x \in V(X)$, $\quad i_X(\eta_X(x))(x) = 1_\mathcal{L}$.

The following theorem is the principal result of the paper.

Theorem 3.1. *Let* $\mathcal{K}$ *be a category and let* $\mathbf{T} = (T, \rightarrow, V, \eta)$ *be a* $\mathcal{L}$-*fuzzy type* $CSLAT(\vee)$-*powerset theory in* $\mathcal{K}$. *Then, the functor* T *is the* $\mathcal{L}$-*fuzzy type F-transform.*

Proof. Let $u : X \twoheadrightarrow Y$ be an epimorphism in $\mathbf{K}$. From the natural transformations $\eta : V \rightarrow T$, $i : T^{op} \rightarrow Z^{op}V$ and $i : T \rightarrow Z.V$, it follows that the diagrams commute:

$$
\begin{array}{ccc}
V(X) \xrightarrow{\eta_X} T(X) \xrightarrow{i_X} \mathcal{L}^{V(X)}, & \qquad & T(X) \xrightarrow{i_X} \mathcal{L}^{V(X)} \\
V(u) \downarrow \qquad \downarrow u_{\overrightarrow{T}} \qquad \downarrow V(u)_{\overrightarrow{Z}} & & u_{\overleftarrow{T}} \uparrow \qquad \uparrow V(u)_{\overleftarrow{Z}} \\
V(Y) \xrightarrow{\eta_Y} T(Y) \xrightarrow{i_Y} \mathcal{L}^{V(Y)} & & T(Y) \xrightarrow{i_Y} \mathcal{L}^{V(Y)}.
\end{array}
$$

Because V is a concrete functor, $V(u) : V(X) \rightarrow V(Y)$ is a surjective map. Let $y \in V(Y)$ and let $a \in V(X)$ be such that $V(u)(a) = y$. Then we set

$$
A_y = u_{\overleftarrow{T}} u_{\overrightarrow{T}} (\eta_X(a)) \in T(X).
$$

The definition of A_y does not depend on the choice of the element a. In fact, let $b \in V(X)$ be such that $V(u)(b) = V(u)(a) = y$. Then we have

$$
u_{\overrightarrow{T}}(\eta_X(a)) = \eta_Y(V(u)(a)) = \eta_Y(V(u)(b)) = u_{\overrightarrow{T}}(\eta_X(b)).
$$

We prove that $\mathcal{A} = \{A_y : y \in V(Y)\}$ is the $\mathcal{L}$-fuzzy type partition of $T(X)$. Hence, we need to prove that for any $x \in V(X), y = V(u)(x)$, $i_X(A_y)(x) = 1_L$ holds. According to Definition 3.4, (2c), we have

$$
\begin{aligned}
i_X(A_y)(x) &= i_X.u_{\overleftarrow{T}}.u_{\overrightarrow{T}}(\eta_X(x))(x) = V(u)_{\overleftarrow{Z}}.i_Y.u_{\overrightarrow{T}}(\eta_X(x))(x) \\
&= V(u)_{\overleftarrow{Z}}.V(u)_{\overrightarrow{Z}}.i_X(\eta_X(x))(x) = V(u)_{\overleftarrow{Z}}.i_X(\eta_X(x))(V(u)(x)) \\
&= \bigvee_{z \in V(X), V(u)(z) = V(u)(x)} i_X(\eta_X(x))(z) \geq i_X(\eta_X(x))(x) = 1_L.
\end{aligned}
$$

We prove that the diagram from the theorem commutes. Let $f \in T(X), y \in V(Y)$. Then, according to Definition 3.4, parts (a), (b) Definition 3.2 and the relation (2), we obtain

$$i_Y.u_T^{\rightarrow}(f)(y) = V(u)_Z^{\leftarrow}.i_X(f)(y) = V(u)_Z^{\leftarrow}.i_X(\bigvee_{x \in X} i_X(f)(x)_X \star \eta_X(x))(y)$$

$$= V(u)_Z^{\leftarrow}(\bigvee_{x \in X} \underline{i_X(f)(x)}_{V(X)} \otimes i_X(\eta_X(x)))(y)$$

$$= \bigvee_{x \in X} V(u)_Z^{\rightarrow}(i_X(f)(x) \otimes i_X(\eta_X(x)))(y)$$

$$= \bigvee_{x \in X} i_X(f)(x) \otimes V(u)_Z^{\rightarrow}.i_X(\eta_X(x))(y)$$

$$= \bigvee_{x \in X} i_X(f)(x) \otimes i_Y.\eta_Y(V(u)(x))(V(u)(a))$$

$$= \bigvee_{x \in X} i_X(f)(x) \otimes i_Y.\eta_Y(V(u)(a))(V(u)(x))$$

$$= \bigvee_{x \in X} i_X(f)(x) \otimes i_Y.u_T^{\rightarrow}(\eta_X(a))(V(u)(x))$$

$$= \bigvee_{x \in X} i_X(f)(x) \otimes V(u)_Z^{\leftarrow}.i_Y.u_T^{\rightarrow}(\eta_X(a))(x)$$

$$= \bigvee_{x \in X} i_X(f)(x) \otimes i_X u_T^{\leftarrow} u_T^{\rightarrow}(\eta_X(a))(x) = \bigvee_{x \in X} i_X(f)(x) \otimes i_X(A_y)(x)$$

$$= F_{V(X),i_X(A)}^{\uparrow}(i_X(f))(y).$$

Hence, the diagram commutes.   □

4   Examples

In this section we show several examples of powerset theories which are used in fuzzy sets and which were introduced by Rodabaugh [17], Höhle [4], Solovyov [18] and Močkoř [5,7]. We show that all these powerset theories are, in fact, fuzzy type F-transforms.

Example 4.1. $CSLAT(\vee)$-*Powerset theory* $\mathcal{P} = (P, \rightarrow, id, \eta)$ *in the category* **Set**, *where*

1. $P : |\textbf{Set}| \rightarrow |CSLAT(\vee)|$ *is defined by* $P(X) = (2^X, \subseteq)$, *and any element* S *of* $P(X)$ *is identified with the characteristic function* χ_S^X *of a subset* $S \subseteq X$ *in* X.
2. *for each* $f : X \rightarrow Y$ *in* **Set**, $f_P^{\rightarrow} : P(X) \rightarrow P(Y)$ *is defined by* $f_P^{\rightarrow}(\chi_S^X) = \chi_{f(S)}^Y$,
3. *for each* $X \in$ **Set**, $\eta_X : X \rightarrow P(X)$ *is the characteristic function* $\chi_{\{x\}}^X$ *of a subset* $\{x\}$ *defined in* X.

Example 4.2. $CSLAT(\vee)$-*Powerset theory* $\mathcal{Z} = (Z, \rightarrow, id, \chi)$ *in the category* **Set**, *where*

1. $Z : |\textbf{Set}| \rightarrow |CSLAT(\vee)|$ *is defined by* $Z(X) = \mathcal{L}^X$,
2. *for each* $f : X \rightarrow Y$ *in* **Set**, $f_Z^{\rightarrow} : \mathcal{L}^X \rightarrow \mathcal{L}^Y$ *is defined by* $f_Z^{\rightarrow}(s)(y) = \bigvee_{x \in X, f(x)=y} s(x)$,
3. *for each* $X \in \textbf{Set}$, $\chi^X : X \rightarrow \mathcal{L}^X$ *is defined by* $\chi^X(a) = \chi_{\{a\}}^X$, *for* $a \in X$.

Example 4.3. $CSLAT(\vee)$-*Powerset theory* $\mathcal{D} = (D, \rightarrow, id, \rho)$ *in the category* **Set**, *where*

1. $D : |\textbf{Set}| \rightarrow |CSLAT(\vee)|$ *is defined by* $D(X) = $ *the set of all cuts* $(C_\alpha)_{\alpha \in L}$ *in a set* X, *naturally ordered by inclusion*,
2. *for each* $f : X \rightarrow Y$ *in* **Set**, $f_D^{\rightarrow} : D(X) \rightarrow D(Y)$ *is defined by* $f_D^{\rightarrow}((C_\alpha)_\alpha) = (\overline{f(C_\alpha)})_\alpha \in D(Y)$, *where the closure* $(\overline{S_\alpha})_\alpha$ *in a set* Y *is defined by* $\overline{S_\alpha} = \{a \in Y : \bigvee_{\beta : a \in C_\beta} \beta \geq \alpha\}$,
3. *for each* $X \in \textbf{Set}$, $\rho_X : X \rightarrow D(X)$ *and* $\rho_X(x)$ *is defined as the constant cut* $(\{x\})_\alpha$.

Example 4.4. $CSLAT(\vee)$-*Powerset theory* $\mathcal{E} = (E, \rightarrow, V, \widehat{\chi})$ *in the category* $\textbf{Set}(\mathcal{L})$, *where*

1. $E : |\textbf{Set}(\mathcal{L})| \rightarrow |CSLAT(\vee)|$, *where* $E(X, \delta)$ *is the set of all functions* $f \in \mathcal{L}^X$ *extensional with respect to the similarity relation* δ, *ordered point-wise*,
2. *for each morphism* $f : (X, \delta) \rightarrow (Y, \gamma)$ *in* $\textbf{Set}(\mathcal{L})$, $f_E^{\rightarrow} : E(X, \delta) \rightarrow E(Y, \gamma)$ *is defined by* $f_E^{\rightarrow}(s)(y) = \bigvee_{x \in X} s(x) \otimes \gamma(f(x), y)$,
3. $V : \textbf{Set}(\mathcal{L}) \rightarrow \textbf{Set}$ *is the concretel functor*,
4. *for each* $(X, \delta) \in \textbf{Set}(\mathcal{L})$, $\widehat{\chi}_{(X,\delta)} : X \rightarrow E(X, \delta)$ *is defined by* $\widehat{\chi}_{(X,\delta)}(a)(x) = \delta(a, x)$, *for* $a, x \in X$.

Example 4.5. $CSLAT(\vee)$-*Powerset theory* $\mathcal{C} = (C, \rightarrow, V, \overline{\chi})$ *in the category* $\textbf{Set}(\mathcal{L})$, *where*

1. $C : |\textbf{Set}(\mathcal{L})| \rightarrow |CSLAT(\vee)|$ *is defined by* $C(X, \delta) = $ *set of all f-cuts in* (A, δ) *in the category* $\textbf{Set}(\mathcal{L})$, *naturally ordered by inclusion*,
2. *for each morphism* $f : (X, \delta) \rightarrow (Y, \gamma)$ *in* $\textbf{Set}(\mathcal{L})$, $f_C^{\rightarrow} : C(X, \delta) \rightarrow C(Y, \gamma)$ *is defined by*

$$f_C^{\rightarrow}((E_\alpha)_\alpha) = (\overline{f(E_\alpha)})_\alpha, \quad \overline{f(E_\alpha)} = \{b \in Y : \bigvee_{(y,\beta): y \in f(E_\beta)} \beta \otimes \gamma(b, y) \geq \alpha\},$$

3. $V : \textbf{Set}(\mathcal{L}) \rightarrow \textbf{Set}$ *is the concretel functor*,
4. *for each* $(X, \delta) \in \textbf{Set}(\mathcal{L})$, $\overline{\chi}_{(X,\delta)} : X \rightarrow C(X, \delta)$ *is defined by* $\overline{\chi}_{(X,\delta)}(a) = (\overline{\{a\}})_\alpha$, *where* $\overline{\{a\}}_\alpha = \{b \in X : \delta(a, b) \geq \alpha\}$.

Example 4.6. $CSLAT(\vee)$-*powerset theory* $\mathcal{F} = (F, \rightarrow, W, \vartheta)$ *in the category* **SpaceFP** *is defined by*

(1) $F : |\textbf{SpaceFP}| \rightarrow |CSLAT(\vee)|$, *defined by*

$$F(X, \mathcal{A}) = \{f | f : X \rightarrow L \text{ is extensional in } (X, \mathcal{A})\},$$

ordered pointwise.
(2) For each $(f, u) : (X, \mathcal{A}) \rightarrow (Y, \mathcal{B})$ *in* **SpaceFP**, $(f, u)_F^{\rightarrow} : F(X, \mathcal{A}) \rightarrow F(Y, \mathcal{B})$ *is defined by*

$$g \in F(X, \mathcal{A}), y \in Y, \quad (f, u)_F^{\rightarrow}(g)(y) = \bigvee_{x \in X} g(x) \otimes \delta_{Y, \mathcal{B}}(f(x), y),$$

where $\delta_{Y, \mathcal{B}}$ *is the characteristic similarity relation in* Y *in a space with a fuzzy partition* $(Y, \mathcal{B})$.
(3) $V : \textbf{SpaceFP} \rightarrow \textbf{Set}$ *is the concretel functor,* $V(X, \mathcal{A}) = X$,
(4) For each $(X, \mathcal{A})$ *in* **SpaceFP**, $\vartheta_{(X, \mathcal{A})} : V(X, \mathcal{A}) \rightarrow F(X, \mathcal{A})$, $\vartheta_{(X, \mathcal{A})}(a)(x) = \delta_{X, \mathcal{A}}(a, x)$, *for each* $a, x \in X$.

Proposition 4.1. *Let us consider the powerset theories from the Examples 4.1–4.6.*

1. *Functors from Examples 4.1–4.6 are of the* $\mathcal{L}$-*fuzzy type.*
2. $CSLAT(\vee)$-*powerset theories from Examples 4.1–4.6 are* $\mathcal{L}$-*fuzzy type powerset theories.*
3. *Functors from Examples 4.1–4.6 are* $\mathcal{L}$-*fuzzy type F-transforms.*

Instead of the proof, for illustration we show only how the natural transformations i and operations $\star$ from the Definition 3.2 are defined for functors E and C. The rest of technical proof will be published elsewhere.

(1) Example 4.4: For $(X, \delta) \in \textbf{Set}(\mathcal{L})$, $i_{(X, \delta)} : E(X, \delta) \subseteq \mathcal{L}^X$ is the embedding, and for $\alpha \in \mathcal{L}, f \in E(X, \delta)$, $\alpha \star f := \alpha \otimes f$.

(2) Example 4.5: For $(X, \delta) \in \textbf{Set}(\mathcal{L})$, $i_{(X, \delta)} : C(X, \delta) \rightarrowtail \mathcal{L}^X$, and the operation $\star$ are defined by

$$x \in X, \quad i_{(X, \delta)}((C_\alpha)_\alpha)(x) = \bigvee_{\{\beta : x \in C_\beta\}} \beta,$$

$$\lambda \in \mathcal{L}, (C_\alpha)_\alpha \in C(X, \delta), \quad \lambda \star (C_\alpha)_\alpha := (G_\alpha)_\alpha, G_\alpha = \{x \in X : \lambda \otimes \bigvee_{\beta, x \in C_\beta} \beta \geq \alpha\}.$$

5   Conclusions

We introduce two types of aggregation operators for lattice-valued fuzzy sets, called *fuzzy type powerset operators* and *fuzzy type F-transforms*, which are derived from classical powerset operators and F-transforms, respectively, and we proved that any $\mathcal{L}$-fuzzy type powerset theory is also $\mathcal{L}$-fuzzy type F-transform. This result makes it possible to use some theoretical tools from powerset theory (e.g., theory of monads) also in F-transform theory.

References

1. Calvo, T., Kolesárová, A., Komorníková, M., Mesiar, R.: Aggregation operators: properties, classes and construction methods. In: Calvo, T., Mayor, G., Mesiar, R. (eds.) Aggregation Operators, pp. 3–107. Physica-Verlag, Heidelberg (2002)
2. Dubois, E., Prade, H.: A review of fuzzy set aggregation connectives. Inf. Sci. **36**, 85–121 (1985)
3. Grabisch, M., Marichal, J.-L., Mesiar, R., Pap, E.: Aggregation Functions. Cambridge University Press, Cambridge (2009)
4. Höhle, U.: Many Valued Topology and its Applications. Kluwer Academic Publishers, Boston (2001)
5. Močkoř, J.: Closure theories of powerset theories. Tatra Mountains Math. Publ. **64**, 101–126 (2015)
6. Močkoř, J.: Cut systems in sets with similarity relations. Fuzzy Sets Syst. **161**, 3127–3140 (2010)
7. Močkoř, J.: Powerset operators of extensional fuzzy sets. Iran. J. Fuzzy Syst. **15**(2), 143–163 (2017)
8. Močkoř, J., Holčapek, M.: Fuzzy objects in spaces with fuzzy partitions. Soft Comput. **21**(24), 7269–7284 (2017)
9. Močkoř, J.: Spaces with fuzzy partitions and fuzzy transform. Soft Comput. **21**, 3479–3492 (2017)
10. Novák, V., Perfilieva, I., Močkoř, J.: Mathematical Principles of Fuzzy Logic. Kluwer Academic Publishers, Boston (1999)
11. Perfilieva, I.: Fuzzy transforms and their applications to image compression. In: Bloch, I., Petrosino, A., Tettamanzi, A.G.B. (eds.) WILF 2005. LNCS (LNAI), vol. 3849, pp. 19–31. Springer, Heidelberg (2006). https://doi.org/10.1007/11676935_3
12. Perfilieva, I.: Fuzzy transforms: a challenge to conventional transform. In: Hawkes, P.W. (ed.) Advances in Image and Electron Physics, vol. 147, pp. 137–196. Elsevies Acad. Press, San Diego (2007)
13. Perfilieva, I., Novak, V., Dvořak, A.: Fuzzy transforms in the analysis of data. Int. J. Approximate Reasoning **48**, 36–46 (2008)
14. Perfilieva, I.: Fuzzy transforms: theory and applications. Fuzzy Sets Syst. **157**, 993–1023 (2006)
15. Perfilieva, I., Singh, A.P., Tiwari, S.P.: On the relationship among F-transform, fuzzy rough set and fuzzy topology. In: Proceedings of IFSA-EUSFLAT, pp. 1324–1330. Atlantis Press, Amsterdam (2015)
16. Rodabaugh, S.E.: Powerset operator foundation for poslat fuzzy SST theories and topologies. In: Höhle, U., Rodabaugh, S.E. (eds.) Mathematics of Fuzzy Sets: Logic, Topology and Measure Theory, The Handbook of Fuzzy Sets Series, vol. 3, pp. 91–116. Kluwer Academic Publishers, Boston (1999)
17. Rodabaugh, S.E.: Relationship of algebraic theories to powerset theories and fuzzy topological theories for lattice-valued mathematics. Int. J. Math. Math. Sci. **2007**, 1–71 (2007)
18. Solovyov, S.A.: Powerset oeprator foundations for catalg fuzzy set theories. Iran. J. Fuzzy Syst. **8**(2), 1–46 (2001)
19. Takači, A.: General aggregation operators acting on fuzzy numbers induced by ordinary aggregation operators. Novi Sad J. Math. **33**(2), 67–76 (2003)
20. Zadeh, L.A.: Fuzzy sets. Inf. Control **8**, 338–353 (1965)

Implicative Weights as Importance Quantifiers in Evaluation Criteria

Jozo Dujmović[(✉)]

San Francisco State University, San Francisco, CA 94132, USA
jozo@sfsu.edu

Abstract. This paper investigates properties of implicative weights and the use of implicative weights in evaluation criteria. We analyze and compare twelve different forms of implication and compare them with multiplicative weights and exponential weights that are also used in evaluation criteria. Since weighted conjunction is based on implicative weights, we also investigate the usability of weighted conjunction in evaluation criteria.

Keywords: Graded logic · Importance · Implicative weights · GCD
Evaluation · Logic aggregation

1 Introduction

All logic aggregators used in evaluation criteria aggregate degrees of truth (or degrees of fuzzy membership). The degrees of truth quantify the truth of value statements. Each value statement specifies the degree of contribution of specific attribute to the overall suitability of evaluated object. Consequently, the degrees of truth are not anonymous real numbers, because they have precisely defined role and meaning. In other words, the degrees of truth (and degrees of fuzzy membership) have semantic identity.

The individual attribute contributions to attainment of stakeholder's goals are generally different, yielding different degrees of attribute importance. Obviously, more important are those attributes that more contribute to the overall suitability of evaluated object. The degrees of importance are human percepts that are quantified using weights. Consequently, the weights are interpreted as semantic components of evaluation criteria. Logic aggregators combine semantic and formal logic aspects of aggregation. Undeniably, the purpose of logic aggregation is insight, not numbers.

A natural way to quantify weights is to use real numbers from a fixed interval, and to verbalize those using rating scales. In the context of evaluation, we assume that weights in direct proportion express the degree of importance, which is always positive. Indeed, in the area of evaluation, weights cannot be zero. A zero weight obviously denotes the total insignificance of an input (elementary or compound attribute), and such inputs are justifiably excluded from consideration. Rational thinkers ignore insignificant inputs and focus on reasoning based on the "first things first" concept. So, in graded evaluation logic, weights are always positive.

Similarly to zero-sum games, increasing the relative importance (impact) of an input in a group of inputs automatically means decreasing the relative importance

---

The original version of this chapter has been revised: Minor errors in the text have been corrected. The correction to this chapter is available at https://doi.org/10.1007/978-3-030-00202-2_26

V. Torra et al. (Eds.): MDAI 2018, LNAI 11144, pp. 193–205, 2018.
https://doi.org/10.1007/978-3-030-00202-2_16

(impact) of all other inputs in the group. Therefore, the weights must be normalized. Following are three fundamental normalization methods:

1. *Sum-normalized weights* (constant sum of weights equal 1):

$$\mathbf{W} = (W_1, \ldots, W_n), \quad W_1 + \cdots + W_n = 1, \quad 0 < W_i < 1, \quad i = 1, \ldots, n; \quad n \geq 2.$$

2. *Max-normalized weights* (maximum weight equal 1):

$$\mathbf{v} = (v_1, \ldots, v_n), \quad \max(v_1, \ldots, v_n) = 1, \quad 0 < v_i \leq 1, \quad i = 1, \ldots, n.$$

3. *Count-normalized weights* (sum of weights equals the number of inputs n):

$$\mathbf{p} = (p_1, \ldots, p_n), \quad p_1 + \cdots + p_n = n, \quad 0 < p_i < n, \quad i = 1, \ldots, n.$$

These weights are used in different contexts. Assuming that they express the same importance degrees, it is easy to transform one of them into another by keeping them proportional as follows:

$$W_i = v_i/(v_1 + \cdots + v_n) = p_i/n,$$
$$v_i = W_i/\max(W_1, \ldots, W_n) = p_i/\max(p_1, \ldots, p_n),$$
$$p_i = nW_i = nv_i/(v_1 + \cdots + v_n), \quad i = 1, \ldots, n.$$

In the special case of equal importance and equal weights, the values of weights become $W_i = 1/n$, $v_i = 1$, $p_i = 1$, $i = 1, \ldots, n$.

All forms of weights support the same "implication concept:" *it is not acceptable that an important input (requirement) is insufficiently satisfied.* If x denotes an input argument and W, v, p are its weights (degrees of importance), then the implication concept emerges in three characteristic forms:

- Multiplicative: $W \cdot x$
- Implicative: $v \to x = \bar{v} \lor x = \overline{v \land \bar{x}} = 1 - [v \land (1 - x)]$
- Exponential: x^p

All these forms support the same idea: if the weight is large, then x should also be large in order to provide a high contribution to the overall score. Implicative weights are particularly interesting because only implicative weights belong to [0, 1] without additive restrictions. Thus, only implicative weights have the status of *independent logic variables* and can be interpreted as degrees of truth. Multiplicative weights and exponential weights do not have the interesting property that semantic aspects of evaluation (degrees of importance) can be interpreted in the same way as the degrees of satisfaction. In the case of implicative weights evaluation models can be developed strictly inside the domain of formal logic. This is the motivation for the analysis presented in this paper. The context of our analysis is evaluation: we investigate the suitability of implicative weights as importance quantifiers in evaluation criteria.

Evaluation criteria use logic aggregators to compute an overall suitability indicator using n arguments that represent suitability indicators of selected input attributes. Such forms of logic aggregators were proposed in the context of approximate reasoning and information retrieval [1, 2], and general aggregation models [3, 4]. Our goal is to evaluate the applicability of such models in the area of evaluation.

2 Implicative Weights in Conjunctive Aggregators

The motivation for using implicative weights $v \rightarrow x = \overline{v \wedge \bar{x}}$ comes directly from the verbal interpretation of this relationship between the degree of importance (weight) v and the degree of satisfaction/truth x: "it is not acceptable if input x is important and it is not (sufficiently) satisfied" or "important input must be highly satisfied." If x is very important (v is close to 1) and x is very much satisfied (x is close to 1) then $v \rightarrow x \approx 1 \rightarrow 1 = 1$. This is a correct conclusion: we are completely satisfied with the ideal combination of importance and satisfaction. The next step in that direction is the case of multiple inputs $x_1, \ldots, x_n$ and the criterion "all important inputs should be simultaneously satisfied" which directly yields conjunctive aggregators based on multiplicative, implicative and exponential forms:

$$y = (W_1 x_1^r + \cdots + W_n x_n^r)^{1/r}, \quad r < 1,$$
$$y = (v_1 \rightarrow x_1) \wedge \cdots \wedge (v_n \rightarrow x_n) = (\bar{v}_1 \vee x_1) \wedge \cdots \wedge (\bar{v}_n \vee x_n),$$
$$y = x_1^{p_1} \times \cdots \times x_n^{p_n}, \quad n \geq 2.$$

Multiplicative weights are used in additive aggregation forms and exponential weights are used in multiplicative aggregation forms. So, there is a natural interest to compare aggregators with implicative weights and other aggregators. Initial steps in that direction can be found in [4, 5].

If implication is modeled as $v \rightarrow x = \max(1 - v, x)$ then it is easy to see that, from the standpoint of logic aggregators used in evaluation criteria, this function has acceptable, questionable and unacceptable properties shown in Table 1. To find more convenient properties, we must investigate other forms of implication function.

3 Implicative Weight Functions

In the case of implicative weights, there are various ways to implement the necessary negation and conjunction, yielding a variety of implication functions. Presentations and analyses of various forms of implication can be found in [6–10]. The most popular forms of implication include the following:

Dienes: $v \rightarrow x = \max(1 - v, x)$

Reichenbach: $v \rightarrow x = 1 - v(1 - x)$

Gödel: $v \rightarrow x = \begin{cases} 1 & \text{if } v \leq x \\ x & \text{otherwise} \end{cases}$

Yager: $v \rightarrow x = x^v$

Lukasiewicz: $v \rightarrow x = \min(1 - v + x, 1)$

Table 1. Interpretation of properties of implication $\bar{v} \vee x$, $v \in [0, 1]$, $x \in [0, 1]$

v	x	$\bar{v} \vee x$	Interpretation	Evaluation
0	0	1	Insignificant input can be unsatisfied without consequences	Acceptable
0	1	1	Degree of satisfaction of insignificant input makes no visible impact	Acceptable
1	0	0	It is unacceptable to not satisfy an extremely important requirement	Acceptable
1	1	1	We are fully satisfied if a very important input is completely satisfied	Acceptable
0	1	1	The impact of the fully satisfied input is the same regardless of its degree of importance	Questionable
1	1	1		
¾	0	¼	Penalty for unsatisfied inputs equals their degree of importance. Thus, unsatisfied inputs have the power of annihilation only if they have the maximum degree of importance	Unacceptable
½	0	½		
¼	0	¾		
¼	0	¾	Insensitivity to the degree of satisfaction of x as long as $x \leq 1 - v$	Unacceptable
¼	¾	¾		

$$\text{Drastic: } v \rightarrow x = \begin{cases} 1 - v & \text{if } x = 0 \\ x & \text{if } v = 1 \\ 1 & \text{otherwise} \end{cases}$$

$$\text{Einstein: } v \rightarrow x = 1 - \frac{v(1-x)}{1 + (1-v)x}$$

$$\text{Hamacher: } v \rightarrow x = 1 - \frac{v(1-x)}{1 - (1-v)x}$$

$$\text{Goguen: } v \rightarrow x = \begin{cases} 1 & \text{if } v \leq x \\ x/v & \text{otherwise} \end{cases}$$

$$\text{Rescher: } v \rightarrow x = \begin{cases} 1 & \text{if } v \leq x \\ 0 & \text{if } v > x \end{cases}$$

$$\text{Zadeh: } v \rightarrow x = \max(1 - v, \min(v, x))$$

$$\text{Fodor: } v \rightarrow x = \begin{cases} 1 & \text{if } v \leq x \\ \max(1 - v, x) & \text{if } v > x \end{cases}$$

Multiplicative weights and additive aggregation forms provide both soft and hard models of simultaneity and substitutability [11]. As opposed to that, multiplicative models with exponential weights are always hard. In the case of Dienes type of implicative weights and weighted conjunction, if $1 = v_1 > v_2 \geq \cdots \geq v_n > 0$, then the weighted conjunction $y = x_1 \wedge (\bar{v}_2 \vee x_2) \wedge \cdots \wedge (\bar{v}_n \vee x_n)$ is only hard with respect to x_1 (and all other inputs that might have the maximum importance 1), but soft with respect to all other inputs. This is a typical property of partial absorption [12] but not the property of partial conjunction. This inconvenient property also indicates that it is necessary to investigate other forms of implication from the standpoint of their suitability for combining weights and degrees of truth in logic aggregators.

4 GCD and Monotonicity with Respect to Andness and Orness

In graded logic, we are primarily interested in models of simultaneity, substitutability, and negation. Assuming the use of standard negation $N(x) = 1 - x$, the fundamental logic function is the Graded Conjunction/Disjunction (GCD) [5, 11]. Let $\mathbf{x}$ denote an array of n degrees of truth and let $\mathbf{W}$ denote the corresponding array of n weights. GCD aggregators provide continuous transition from the drastic conjunction $y = \lfloor x_1 \cdots x_n \rfloor$ to the drastic disjunction $y = 1 - \lfloor (1 - x_1) \cdots (1 - x_n) \rfloor$ controlled by the global andness α, or the global orness $\omega = 1 - \alpha$. Consequently, GCD is an andness-directed aggregator defined in the full range $-1/(n-1) \leq \alpha \leq n/(n-1)$ [13]. A general interpolative GCD aggregator [11] that uses weighted power mean with exponent $r_{wpm}(\alpha)$, the adjustable threshold andness α_θ, and t-norm/conorm, can be written as follows [13]:

$$z = A(\mathbf{x}; \mathbf{W}, \alpha)$$

$$= \begin{cases} F(\mathbf{x}; \mathbf{W}, \alpha) = \begin{cases} \left\lfloor \prod_{i=1}^{n} x_i \right\rfloor, & \alpha = \alpha_{\max} = n/(n-1). \\[2mm] \left(\prod_{i=1}^{n} x_i \right)^{\{(n+1)/[n-(n-1)\alpha]\}^{1/n} - 1}, & \alpha_t < \alpha < \alpha_{\max}. \\[2mm] \prod_{i=1}^{n} x_i, & \alpha = \alpha_t = \frac{n2^n - n - 1}{(n-1)2^n}. \\[2mm] \frac{\alpha_t - \alpha}{\alpha_t - 1} \min(\mathbf{x}) + \frac{\alpha - 1}{\alpha_t - 1} \prod_{i=1}^{n} x_i, & 1 < \alpha < \alpha_t. \\[2mm] \min(x_1, \ldots, x_n), & \alpha = 1. \\[2mm] \left(\sum_{i=1}^{n} W_i x_i^{r_{wpm}(\alpha)} \right)^{1/r_{wpm}(\alpha)}, & \alpha_\theta \leq \alpha < 1. \\[2mm] \frac{\alpha_\theta - \alpha}{\alpha_\theta - 1/2} \left(\sum_{i=1}^{n} W_i x_i \right) + \frac{\alpha - 1/2}{\alpha_\theta - 1/2} \left(\sum_{i=1}^{n} W_i x_i^R \right)^{1/R}, & 1/2 < \alpha < \alpha_\theta. \\[2mm] \sum_{i=1}^{n} W_i x_i, & \alpha = \omega = 1/2. \end{cases} \\[2mm] 1 - F(1 - \mathbf{x}; \mathbf{W}, 1 - \alpha), & \alpha_{\min} = -1/(n-1) \leq \alpha < 0.5. \end{cases}$$

$$n > 1, \quad \mathbf{x} = (x_1, \ldots, x_n), \quad 1 - \mathbf{x} = (1 - x_1, \ldots, 1 - x_n),$$

$$\mathbf{W} = (W_1, \ldots, W_n), \quad 0 < W_i < 1, \quad i = 1, \ldots, n, \quad \sum_{i=1}^{n} W_i = 1.$$

$$R = r_{wpm}(\alpha_\theta); \frac{2}{3} \leq \frac{n}{n-1} - \frac{n+1}{n-1} \left(\frac{n}{n-1} \right)^n \leq \alpha_\theta < 1; \text{ (frequently } \alpha_\theta = 3/4).$$

The properties of the GCD aggregator are selected by selecting the desired andness and desired weights. The presented version of GCD uses multiplicative weights only in the region from pure conjunction to pure disjunction $(0 \leq \alpha \leq 1)$; the reasons for this decision are presented in the next section.

The GCD aggregator is andness-monotonic: $\alpha_1 < \alpha_2 \Rightarrow A(\mathbf{x}; \mathbf{W}, \alpha_1) \geq A(\mathbf{x}; \mathbf{W}, \alpha_2)$, $\omega_1 < \omega_2 \Rightarrow A(\mathbf{x}; \mathbf{W}, 1 - \omega_1) \leq A(\mathbf{x}; \mathbf{W}, 1 - \omega_2)$ (the monotonicity is strict if arguments are different from annihilators and non-identical). This property is consistent with observations of human reasoning: as the andness increases, decision maker requires more simultaneity. The percept of the importance of inputs automatically increases, and for hard GCD all inputs must be satisfied regardless to differences in their relative importance. Ultimately, the aggregated degree of truth approaches the minimum input (i.e., the weakest link in the chain of degrees of truth), yielding the weight-independent pure conjunction $A(\mathbf{x}; \mathbf{W}, 1) = x_1 \wedge \cdots \wedge x_n$. Therefore, if arguments are not identical, or annihilators, GCD must be strictly decreasing in α and strictly increasing in ω. This is the fundamental property of conjunctive and disjunctive logic aggregators.

5   Andness-Domination Versus Weight-Domination

The percept of overall importance of each GCD input depends on two components: the relative importance (weight) and andness/orness. Any increase of the weight of an input causes an increase of the percept of the input's overall importance. The same effect is caused by high values of andness (or orness): the percept of importance is an increasing function of $\max(\alpha, \omega)$. Indeed, in the case of conjunctive aggregators, more simultaneity we need, more important becomes each input, because its nonsatisfaction cannot be tolerated. Similarly, for disjunctive aggregators, increasing substitutability and orness gives more power to any input to fully satisfy the disjunctive criterion; that automatically increases the overall percept of its importance.

The modeling of process how humans combine the impacts of weights and the impact of andness/orness to produce the percept of overall importance is certainly a delicate question. An answer to this question, in the form of methods for decomposing the percept of the overall importance of inputs into the most appropriate weight and the most appropriate andness can be found in [11].

Observations of human reasoning indicate that, except for identical inputs, and inputs that are annihilators, GCD aggregators must satisfy $\partial A(\mathbf{x}; \mathbf{W}, \alpha)/\partial \alpha < 0$ and also $\partial A(\mathbf{x}; \mathbf{W}, 1 - \omega)/\partial \omega > 0$ in the whole range of andness, from drastic conjunction to drastic disjunction. For conjunctive aggregators, the impact of weights is a decreasing function of andness. Similarly, for disjunctive aggregators, the impact of weights is a decreasing function of orness. To prove this claim, suppose that we want simultaneous satisfaction of multiple inputs that have different degrees of relative importance. The degree of simultaneity is defined as the global andness α. For low values of α, (e.g. $\alpha \approx 1/2$) the impact of important inputs (those with higher weights) is visibly higher than the impact of inputs that have lower weights. As we increase the degree of simultaneity the percept of importance of inputs increases because it becomes less and less acceptable that any input is insufficiently satisfied. For values $\alpha \geq \alpha_\theta$, GCD becomes hard, and all inputs become important because each of them models a mandatory requirement and must be satisfied. In such a situation, for conjunctive aggregators, the highest impact comes from the smallest inputs and not from the most important inputs. For high values of α, (e.g. $\alpha \approx 1$) all inputs becomes equally important because no input can be unsatisfied regardless the level of its initial relative importance. Consequently, for high andness all inputs become equally important. We call this property *andness-domination*, because the impact of different weights is first diminished and then annihilated at high degrees of andness. The impact of andness dominates the impact of weights and in the case of full conjunction $y = x_1 \wedge \cdots \wedge x_n$ the impact of different weights completely disappears, as in Boolean logic. Based on this reasoning, our model of GCD has no weights in the area of hyperconjunction (the range from the full conjunction to the drastic conjunction). For disjunctive aggregators, due to De Morgan duality, the situation is completely symmetric to the conjunctive case.

Let us now investigate the following weighted conjunction, which is based on Dienes type of implicative weights:

$$y(x_1, \ldots, x_n) = (v_1 \rightarrow x_1) \wedge \cdots \wedge (v_n \rightarrow x_n) = (\bar{v}_1 \vee x_1) \wedge \cdots \wedge (\bar{v}_n \vee x_n).$$

In this model, each argument is shielded by negated weight, in the sense that if $\bar{v}_i \geq x_i$ then $\bar{v}_i \vee x_i = \bar{v}_i$. In the case of sorted weights, if $1 = v_1 > v_2 \geq \cdots \geq v_n > 0$, then this model of weighted simultaneity has the following properties:

$$y(0, 1, \ldots, 1) = 0; \quad y(1, 0, \ldots, 0) = \bar{v}_2 = 1 - \max(v_2, \ldots, v_n) > 0,$$

$$y(1, 0, 0, \ldots, 0) = y(\bar{v}_2, 0, 0, \ldots, 0) = \bar{v}_2,$$

$$y(\bar{v}_2, 0, 0, \ldots, 0) = y(\bar{v}_2, \bar{v}_2, \ldots \bar{v}_2) = y(1, 0, 1, \ldots, 1) = y(1, \bar{v}_2, 1, \ldots, 1) = \bar{v}_2$$

It is easy to see that these properties are generally very questionable and unacceptable in the area of evaluation criteria. The first formula shows that this model provides soft aggregation for all inputs except the most important input (which is hard); this is not the property of either hard or soft partial conjunction, where all inputs must behave in a homogeneous way. The second formula shows that this form of implicative weights is insensitive to variation of the most important input in the range $\bar{v}_2 \leq x_1 \leq 1$, regardless the fact that only x_1 is satisfied, and for all $0 \leq x_i \leq \bar{v}_2$, $i = 2, \ldots, n$; there is no logic justification for this property. The third formula shows the unjustifiable insensitivity to large variations of inputs as long as the second input is not greater than $\bar{v}_2$.

In all presented examples the result of aggregation is the weight $\bar{v}_2$. The weights are an expression of stakeholder's goals and requirements and not an expression of properties of an evaluated object. We call this property of highly conjunctive aggregators the *weight domination,* to contrast it to the previous more justifiable *andness domination.* Of course, the presented example of output equal to weight occurs only for specific set of distributions of input suitability degrees. Consequently, it indirectly depends on inputs, but nevertheless represents a feature that cannot be tolerated in evaluation models and disqualifies the use of the most natural Dienes implication in evaluation criteria. On the other hand, this opens the question whether other forms of implication yield more desirable logic aggregation models.

6 Experimental Comparison of Implicative Weight Models

The goal of experimental comparison of various forms of implication in weighted simultaneity models is to select the form of implication that is the most suitable for solving practical evaluation problems. We are going to analyze classic implications presented in Sect. 3. Of course, there are more implication functions [9, 10] and they should be analyzed in a similar way as a part of future work.

Suppose that a homebuyer reduced the decision about the overall home suitability (H) to two fundamental final components: the suitability of home location (L) and the quality of home (Q, evaluated without taking into account its cost and location). Therefore, $H = f(L, Q)$ and there is no doubt that the function f is conjunctive because everybody *simultaneously* wants a good home and a nice location. In the decision process, the cost will later be compared to H.

Let the location be two times more important than the home quality (the owner can improve the quality of home, while the quality of location usually remains unchanged). If we use the basic Dienes implication to make the weighted conjunction, then we have the following:

$$H_{wc}(L, Q) = (\bar{v}_L \vee L) \wedge (\bar{v}_Q \vee Q)$$
$$v_L = 1, \quad v_Q = 0.5 \quad \Rightarrow H_{wc}(L, Q) = L \wedge (0.5 \vee Q)$$

This criterion yields the following completely meaningless results:

$H_{wc}(0.5, 0) = 0.5$: An unacceptable home in an average location makes the homebuyer 50% satisfied.

$\forall Q \in [0, 1] \Rightarrow H_{wc}(0.5, Q) = 0.5$: In an average location homebuyer's satisfaction is always 50% and it does not depend on home quality.

$H_{wc}(0.5, 0.5) = H_{wc}(1, 0.5) = 0.5$: An average home in an average location and an average home in an ideal location are equally desirable.

The problem we face with this form of weighted conjunction is the weight-domination problem that is clearly visible in the general case of weights $v_L = 1, 0 < v_Q < 1$, as follows:

$$\forall L \geq \bar{v}_Q, \quad \forall Q \leq \bar{v}_Q \quad \Rightarrow H_{wc}(L, Q) = L \wedge (\bar{v}_Q \vee Q) = \bar{v}_Q = const.$$

Thus, this criterion absurdly claims that in the large range of values of input arguments L and Q the resulting suitability does not depend on arguments at all, but it is equal to the weight complement $\bar{v}_Q$. However, $\bar{v}_Q$ is a parameter that reflects the interests of homebuyer and has nothing in common with the properties of the evaluated object, while the properties of evaluated object (L and Q) are made irrelevant and completely neglected in a wide range of their values.

In graded logic aggregation the proper solution of the homebuyer criterion problem is any idempotent *hard partial conjunction*. For example, let us take a simple weighted harmonic mean:

$$H_{har}(L,Q) = \frac{1}{W_L/L + W_Q/Q} = \frac{LQ}{QW_L + LW_Q}$$

$$W_L = 2/3, \quad W_Q = 1/3 \quad \Rightarrow \quad H_{har}(L,Q) = \frac{LQ}{2Q/3 + L/3} = \frac{3LQ}{L + 2Q}$$

Let us again suppose that the quality of home location is two times more important than the quality of home. Following are four characteristic examples of results generated by this aggregator:

$H_{har}(0.5, 0) = 0$ (an unacceptable home is rejected in any location)
$H_{har}(0.5, 0.5) = 0.5$ (an average home in an average location gives average satisfaction)
$H_{har}(0.5, 1) = 0.6$ (in an average location an ideal home satisfies 60% of requirements)
$H_{har}(1, 0.5) = 0.75$ (in an ideal location an average home satisfies 75% of requirements)

These results are intuitively acceptable and we can ask the obvious question: is it possible to achieve such results using weighted conjunction based on an appropriate implication function? Since there are many forms of implication and many models of simultaneity, some combinations can be more suitable and some combinations are less suitable. A positive answer to this question can be found in [13].

Let us compute $H_{har}(L,Q)$ and $H_{wc}^{[imp]}(L,Q)$ for various implications [*imp*] and for nine combinations of arguments L and Q belonging to set $\{0, 0.5, 1\}$. For comparison of implicative weighted conjunction and the harmonic mean we use the weighted absolute error $E^{[imp]} = 100 \, \Sigma_{i=1}^{9} \, W_i \left| H_{wc}^{[imp]}(L,Q) - H_{har}(L,Q) \right| \Big/ \Sigma_{i=1}^{9} W_i$ [%] defined using weights W_i that reflect the idea that some errors can be easier tolerated than other errors. We use the following weights: 4 (for idempotency errors), 3 (for Boolean implication compatibility errors), 2 (for annihilation errors) and 1 (for all other errors). For $W_L = 2/3$, $W_Q = 1/3$, $v_L = 1$, $v_Q = 0.5$. Following is the ranking of implications according to decreasing error:

$$E^{[Yag]} = 0.595\%, \quad E^{[God]} = 1.46\%, \quad E^{[Gog]} = 1.46\%, \quad E^{[Rei]} = 10.8\%,$$
$$E^{[Ein]} = 11\%, \quad E^{[Ham]} = 11.2\%, \quad E^{[Die]} = 11.9\%, \quad E^{[Luk]} = 11.9\%,$$
$$E^{[Dra]} = 11.9\%, \quad E^{[Res]} = 11.9\%, \quad E^{[Fod]} = 11.9\%, \quad E^{[Zad]} = 20.2\%.$$

The most convenient form of implication for implicative evaluation criteria is Yager's implication, followed by implications proposed by Gödel and Goguen. Using Yager's implication the weighted conjunction aggregator with implicative weights becomes $(v_L \rightarrow L) \wedge (v_Q \rightarrow Q) = L^{v_L} \wedge Q^{v_Q} = L \wedge Q^{v_Q}$. The differences between various weighted conjunction formulas depend on v_L and v_Q. In the case $v_L = v_Q$ all errors (except for Rescher) become small and equal: 1.39%. Therefore, some forms of implication can be used to provide implicative weights in evaluation criteria.

Weighted conjunctions can be organized with various models of conjunction. In addition, weighted conjunction and weighted disjunction can be natural limit cases of aggregators that perform continuous andness-directed transition from the weighted conjunction to the weighted disjunction. Such an aggregator was proposed by Henrik Legind Larsen [3–5]. Larsen developed a graded conjunction/disjunction aggregator based on Reichenbach implication, called AIWA [3]. AIWA can be interpreted as a form of GCD where the weighted conjunction $y_\wedge$ and the weighted disjunction $y_\vee$ (based on Reichenbach implication) are special cases of AIWA for andness 1 and 0 respectively. The AIWA aggregator uses the maximum-normalized weights $\mathbf{v} = (v_1, \ldots, v_n)$, $\max(v_1, \ldots, v_n) = 1$. The logic properties of AIWA are adjusted using the AIWA andness $a \in [0, 1]$, as follows:

$$h(\mathbf{x}; \mathbf{v}, a) = \begin{cases} \max(v_1 x_1, \ldots, v_n x_n), & a = 0 \\ \left(\dfrac{\sum_{i=1}^n (v_i x_i)^r}{\sum_{i=1}^n v_i^r} \right)^{1/r}, & 0 < a \le \frac{1}{2} \\ \left(1 - \dfrac{\sum_{i=1}^n (v_i(1-x_i))^{1/r}}{\sum_{i=1}^n v_i^{1/r}} \right)^r, & \frac{1}{2} \le a < 1 \\ \min[1 - v_1(1-x_1), \ldots, 1 - v_n(1-x_n)], & a = 1 \end{cases}$$

$$r = \frac{1}{a} - 1 = \frac{1-a}{a} = \frac{\bar{a}}{a}, \qquad \max(v_1, \ldots, v_n) = 1$$

$$\min[1 - v_1(1-x_1), \ldots, 1 - v_n(1-x_n)] = 1 - \max[v_1(1-x_1), \ldots, v_n(1-x_n)]$$

At andness 1 and 0, AIWA becomes the implicative weighted conjunction and weighted disjunction respectively, based on Reichenbach implication. At andness ½, the AIWA aggregator becomes the weighted arithmetic mean. In the case of equal weights $v_1 = \cdots = v_n = 1$ the AIWA aggregator becomes the power mean. AIWA satisfies De Morgan duality and can be written as follows:

$$h(\mathbf{x}; \mathbf{v}, a) = \begin{cases} \lim_{r \to (1-a)/a} \left(\dfrac{\sum_{i=1}^n (v_i x_i)^r}{\sum_{i=1}^n v_i^r} \right)^{1/r}, & 0 \le a \le \frac{1}{2} \\ 1 - h(\mathbf{1} - \mathbf{x}; \mathbf{v}, 1 - a), & \frac{1}{2} \le a \le 1 \end{cases}$$

Since $r = a^{-1} - 1 \ge 1$, numeric computations are always performed for partial disjunction and $r \ge 1$. Similarly to the exponential mean, and OWA, the AIWA partial disjunction is a *soft aggregator*, and because of De Morgan duality the corresponding partial conjunction is also soft. Therefore, AIWA cannot be used in cases where we need hard aggregators. In addition, the implicative weighted conjunction and weighted disjunction based on Reichenbach implication are also soft. More precisely, they are hard in the case of the most important input and soft in the case of all less important inputs. In that sense they seem to be more related to partial absorption than to conjunction or disjunction. It is not an observable human property that at the highest level of simultaneity or substitutability the slightest inequality of weights deprives an input of the right to be mandatory or sufficient. This AIWA property is inconsistent with the fact that the hard partial conjunction/disjunction can be observed in human reasoning already at the level of andness/orness below 75% [11].

In the case of logic aggregators, the output of GCD must be a monotonically decreasing function of andness. Indeed, any increase of andness increases the penalty for insufficient simultaneity and the output suitability should monotonically decrease towards the minimum input value. However, as shown in [13], for some combinations of inputs and weights AIWA output monotonically decreases when we increase andness, but for other combinations that is not the case. Consequently, in a general case, AIWA does not have the formal status of logic aggregator.

7 Conclusions

Implicative weights have attractive property that weights can be interpreted as *independent logic variables*. Consequently, both semantic and formal logic components of evaluation criteria can be expressed using soft computing logic functions. Implicative weights clearly support the requirement that it is not acceptable if an input argument is important and insufficiently satisfied.

The presented analysis of implicative weights and weighted conjunction shows that implicative weights and weighted conjunction/disjunction currently provide a limited applicability in graded evaluation logic. The main problems we found with implicative weights and related aggregators in applications based on graded logic criteria are the following:

(1) The weight-dominated aggregation where, at the extreme levels of andness, instead of fading out, the weights can become a dominant contributor to the result of aggregation. This property prevents compatibility with classic Boolean logic and interpolative connection with hyperconjunction/hyperdisjunction.
(2) There are significant differences of properties of aggregators in the case of equal weights, and the same aggregators in the case of different weights.
(3) Significant regions of insensitivity to variations of input values reduce both the compensativeness of aggregators and the desirable (strict) monotonicity.
(4) The absence of homogeneous annihilators: regardless of andness, the power of annihilation is reserved only for the most important input.
(5) Insufficient monotonicity of aggregation results with respect to andness/orness.
(6) Predominantly soft aggregation, insufficient for modeling all observable properties of human evaluation reasoning.

Implicative weights and related aggregators seem to currently have a modest potential for applicability in modeling evaluation criteria. Taking into account the variety of existing implication functions, and the variety of simultaneity models, it is always necessary to match these two components to find the most convenient combination for each particular application area. We presented an example of good match, but generally, that can be a significant effort.

The impact of weights as indicators of relative importance is the highest in the vicinity of the medium andness $\alpha = \frac{1}{2}$, which corresponds to the arithmetic mean. If the andness increases and approaches 1, the impact of weights decreases. Sooner or later, it must disappear. If a high andness $(\alpha \geq 1)$ eliminates the impact of weights, it is reasonable to claim that such aggregators are andness-dominated. This is the property of human reasoning and the property of multiplicative weights. In the case of implicative weights the situation is different: the models of simultaneity are weight-dominated.

The results of this paper indicate the direction of future work. First, we need a systematic analysis of various families of implication functions as providers of implicative weights in models of simultaneity and substitutability used in evaluation criteria. Second, we need additional study of relationships between andness-dominance and weight-dominance. Third, the goal of logic is to provide justifiable models of human reasoning, and the validity of such models is not a mathematical problem, but a problem that can only be solved using experiments with human subjects. At this time, that area of research seems to be fully neglected.

References

1. Yager, R.R.: An approach to inference in approximate reasoning. Int. J. Man-Mach. Stud. **12**, 323–338 (1980)
2. Yager, R.R.: On some new classes of implication operators and their role in approximate reasoning. Inf. Sci. **167**, 193–216 (2004)
3. Larsen, H.L.: Efficient andness-directed importance weighted averaging operators. Int. J. Uncertain. Fuzziness Knowl. Based Syst. **12**(Suppl.), 67–82 (2003)
4. Larsen, H.L.: Multiplicative and implicative importance weighted averaging aggregation operators with accurate andness direction. In: Carvalho, J.P., Dubois, D., Kaymak, U., da Costa Sousa, J.M. (eds.) Proceedings of the Joint 2009 International Fuzzy Systems Association World Congress and 2009 European Society of Fuzzy Logic and Technology Conference, Lisbon, Portugal, pp. 402–407 (2009)
5. Dujmović, J., Larsen, H.L.: Generalized conjunction/disjunction. Int. J. Approx. Reason. **46**, 423–446 (2007)
6. Fodor, J., Roubens, M.: Fuzzy Preference Modelling and Multicriteria Decision Support. Kluwer Academic Publishers, Dordrecht (1994)
7. Baczyński, M., Jayaram, B.: Fuzzy Implications. Studies in Fuzziness and Soft Computing, vol. 231. Springer, Heidelberg (2008). https://doi.org/10.1007/978-3-540-69082-5
8. Baczynski, M., Jayaram, B.: Yager's classes of fuzzy implications: some properties and intersections. Kybernetika **43**(2), 157–182 (2007)
9. Massanet, S., Pradera, A., Ruiz-Aguilera, D., Torrens, J.: From three to one: equivalence and characterization of material implications derived from co-copulas, probabilistic S-implications and survival S-implications. Fuzzy Sets Syst. **323**, 103–116 (2017)
10. Baczynski, M., Jayaram, B., Massanet, S., Torrens, J.: Fuzzy implications: past, present, and future. In: Kacprzyk, J., Pedrycz, W. (eds.) Springer Handbook of Computational Intelligence, pp. 183–202. Springer, Heidelberg (2015). https://doi.org/10.1007/978-3-662-43505-2_12

11. Dujmović, J.: Weighted compensative logic with adjustable threshold andness and orness. IEEE Trans. Fuzzy Syst. **23**(2), 270–290 (2015)
12. Dujmović, J.: Partial absorption function. J. Univ. Belgrade EE Dept. Ser. Math. Phys. **659**, 156–163 (1979)
13. Dujmović, J.: Soft Computing Evaluation Logic. Wiley/IEEE (2018)

Balancing Assembly Lines and Matching Demand Through Worker Reallocations

Randall Mauricio Pérez-Wheelock$^{(\boxtimes)}$ and Van-Nam Huynh

Japan Advanced Institute of Science and Technology,
Nomi, Ishikawa 923-1292, Japan
{randall.m.perez,huynh}@jaist.ac.jp

Abstract. Assembly lines are of great importance in most actual production systems and thus continue attracting strong research interest. We address a real industry scenario where the aim of the line is to target a production output that meets, as much as possible, a given demand forecast. To the best of our knowledge, the existing literature has not tackled this problem, and we named it *the demand-driven assembly line (re)balancing problem*. A mixed integer programming model is developed, solved using genetic algorithm, and tested in the straight assembly line, providing useful insights about the dynamics of worker reallocations.

1 Introduction

The extensive literature on assembly line balancing (ALB) has focused on maximizing line efficiency, overlooking strategic use or neglecting the organization's overall operations effectiveness [21]. Wilson [19,20] argues that Ford's assembly lines were optimized both 'locally' as individual production systems; and also 'globally' as constituent sub-systems of Ford's larger, vertically integrated supply chain system. Wilson [21] also reveals with data that, in fact, Ford's operations were adaptable to strongly increasing and highly variable demand.

Needless to say, the importance of matching supply with demand is universally recognized. However, to our surprise, demand fluctuations still have not been explicitly considered to perform task assignment and/or worker allocation (to stations) in the assembly line balancing problem (ALBP).

This paper introduces *the demand-driven assembly line (re)balancing problem* (DDALBP). The proposed model aims to balance and rebalance an assembly system over a planning period, adjusting the production output of the line as much as possible to the forecast market demand, by means of worker allocation and reallocation. The model also aims to achieve smooth production flow and considers learning and forgetting effects.

In 2007, Miralles et al. [8] introduced *the assembly line worker assignment and balancing problem* (ALWABP). The ALWABP appears in real assembly lines when not all workers can execute all tasks, and the operation time of each task is different depending upon who executes the task. Traditionally, the aim in the ALBP has been the assignment of tasks to stations. In the ALWABP there is a

© Springer Nature Switzerland AG 2018
V. Torra et al. (Eds.): MDAI 2018, LNAI 11144, pp. 206–217, 2018.
https://doi.org/10.1007/978-3-030-00202-2_17

double assignment: (1) tasks to stations and (2) workers to stations. The authors presented the mathematical model for the ALWABP and a case study based on a Spanish sheltered workcenter for disabled workers.

Different methods have been proposed for solving the ALWABP; for instance: hybrid method Clustering Search [4,5], branch-and-bound [3,9,18], beam search [2], a constructive heuristic framework with priority rules [10], an iterative genetic algorithm [13], a multi-objective evolutionary algorithm [22].

In 2015, Moreira et al. [11] introduced *the assembly line worker integration and balancing problem* (ALWIBP). The scenario seen in the ALWIBP is similar to the ordinary company, where only few disabled workers have to be integrated. The authors presented mathematical models and heuristic methodologies to solve the problem. Moreira et al. [12] proposed the use of Miltenburg's regularity criterion to evenly distribute workers with special characteristics along the line, as well as two fast heuristics for the assignment of tasks and workers to stations.

More recently, Stall-Sikora et al. [16] introduced *the traveling worker assembly line (re)balancing problem* (TWALBP). This new problem variation arises if workers can be assigned to more than one station. Hence, workers are able to move between stations, allowing them to perform tasks from different regions of the precedence diagram. Each worker limits the cycle time by the sum of the processing times of the tasks assigned to him/her and his/her movement times between stations. The authors presented a mixed integer programming (MIP) model with a traveling salesman problem (TSP) formulation integrated in the balancing model to solve the problem.

All of these works deal interestingly and cleverly with the allocation of workers and tasks to stations. They aim mainly at minimizing the cycle time (i.e., maximizing throughput), regardless of the production output that is actually required. Also, task times are deterministic and differ depending on who executes the task. In our modest effort, we intend to offer a balancing model that differs in two fundamental ways: (1) The allocation of workers is driven by a forecast market demand. (2) Task times are dynamic since our model considers learning and forgetting effects.

The rest of this paper is organized as follows: Sect. 2 describes our problem and formulates it mathematically; numerical experiments are presented and discussed in Sect. 3; and Sect. 4 contains some brief conclusions.

2 Mathematical Formulation

2.1 Input Description

We consider an assembly line (AL) composed of a specific number of stations $j = 1, 2, 3, \ldots, J$, organized in straight layout. Each station is equipped with specific tooling and equipment in order to provide the station the required functionality to execute the particular subset of tasks assigned to it. Progressively, on each station, the bill-of-material parts and components are attached to the jobs or workpieces, which become finished products at the end of the line.

In the factory, there is a fixed number of workers, K, which may (or may not) have an initial skill inventory, $S_{jk}^{initial}$, the number of units that worker k would be able to process (in one working period) in his/her first assignment to station j. This capacity improves as long as the worker continues performing on the same station; otherwise, the skill level deteriorates.

The jobs processed at one station are put on a buffer, and these jobs become input for the next station. Each station must guarantee a minimum work-in-process inventory, WIP_{min}, for its downstream station at the end of each working period in order to ensure smooth production flow at the beginning of the next working period (i.e., avoid the waiting time of feeding the line).

By assigning workers to stations, the AL should be balanced in the best possible way on each planning period $\ell = 0, 1, 2, \ldots, L-1$ to satisfy $D(\ell + 1)$, the forecast market demand of the next period, $\ell + 1$.

2.2   Incorporating Learning and Forgetting Effects

We incorporate the formulations for skill improvement and skill deterioration proposed by Azizi et al. [1]. On the one hand, when a worker is assigned to a station, his/her skill improves as he/she performs in the same station. Skill improvement can be modeled with:

$$S_{jk\ell} = S_j^{max} - \left(S_j^{max} - S_{jka}^{rem}\right) e^{\beta_k (\ell - a)} \tag{1}$$

where $S_{jk\ell}$ is the skill level of worker k in station j on period ℓ, S_j^{max} is the theoretical maximum level of skill at station j, S_{jka}^{rem} is the skill level that worker k had in station j when he/she was assigned to that station (on period a). At time zero, however, S_{jk0}^{rem} corresponds to the worker's initial skill level, $S_{jk0}^{rem} = S_{jk}^{initial}$. If a worker is assigned to a station by the first time, then S_{jka}^{rem} in (1) must be substituted by $S_{jk}^{initial}$. β_k is the learning slope of worker k given by $\beta_k = (log\ r_k)/(log\ 2)$, where r_k is the learning coefficient of worker k.

On the other hand, as the worker continues to learn the new skill, his/her previously gained skill decays as a result of the forgetting phenomenon. The corresponding skill deterioration formula is:

$$S_{jk\ell}^{rem} = S_j^{min} + \left(S_{jkd} - S_j^{min}\right) e^{\gamma_k (\ell - d)} \tag{2}$$

where $S_{jk\ell}^{rem}$ is the remnant skill of worker k in station j on period ℓ; S_j^{min} is the theoretical minimum level of skill at station j; S_{jkd} is the skill level that worker k had in station j when he/she departed last time (on period d) from that station; and γ_k is the forgetting slope of worker k given by $\gamma_k = (log\ f_k)/(log\ 2)$, where f_k is the forgetting coefficient of worker k.

According to (1) and (2), at infinite time, the skill level reaches, respectively, the maximum and the minimum levels: $S_{jk\infty} = S_j^{max}$ and $S_{jk\infty}^{rem} = S_j^{min}$. However, achieving the maximum or the minimum level of skill in infinite time is unrealistic. Therefore, the concepts *skill upper bound* and *skill lower bound* are introduced: S_j^{UB} and S_j^{LB}. Their relationship with S_j^{max} and S_j^{min} is $S_j^{UB} = S_j^{max} - \delta_j$ and $S_j^{LB} = S_j^{min} + \epsilon_j$, respectively, where δ_j and ϵ_j are the skill upper bound threshold value and the skill lower bound threshold value for station j.

2.3   Objective Functions

The number of units that a worker can process is given by $S_{jk\ell}$, as descried in the previous section. The *theoretical* number of units that station j can process on period ℓ, $P(j, \ell)$, depends on the worker who will be assigned to that station; i.e., $P(j, \ell) = x_{jk\ell} \cdot S_{jk\ell}$, where $x_{jk\ell}$ is a binary decision variable that equals 1 if worker k is assigned to station j on period ℓ; otherwise it equals 0. $Q(j, \ell)$ is the *actual* number of units processed by station j on period ℓ based (not only on the worker assigned there, but also) on the number of units that the upstream station is feeding, and the station's available WIP inventory.

The essential goal in the DDALBP is to match the production output of the AL to the forecast market demand. Therefore, in the ideal case, the actual number of units produced by the last station on any given period should match the forecast demand of the next period, $Q(J, \ell) = D(\ell + 1)$. Therefore, our first objective function (OF) can be expressed as:

$$\text{Min } Z_1 = \sum_{\ell=0}^{L-1} |D(\ell + 1) - Q(J, \ell)| \tag{3}$$

If we distinguish between loosing sales, $D(\ell + 1) > Q(J, \ell)$, and building inventory, $Q(J, \ell) > D(\ell + 1)$, this objective function can be reformulated as:

$$\text{Min } Z_1 = g \sum_{\ell=0}^{L-1} \text{Max}\,\{0, D(\ell+1) - Q(J, \ell)\} + h \sum_{\ell=0}^{L-1} \text{Max}\,\{0, Q(J, \ell) - D(\ell+1)\} \tag{4}$$

where g is the unit cost of lost sales and h is the unit cost of holding inventory.

In addition to meet demand, the solution to the DDALBP also aims to achieve the smoothest possible production flow. The following two proposed OFs are modified from Song et al. [15]. If $\overline{Q}$ represents the average number of units processed among all stations, then, our second OF can be written as:

$$\text{Min } Z_2 = \sum_{\ell=0}^{L-1} \sqrt{\frac{1}{J} \sum_{j=1}^{J} \left[Q(j, \ell) - \overline{Q}(\ell) \right]^2} \tag{5}$$

The number of units produced by the bottleneck station is $Q(bn) = min\,\{Q(1), \ldots, Q(J)\}$. The production waste of station j on period ℓ is defined as $Q_w(j, \ell) = P(j, \ell) - Q(bn, \ell)$. The total production waste (of the whole AL) on period ℓ is $Q_w(\ell) = \sum_{j=1}^{J} [P(j, \ell) - Q(bn, \ell)]$, and our third OF can be written as:

$$\text{Min } Z_3 = \sum_{\ell=0}^{L-1} \sum_{j=1}^{J} [P(j, \ell) - Q(bn, \ell)] \tag{6}$$

2.4   Restrictions

The following restrictions are related to the number of workers: Constraint (7) indicates that on each period, every worker is assigned to one station. Constraint

(8) indicates that on each period, each station receives exactly one worker. Constraint (9) indicates that on each period, the sum of workers assigned along the different stations cannot exceed the number of workers available in the factory.

$$\sum_{j=1}^{J} x_{jk\ell} = 1 \quad \forall \, k \in K, \; \ell = 0, 1, \ldots, L-1 \tag{7}$$

$$\sum_{k=1}^{K} x_{jk\ell} = 1 \quad \forall \, j \in J, \; \ell = 0, 1, \ldots, L-1 \tag{8}$$

$$\sum_{j=1}^{J} \sum_{k=1}^{K} x_{jk\ell} \leq K \quad \ell = 0, 1, \ldots, L-1 \tag{9}$$

Constraint (10) links the worker allocation decision to the theoretical number of units produced.

$$P(j, \ell) = \sum_{k=1}^{K} (x_{jk\ell} \cdot S_{jk\ell}) \quad \forall \, j \in J, \; \ell = 0, 1, \ldots, L-1 \tag{10}$$

The following restrictions regulate the actual number of units processed by the stations: Constraint (11) indicates that the actual number of units processed by station 1 equals its own theoretical number of units processed. Constraints (12) and (13) compute the number of units processed by all other stations on period 0 (when there is no WIP inventory), and on subsequent periods (when there may exist WIP inventory), respectively. Constraint (14) stipulates that, on any given period, the actual number of units processed by the last station (J) must satisfy the forecast market demand of the next period.

$$Q(j, \ell) = P(j, \ell) \quad j = 1, \; \ell = 0, \ldots, L-1 \tag{11}$$

$$Q(j, \ell) = \min \{P(j, \ell), Q(j-1, \ell)\} \quad j = 2, \ldots, J, \; \ell = 0 \tag{12}$$

$$Q(j, \ell) = \min \{P(j, \ell), Q(j-1, \ell) + WIP(j, \ell-1)\} \quad j = 2, \ldots, J, \; \ell = 1, \ldots, L-1 \tag{13}$$

$$Q(J, \ell) \geq D(\ell+1) \quad \ell = 0, \ldots, L-1 \tag{14}$$

The following restrictions control the amount of WIP inventory: Constraints (15) and (16) compute the WIP inventory that remains at each station at the end of period 0, and at the end of subsequent periods, respectively. Constraint (17) requires the WIP inventory at the stations to be at least the minimum necessary to ensure immediate work at the beginning of each working period. Period 0 is excluded from this WIP constraint because on period 0 the AL is empty; there is no WIP inventory at all. Station 1 is excluded from these three WIP constraints because it is not fed by WIP inventory from a previous station; instead, it is fed by raw materials. Finally, (18) shows the binary restriction of the decision variables.

$$WIP(j,\ell) = Q(j-1,\ell) - Q(j,\ell) \quad j = 2,\ldots,J, \ \ell = 0 \tag{15}$$

$$WIP(j,\ell) = WIP(j,\ell-1) + Q(j-1,\ell) - Q(j,\ell) \quad j = 2,\ldots,J, \ \ell = 1,\ldots,L-1 \tag{16}$$

$$WIP(j,\ell) \geq WIP_{min} \quad j = 2,\ldots,J, \ \ell = 1,\ldots,L-1 \tag{17}$$

$$x_{jk\ell} \in \{0,1\} \tag{18}$$

3 Numerical Experiments

We now demonstrate the use of the developed MIP formulation, solved with genetic algorithm, in the Jackson problem. The data of this problem comes from a benchmark data set, a collection of *simple assembly line balancing problems* (SALBP) that appears in Scholl [14].

3.1 Experimental Design

The Jackson problem consists of 11 tasks. For the purpose of assuming an installed AL, we considered a hypothetical scenario in which 30 units need to be produced each day in 480 min of available productive time per day. Hence, we designed 3 stations arranged in straight layout: The cycle time is $CT = 480$ min $\div$ 30 units $= 16$ min/unit. The minimum number of stations is computed by dividing the total task time (46 min) by CT, $\lceil 46 \div 16 \rceil = 3$ stations.

The distribution of the 11 tasks into the 3 stations is shown in the first two columns of Table 1. The third column is the station load (SL), the sum of the times of the tasks assigned to each station. The fourth column is the station utilization, $SU = SL \div CT$. The fifth column computes the theoretical maximum number of units that can be processed at each station within the available productive time. For instance, for a station load of 16 min, 30 units can be processed within the available time (480 min). These figures in the fifth column correspond to the definition of S_j^{max} discussed in Sect. 2.2. The last column shows the S_j^{min} values, which were set at 0 for the 3 stations.

A total of 32 experiments were designed to test our MIP formulation in the straight AL layout designed for the Jackson problem. Four cases were obtained by combining low and high values of the learning and forgetting coefficients. These four cases were run under two scenarios: (a) workers have some initial skill inventory ($S^{initial}$ values were obtained randomly in $unif(4,9)$), and (b) workers have no initial skill inventory ($S^{initial} = 0$ for all workers). These two scenarios already generate 8 cases, which were solved with two different targets: (i) Match demand on individual periods: this is the case when g and h have equal values (Eq. 3). The model run was (3), (5)–(18). (ii) Match demand over the whole planning horizon: g and h are differentiated (Eq. 4). We assumed that

Table 1. Pre-processing of the Jackson problem

Station	Tasks	SL (min)	SU (%)	S_j^{max}	S_j^{min}
$j = 1$	1, 2, 4, 5	16	100	30.00	0
$j = 2$	3, 6, 7, 8	16	100	30.00	0
$j = 3$	9, 10, 11	14	88	34.29	0

Table 2. Deployment of cases run

Group 1: Increasing demand pattern				
Case	Target	Skill inventory	r	f
1	Match demand	Workers have	low	high
2	on individual	some initial	low	low
3	periods.	skill inventory.	high	high
4			high	low
5		$S^{initial} = 0$ for	low	high
6		all workers.	low	low
7			high	high
8			high	low
9	Match demand	Workers have	low	high
10	over the whole	some initial	low	low
11	planning horizon.	skill inventory.	high	high
12			high	low
13		$S^{initial} = 0$ for	low	high
14		all workers.	low	low
15			high	high
16			high	low

the opportunity cost is more costly than holding inventory. The model run was (4)–(18). With these two targets under consideration, 16 cases are generated so far. This deployment description of the cases run is illustrated in Table 2 for the first main group of cases, *increasing demand pattern*. The same deployment was used for the second main group of cases, *erratic demand pattern*.

3.2 Results

The matching results of the AL production output to the forecast market demand, for the first main group (cases 1 to 16), are shown in Fig. 1. The orange trendline shows the increasing demand forecast on period $\ell + 1$ (same forecast for all 16 cases), and the lightblue vertical bar, the production output achieved on period ℓ. So, in these graphs, the production output $Q(J, \ell)$ is plotted against the forecast demand of the next period, $D(\ell + 1)$.

From the figure we can see that demand is well matched in cases 1 and 2, where workers have fast learning (low r values). However, in cases 3 and 4, it was not possible to fully match the demand forecast of periods 8, 9, and 10 (in spite of keeping the same worker allocation since periods 4 and 3, respectively) due to slow learning (high r values). We may think of the possibility of having the same worker allocation since earlier periods (e.g., since period 2), in order to take advantage of the learning effect, and being able to better match the demand of periods 8, 9, and 10. However, this action would have led to building inventory in earlier periods, which is penalized by Z_1; and the sum of its both components (penalties for building inventory in earlier periods plus penalties for loosing sales in later periods) yield a worse value for Z_1.

Contrasting cases 3 and 4 vs. cases 11 and 12 lead to an interesting observation. Cases 11 and 12 are identical to cases 3 and 4, expect for the fact that loosing sales is more highly penalized than building inventory. Hence, cases 11 and 12 do keep the same worker allocation since earlier periods and build inventory (with small penalty) in order to better match the increasing demand of later periods (because not achieving the production target has higher penalty).

Demand is better matched in cases 9 and 10 (fast learning) than in cases 11 and 12 (slow learning).

Cases 5 to 8 are identical to cases 1 to 4, except for the fact that workers have no initial skill inventory. Due to the increasing demand, and consequently, the need to speed up production, less worker reallocation takes place. Worker reallocation is measured as the number of times that a worker changes station along a planning horizon. For instance, in Table 3, which shows the worker allo-

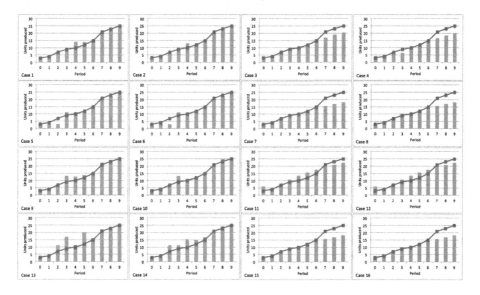

Fig. 1. Increasing demand pattern (orange) vs. production output achieved (lightblue). (Color figure online)

Table 3. Worker allocation and reallocation over the planning period, case 1

Worker	0	1	2	3	4	5	6	7	8	9	Number of reallocations
$k = 1$	3	1	1	2	2	2	3	1	2	2	5
$k = 2$	1	3	2	1	1	3	2	2	1	1	6
$k = 3$	2	2	3	3	3	1	1	3	3	3	3

cation solution for case 1, we can see that worker 3 changed station 3 times (on period 2 he/she changed to station 3; then, on period 5, to station 1; and on period 7, to station 3; totaling 3 reassignments).

Cases 13 to 16 are the last set of cases in this first main group. Similar to cases 5 to 8, $S^{initial} = 0$. Also, g is higher than h, like in cases 9 to 12. Because there is no initial skill inventory, workers tend to remain in the same station. Taking advantage of the learning effect is the preferable course of action since building inventory is less penalized than loosing sales. In cases 15 and 16, it is not possible to match the increasing demand of the last three periods due to slower learning (compared to cases 13 and 14).

Figure 2 presents the matching results for the second main group (cases 17 to 32). Some cases are discussed: Cases 17 and 18 are interestingly well matched from period 3. Such matching was not possible in cases 19 and 20 due to slower learning and faster forgetting. These four cases (workers have some initial skill inventory) present a higher number of reallocations than the next four cases (no initial skill inventory). In fact, in cases 23 and 24 workers remain in the same station along the whole planning period.

Case 26 shows an interesting phenomenon. On period 4, a different worker allocation could have been devised so as to produce less units (like in case 25), and then, on period 5, return to the previous allocation (same assignment of period 3). This action was not taken because it would have been very costly on period 5. Because in case 26 forgetting occur faster (than in case 25), on period 5 it would had not been possible to match demand. So, it was preferred to accumulate inventory (on period 4) because it cost less than loosing sales (on period 5).

One final point of discussion is the following: Which workers are reallocated more frequently? In our small experiment, with three workers, which we can label as the "best" worker (i.e., fast learning and slow forgetting), the "average" worker, and the worst worker (i.e., slow learning and fast forgetting), swap of workers occurred more frequently between the "best" and the "average" workers. The "worst" worker tended to have less number of reallocations (under both increasing and erratic demand patterns), in order to take advantage of the learning effect.

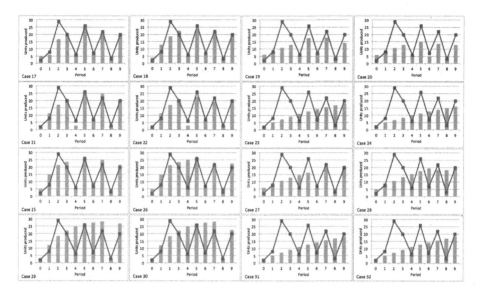

Fig. 2. Erratic demand pattern (orange) vs. production output achieved (lightblue). (Color figure online)

3.3 Further Discussion

Contrary to the well-known principle "divide and conquer", conquering new research frontiers in ALB is exactly about doing the opposite: it is about simultaneously considering branches that are connected to the problem, and about considering more reality. This paper already considered the demand forecast in the ALBP. This is only one connecting branch. Recently, Sternatz [17] introduced *the joint line balancing and material supply problem*. Materials supply to the AL is another connecting branch. So, it looks like a new trend in this research field is to connect inbound and outbound logistics issues to the ALBP. In the past, materials requirement planning, ALBP, and other problems had been addressed separately. Probably, the reason that explains the new trend to formulate a bigger problem is the fact that addressing the problem as a whole yields a better solution, closer to optimum, than dividing the problem into sub-problems and solving them separately and independently. In this regard, future research efforts should be addressed to holistically optimize the work system.

Another way to extend the research frontier in this field is by considering additional reality to improve the accuracy of the models. For instance, the processing rate at which workers process units is affected not only by learning-forgetting effects. Fatigue-recovery and motivation-boredom are realities that certainly affect the processing times. Givi et al. [7] considered fatigue-recovery and learning-forgetting parameters to develop a mathematical model that estimates the human error rate when performing an assembly job. In addition to the error rate, it would be desirable to estimate the processing time. Azizi et al. [1] developed motivation-boredom and skill improvement-deterioration formulations to model job rotation in a manufacturing system. Corominas et al. [6] addressed an important element of reality: if task A and task B are similar, then, when a worker performs task A, he/she also gains experience on task B –even though he/she had not performed task B before. Moreover, linked to our model and the final point of discussion in the previous section, it would be interesting to study if the higher frequency of reallocations of the "best" and "average" workers could lead to the development of more stress, fatigue, and a possible sick leave. Simultaneously considering these realities would lead to more accurate values of the processing times of tasks, crucial data for the ALBP.

4   Conclusions

Different from other methods, which focus on maximizing the efficiency of the lines without regard to their role within the extended supply chain, this paper aimed at balancing the AL in such a way that the production output meets a given demand forecast. In particular, this paper introduced the DDALBP, and proposed a formulation to balance the line, in accordance to some forecast market demand, by means of worker allocation and reallocation. Our formulation was tested in the Jackson problem, arranged in straight layout, obtaining useful insights in regard to the behavior of worker reallocations under different scenarios (low/high learning coefficients, low/high forgetting coefficients, workers with/without initial skill inventory, increasing/erratic demand patterns).

Regardless of the physical layout of the line, the presence of some parallel stations, or the presence of feeder lines, a production system is comprised of J stations, which, independently of their physical location within the factory, the system should be balanced in such a way so as to achieve smooth production flow, and match as much as possible the forecast market demand. Therefore, the review and extension of this model, so as to be capable of being applicable to different line layouts and configurations, is already in our next research plan.

Acknowledgment. The authors thank Ou Wei, PhD candidate at JAIST, for the technical assistance provided to run numerical experiments.

References

1. Azizi, N., Zolfaghari, S., Liang, M.: Modeling job rotation in manufacturing systems: the study of employee's boredom and skill variations. Int. J. Prod. Econ. **123**, 69–85 (2010)
2. Blum, C., Miralles, C.: On solving the assembly line worker assignment and balancing problem via beam search. Comput. Oper. Res. **38**, 328–339 (2011)
3. Borba, L., Ritt, M.: A heuristic and a branch-and-bound algorithm for the assembly line worker assignment and balancing problem. Comput. Oper. Res. **45**, 87–96 (2014)
4. Chaves, A.A., Miralles, C., Nogueira Lorena, L.A.: Clustering search approach for the assembly line worker assignment and balancing problem. In: Elwany, M.H., Eltawil, A.B. (eds.) Proceedings of the 37th International Conference on Computers and Industrial Engineering, pp. 1469–1478 (2007)

5. Chaves, A.A., Nogueira Lorena, L.A., Miralles, C.: Hybrid metaheuristic for the assembly line worker assignment and balancing problem. In: Blesa, M.J., Blum, C., Di Gaspero, L., Roli, A., Sampels, M., Schaerf, A. (eds.) Proceedings of the 6th International Workshop on Hybrid Metaheuristics. LNCS, vol. 5818, pp. 1–14. Springer, Heidelberg (2009). https://doi.org/10.1007/978-3-642-04918-7_1
6. Corominas, A., Olivella, J., Pastor, R.: A model for the assignment of a set of tasks when work performance depends on experience of all tasks involved. Int. J. Prod. Econ. **126**, 335–340 (2010)
7. Givi, Z.S., Jaber, M.Y., Neumann, W.P.: Modelling worker reliability with learning and fatigue. Appl. Math. Model. **39**, 5186–5199 (2015)
8. Miralles, C., García-Sabater, J.P., Andrés, C., Cardós, M.: Advantages of assembly lines in sheltered work centres for disabled: a case study. Int. J. Prod. Econ. **110**, 187–197 (2007)
9. Miralles, C., García-Sabater, J.P., Andrés, C., Cardós, M.: Branch and bound procedures for solving the assembly line worker assignment and balancing problem: application to sheltered work centres for disabled. Discret. Appl. Math. **156**(3), 352–367 (2008)
10. Moreira, M.C.O., Ritt, M., Costa, A.M., Chaves, A.A.: Simple heuristics for the assembly line worker assignment and balancing problem. J. Heuristics **18**, 505–524 (2012)
11. Moreira, M.C.O., Miralles, C., Costa, A.M.: Model and heuristics for the assembly line worker integration and balancing problem. Comput. Oper. Res. **54**, 64–73 (2015)
12. Moreira, M.C.O., Pastor, R., Costa, A.M., Miralles, C.: The multi-objective assembly line worker integration and balancing problem of type-2. Comput. Oper. Res. **82**, 114–125 (2017)
13. Mutlu, Ö., Polat, O., Supciller, A.A.: An iterative genetic algorithm for the assembly line worker assignment and balancing problem of type-II. Comput. Oper. Res. **40**, 418–426 (2013)
14. Scholl, A.: Data of assembly line balancing problems. Technical report 16/93, Schriften zur Quantitativen Betriebswirtschaftslehre, Technische Universität Darmstadt, November 1993
15. Song, B.L., Wong, W.K., Fan, J.T., Chan, S.F.: A recursive operator allocation approach for assembly line-balancing optimization problem with the consideration of operator efficiency. Comput. Ind. Eng. **51**, 585–608 (2006)
16. Stall Sikora, C.G., Cantos Lopes, T., Magatão, L.: Traveling worker assembly line (re)balancing problem: Model, reduction techniques, and real case studies. Eur. J. Oper. Res. **259**, 949–971 (2017)
17. Sternatz, J.: The joint line balancing and material supply problem. Int. J. Prod. Econ. **159**, 304–318 (2015)
18. Vilà, M., Pereira, J.: A branch-and-bound algorithm for assembly line worker assignment and balancing problems. Comput. Oper. Res. **44**, 105–114 (2014)
19. Wilson, J.M.: Henry Ford's just-in-time system. Int. J. Oper. Prod. Manag. **15**(12), 59–75 (1995)
20. Wilson, J.M.: Henry Ford: a just-in-time pioneer. Prod. Inven. Manag. J. **36**(2), 26–31 (1996)
21. Wilson, J.M.: Henry Ford vs. assembly line balancing. Int. J. Prod. Res. **52**(3), 757–765 (2014)
22. Zacharia, P.T., Nearchou, A.C.: A population-based algorithm for the bi-objective assembly line worker assignment and balancing problem. Eng. Appl. Artif. Intell. **49**, 1–9 (2016)

Clustering and Classification

Optimal Clustering with Twofold Memberships

Sadaaki Miyamoto[1]([✉]), Jong Moon Choi[1], Yasunori Endo[1],
and Van Nam Huynh[2]

[1] University of Tsukuba, Tsukuba, Japan
`miyamoto.sadaaki.fu@u.tsukuba.ac.jp`
[2] Japan Advanced Institute of Science and Technology, Nomi, Japan
`huynh@jaist.ac.jp`

Abstract. This paper proposes two clustering algorithms of twofold memberships for each cluster. One uses a membership similar to that in K-means, while another membership is defined for a core of a cluster, which is compared to the lower approximation of a cluster in rough K-means. Two ideas for the lower approximation are proposed in this paper: one uses a neighborhood of a cluster boundary and another uses a simple circle from a cluster center. By using the two memberships, two alternate optimization algorithms are proposed. Numerical examples show the effectiveness of the proposed algorithms.

Keywords: Neighborhood · Clustering · K-means · Rough K-means
Twofold memberships

1 Introduction

With the progress of data mining techniques, various methods of data clustering have been studied, among which the K-means [7] are best-known and its fuzzy version of fuzzy c-means [1] is also popular. Still another method of rough K-means [6] has been proposed. Rough K-means has two memberships of upper and lower approximations of clusters, and its usefulness has been empirically discussed using examples. A drawback of rough K-means is that it is not based on optimization of an objective function, unlike K-means and fuzzy c-means. Kinoshita *et al.* [4] showed difficulty and possibility to handle rough K-means within the framework of optimization. In contrast, a major part of methodological considerations of clustering algorithms is based on the formulation of optimization problems. Hence to study an optimization algorithm similar to rough K-means is methodologically important.

This paper proposes two algorithms of twofold memberships: two memberships are defined to each cluster. One is a membership like that in K-means. Another membership is defined for a core of a cluster, which is similar to lower approximations of clusters. Two ideas for the lower approximation are considered in this paper: one uses neighborhoods and another uses a circle from a cluster

V. Torra et al. (Eds.): MDAI 2018, LNAI 11144, pp. 221–231, 2018.
https://doi.org/10.1007/978-3-030-00202-2_18

center. An upper approximation is not used, since the usefulness of an upper approximation in clustering is not clear. Two alternate optimization algorithms for the two methods are proposed.

The rest of this paper is organized as follows. Section 2 introduces notations and the formulation of rough K-means. Section 3 then proposes new algorithms for the two memberships. Section 4 shows artificial and real examples, and finally Sect. 5 concludes the paper.

2 K-Means and Rough K-Means

Clustering using the concept of rough sets [9] has been studied by several researchers [4,6,11]. A drawback in these studies is that the region of clusters do not have a simple geometric shape, while K-means have Voronoi regions [5,8], and fuzzy c-means have simple classification functions [8].

We begin with notations and a brief description of the rough K-means algorithm [6], while K-means algorithm [1,7] is omitted to save space.

Let us assume that $X = \{x_1, \ldots, x_N\}$ is a set of objects for clustering, and x_k $(k = 1, \ldots, N)$ is a point in $\mathbf{R}^p$. Clusters are denoted by G_i $(i = 1, \ldots, c)$ which are subsets of X. G_i has the upper and lower approximations denoted by $U(G_i)$ and $L(G_i)$, respectively, which are subsets of X. $L(G_i)$ is disjoint:

$$L(G_i) \cap L(G_j) = \emptyset, \quad (i \neq j),$$

but $U(G_i)$ is not necessarily disjoint. Each object x_k belongs to some $U(G_i)$ but not necessarily to an $L(G_i)$:

$$\bigcup_{i=1}^{c} U(G_i) = X; \quad \bigcup_{i=1}^{c} L(G_i) \subseteq X.$$

Given cluster centers $(v_1, \ldots, v_c)$, a Voronoi region $V(v_i)$ is given by

$$V(v_i) = \{x \in \mathbf{R}_i : \|x - v_i\| \leq \|x - v_j\|, \forall v_j \neq v_i\}.$$

Note that $V(v_i)$ is a subset of $\mathbf{R}^p$, while $U(G_i)$ and $L(G_i)$ are subsets of X.

Rough K-means by Lingras [6] use two different weights w_1 and w_2 with $w_1 + w_2 = 1$ and $w_1 > w_2 > 0$ and calculate cluster center v_i as follows:

$$v_i = \frac{w_1}{|L(G_i)|} \sum_{x_k \in L(G_i)} x_k + \frac{w_2}{|B(G_i)|} \sum_{x_l \in B(G_i)} x_l, \tag{1}$$

where $|L(G_i)|$ and $|B(G_i)| = |U(G_i) - L(G_i)|$ are respectively a lower approximation and a rough boundary of cluster G_i; $|L(G_i)|$ is the number of objects in $L(G_i)$. Although the definitions of $L(G_i)$ and $U(G_i)$ [6] are omitted here, the upper and lower approximations are not clearly related to a rough approximation and seems *ad hoc*. Second, the method is not formulated as an alternative optimization, unlike the K-means and fuzzy c-means.

3   Algorithms with Twofold Memberships

The above consideration of a clustering algorithm leads us to an algorithm in which the calculation of cluster center is given by, or similar to the one in rough K-means, while we use the above approximation using the neighborhood

$$N_d(x) = \{y \in \mathbf{R}^p : \|y - x\| \le d\}, \tag{2}$$

assuming the Euclidean norm $\|\cdot\|$. Given a region $A \subset \mathbf{R}^p$, the lower approximation $L(A)$ using the neighborhood is

$$L(A) = A - \bigcup_{y \in \partial(A)} N_d(y), \tag{3}$$

where $\partial(A)$ is the boundary of A.

Note that the boundary of K-means clusters is that of a Voronoi region which consists of a hyperplane: $\|x - v_i\|^2 - \|x - v_j\|^2 = 0$, which is reduced to

$$L(x; v_i, v_j) = 2\langle x, v_i + v_j \rangle + \|v_j\|^2 - \|v_i\|^2 = 0. \tag{4}$$

The distance between a point y and hyperplane $L(x; v_i, v_j)$ by (4) is given by

$$Dist(y, L(x; v_i, v_j)) = \frac{|L(y; v_i, v_j)|}{2\|v_i + v_j\|} = \frac{|2\langle y, v_i + v_j \rangle + \|v_j\|^2 - \|v_i\|^2|}{2\|v_i + v_j\|}. \tag{5}$$

Another point in our algorithm is that we do not use an upper approximation of a cluster, but we use a cluster G_i itself and its lower approximation $L(G_i)$. The reason is as follows.

1. It is not difficult to find a cluster G_i and $L(G_i)$ by a K-means type algorithm, while to find an upper approximation of a cluster is more difficult.
2. An upper approximation does not seem to be useful in clustering, while a lower approximation can be used to distinguish of a point in a cluster is near to a boundary or not. The latter is also useful to make an algorithm to be more robust by changing the weight of contribution of a point.

Note that an upper approximation herein means an overlapping part of clusters, while clustering basically means to divide a set of points into disjoint subsets. Hence it is doubtful if overlapping clusters have real merits.

Therefore we consider a cluster G_i and its lower approximation $L(G_i)$, and not $U(G_i)$. In the following algorithm we find a cluster G_i:

$$G_i = V_i(v_i) \cap X \tag{6}$$

when a cluster center v_i is given. The lower approximation is given by (3):

$$L(V_i(v_i)) = \{x \in V_i(v_i) : Dist(x, L(y; v_i, v_j)) \ge d\}, \quad L(G_i) = L(V_i(v_i)) \cap X. \tag{7}$$

and the next cluster center is calculated by:

$$v_i = \frac{w_1}{|L(G_i)|} \sum_{x_k \in L(G_i)} x_k + \frac{w_2}{|G_i - L(G_i)|} \sum_{x_l \in G_i - L(G_i)} x_l. \tag{8}$$

A proposed algorithm is hence as follows.

AKM Algorithm (Another rough K-Means)
Step 1: Set initial values of v_i $(i = 1, \ldots, c)$
Step 2: Find G_i by (6) and $L(G_i)$ by (7).
Step 3: Calculate new v_i $(i = 1, \ldots, c)$ by (8).
Step 4: If cluster centers are convergent, stop; else go to **Step 2**.
End of AKM.

More specifically, the calculation of G_i is as follows:

$$x_k \in G_i \iff \|x_k - v_i\| \leq \|x_k - v_j\|, \quad \forall j \neq i. \tag{9}$$
$$x \in L(G_i) \iff x_k \in G_i \text{ and } Dist(x, L(y; v_i, v_j)) \geq d \tag{10}$$

Note also that calculation of $U(G_i)$ is not so simple as $L(G_i)$ by (10).

This algorithm actually is equivalent to an alternative optimization. Let us consider the following objective function:

$$J(\mathcal{G}, \mathcal{G}', V) = w_2 J_1(\mathcal{G}, V) + (w_1 - w_2) J_2(\mathcal{G}', V), \tag{11}$$

$$J_1(\mathcal{G}, V) = \sum_{x_k \in G_i} \|x_k - v_i\|^2, \tag{12}$$

$$J_2(\mathcal{G}', V) = \sum_{x_k \in G_i', N_d(x_k) \subseteq V(v_i)} \|x_k - v_i\|^2, \tag{13}$$

where $\mathcal{G} = \{G_1, \ldots, G_c\}$, $\mathcal{G}' = \{G_1', \ldots, G_c'\}$ are partitions of X.

Consider the alternative optimization of $J(\mathcal{G}, \mathcal{G}', V)$: then it is easy to see that we can take $\mathcal{G} = \mathcal{G}'$ and v_i is given by (8), since objects in $V(v_i)$ should be allocated to the same G_i even when J_2 is concerned.

AKM is also called **Method 1** below especially when compared with **Method 2** in the next section.

Note 1. The optimal set $\mathcal{G}'$ is not unique but there is no problem to take $\mathcal{G} = \mathcal{G}'$, since the fundamental method is the allocation to the nearest center even when the lower approximation is concerned. Note

$$L(G') = G' \cap \{x \in X \colon N_d(x_k) \subseteq V(v_i)\}.$$

3.1   Second Method with Core Regions of Circles

Another method is related to the above AKM algorithm by changing the idea of lower approximation. In contrast to AKM algorithm in which the boundary of the lower approximation $L(G_i)$ is piecewise linear, another idea is to define a lower approximation by a sphere with its center v_i:

$$L(G_i) = \{x \in G\colon \|x - v_i\| \le \delta\}, \tag{14}$$

where δ is a positive constant.

This lower approximation of a sphere cannot be justified from the standard theory of rough sets, but another method using (14) can be derived. Note that the upper approximation is not used again.

Consider the next objective function in which two matrix variables $U = (u_{ki})$ ($1 \le k \le N$, $1 \le i \le c$) and $U' = (u'_{ki})$ ($1 \le k \le N$, $0 \le i \le c$) are used:

$$J(U, U', V) = \sum_{k=1}^{N} \sum_{i=1}^{c} (w u_{ki} + u'_{ki}) \|x_k - v_i\|^2 + \sum_{k=1}^{N} u'_{k0} \delta^2. \tag{15}$$

where $V = (v_1, \ldots, v_c)$ and w is a positive weight constant. Note that the size of the two matrices U and U' are different. They have constraints:

$$\sum_{i=1}^{c} u_{ki} = 1, \quad 1 \le k \le N, \quad u_{ki} \ge 0, \quad 1 \le k \le N, \ 1 \le i \le c, \tag{16}$$

$$\sum_{i=0}^{c} u'_{ki} = 1, \quad 1 \le k \le N, \quad u'_{ki} \ge 0, \quad 1 \le k \le N, \ 0 \le i \le c. \tag{17}$$

We hence consider the next alternative optimization:

Step 0. Determine the initial values for U and U'.
Step 1. Minimize $J(U, U', V)$ with respect to V while U and U' are fixed to the last optimal solutions.
Step 2. Minimize $J(U, U', V)$ with respect to U and U' while V is fixed to the last optimal solutions.
Step 3. If the optimal solution (U, U', V) is convergent, stop; else go to **Step 2**.

Note moreover that we can assume that optimal solutions for $J(U, U', V)$ satisfy $u_{ki} = 0$ or $u_{ki} = 1$ for all i, k and $u'_{ki} = 0$ or $u'_{ki} = 1$ for all i, k. This is due to the fundamental theorem of linear programming and the linearity of the objective function with respect to u_{ki} and u'_{ki}.

We thus have the optimal solutions:

$$v_i = \frac{\sum_{k=1}^{c} \{w u_{ki} + u'_{ki}\} x_k}{\sum_{k=1}^{c} \{w u_{ki} + u'_{ki}\}}, \tag{18}$$

$$u_{ki} = 1 \iff \|x_k - v_i\| \le \|x_k - v_j\|, \quad \forall j \ne i, \tag{19}$$

$$u'_{k0} = 1 \iff \|x_k - v_j\| > \delta, \quad \forall 1 \le j \le c, \tag{20}$$

$$u'_{ki} = 1 \iff \|x_k - v_i\| \le \|x_k - v_j\|, \quad \forall j \ne i, \ \|x_k - v_i\| \le \delta. \tag{21}$$

Let us introduce G_i and G'_i by

$$G_i = \{x_k \in X : u_{ki} = 1\}, \qquad G'_i = \{x_k \in X : u'_{ki} = 1\}. \tag{22}$$

Since it is easy to see that if $u'_{ki} = 1$ then $u_{ki} = 1$, we have $G'_i \subseteq G_i$. Thus we can define G'_i as a lower approximation: $L(G_i) = G'_i$. Moreover it is straightforward to observe that G_i's form partition of X. This method also has a weighted calculation for cluster centers (18), or in other form:

$$v_i = \frac{w \sum\limits_{x_k \in G_i} x_k + \sum\limits_{x_l \in G'_i} x_l}{w|G_i| + |G'_i|}. \tag{23}$$

Moreover the following property holds:

$$G_i = V(v_i) \cap X, \qquad L(G_i) = G'_i = V(v_i) \cap B_\delta(v_i) \cap X, \tag{24}$$

where $B_\delta(v_i) = \{x \in \mathbf{R}^p : \|x - v_i\| \leq \delta\}$.

We thus have a second method, called **Method 2**. The ideas of **Method 1** and **2** are compared in Fig. 1.

The last method can be generalized to a fuzzy method using a fuzzifying parameter $m > 1$. The objective function is generalized as follows:

$$J(U, U', V) = \sum_{k=1}^{N} \sum_{i=1}^{c} \{w(u_{ki})^m + (u'_{ki})^m\} \|x_k - v_i\|^2 + \sum_{k=1}^{N} (u'_{k0})^m \delta^2. \tag{25}$$

The function (25) is a linear combination of Bezdek's fuzzy c-means and Dave's noise clustering [2]. Details of fuzzy clustering using (25) is omitted here.

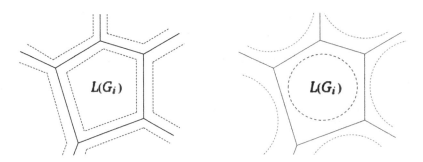

Fig. 1. Method 1 (left) and **Method 2** (right): core regions are shown by dotted lines/circles; Voronoi boundaries are denoted by solid lines.

3.2 Categorical Data

There are studies of clustering handling categorical data, in which a set of attributes is denoted by $\mathcal{A} = (A_1, \ldots, A_p)$. The attribute A_l has its domain

$D_l = \{\gamma_l^q\}$ $(q = 1, \ldots, l_r)$ of symbols. A simplest way to handle such a case is to define a distance $d_l(\gamma, \gamma')$ for A_l:

$$d_l(\gamma, \gamma') = \begin{cases} 1 & (\gamma \neq \gamma'), \\ 0 & (\gamma = \gamma'). \end{cases} \qquad (26)$$

The overall distance $D(x, y)$ is thus given by

$$D(x, y) = \sum_{l=1}^{p} d_l(x^l, y^l), \qquad (27)$$

where x^l and y^l are symbols of lth component of x and y.

The objective function (15) is modified to

$$J(U, U', V) = \sum_{k=1}^{N} \sum_{i=1}^{c} (wu_{ki} + u'_{ki}) D(x, v_i) + \sum_{k=1}^{N} u'_{k0} \delta^2. \qquad (28)$$

Thus the formulation in the case of categorical data is straightforward, but we have to search for optimal v_i in (28), which is not so easy as the case of the Euclidean distance; a simple idea is to use the medoids, the details of which will be discussed in our study in near future.

4   Examples

Two illustrative examples and five real datasets were used to observe the effectiveness of the proposed methods.

4.1   Illustrative Examples

Two artificial datasets on the plane are used. Dataset 1 in Fig. 2 shows large and small circular points. Second data shown in Fig. 3 is called 'synthetic cassini'[1].

Figures 4 and 5 respectively show the results by **Method 1** and **Method 2** for the first example, where squares show points in $L(G_i)$, while × and + show points in $G_i - L(G_i)$. Table 1 shows the values of Rand Index [10] and Adaptive Rand Index [3] for the three examples by the proposed methods and K-means. The two indices were calculated using the Voronoi regions, G_i and not $L(G_i)$, in order to compare the results with those by K-means. The results are almost the same for the three methods in the first example, but the second example shows notable improvement by the proposed method over the K-means.

---

[1] https://clusteval.sdu.dk/1/datasets.

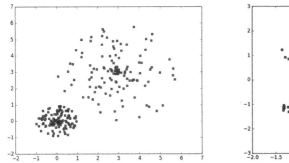

Fig. 2. Dataset 1: Two circles ($N = 250$)   **Fig. 3.** Dataset 2: cassini ($N = 250$)

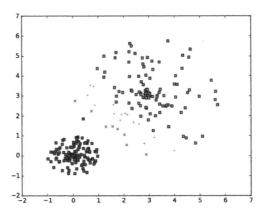

Fig. 4. Dataset1: Method 1; $K = 2, d = 0.7, w_1 = 0.6$

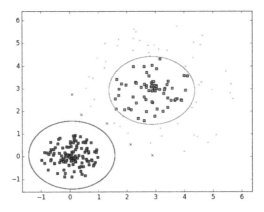

Fig. 5. Dataset1: Method 2; $K = 2, d = 1.5, w = 0.1$

4.2   Real Data

Five datasets of iris, appendicitis, balance, wine, hayes-roth[2] were used to compare results by the proposed two methods and K-means. The results are

Table 1. Values of Rand Index (RI) and Adaptive Rand Index (ARI) by K-means and Methods 1 and 2 for the two data sets

Data	Method	RI	ARI
Dataset1	K-means	0.9303	0.8606
	Method 1	0.9303	0.8606
	Method 2	0.9606	0.9213
Dataset2	K-means	0.7880	0.5402
	Method 1	0.9365	0.8603
	Method 2	0.8988	0.7771

Table 2. RI and ARI by K-means, Methods 1 and 2 for iris

Method	w_1, w	Parameters	RI	ARI
Method 1	0.5	$d = 0.6$	0.8797	0.7302
	0.6	$d = 1.0$	0.8797	0.7312
	0.7	$d = 0.8$	0.8797	0.7302
	0.8	$d = 0.8$	**0.8859**	**0.7455**
Method 2	0.1	$\delta = 0.6$	**0.9124**	**0.8015**
	0.2	$\delta = 0.6$	0.8988	0.7711
	0.3	$\delta = 0.6$	0.8859	0.7437
	0.4	$\delta = 0.8$	0.8859	0.7437
K-means			0.8797	0.7302

Table 3. RI and ARI by K-means, Methods 1 and 2 for appendicitis

Method	w_1, w	Parameters	RI	ARI
Method 1	0.5	$d = 0.5$	0.7973	0.5256
	0.6	$d = 0.6$	**0.7973**	**0.5256**
	0.7	$d = 1.0$	0.7686	0.4337
	0.8	$d = 0.5$	0.7973	0.5256
Method 2	0.1	$\delta = 0.4$	**0.7973**	**0.5256**
	0.2	$\delta = 0.3$	0.7827	0.4966
	0.3	$\delta = 0.4$	0.7686	0.4688
	0.7	$\delta = 0.3$	0.7547	0.4519
K-means			0.6792	0.3141

---

[2] http://sci2s.ugr.es/keel/category.php?cat=clas.

summarized into Tables 2, 3, 4, 5 and 6, where the best values of RI and ARI are shown by bold letters. Thus the proposed methods show better performances over the K-means.

A problem is that good parameters w_i, d, and δ are different on the five different examples. Thus further research is needed to find good or acceptable sets of the parameters.

Table 4. RI and ARI by K-means, Methods 1 and 2 for balance

Method	w_1, w	Parameters	RI	ARI
Method 1	0.5	$d = 13$	0.6216	0.2070
	0.6	$d = 12$	**0.6520**	**0.2787**
	0.7	$d = 14$	0.6363	0.2428
Method 2	0.2	$\delta = 9$	0.5991	0.1605
	0.3	$\delta = 1.6$	0.5999	0.1614
	0.4	$\delta = 2.5$	**0.6098**	**0.1812**
	0.5	$\delta = 8$	0.5991	0.1605
K-means			0.5870	0.1351

Table 5. RI and ARI by K-means, Methods 1 and 2 for wine

Method	w_1, w	Parameters	RI	ARI
Method 1	0.5	$d = 1.1$	0.7204	0.3749
	0.6	$d = 1.1$	0.7204	0.3841
	0.7	$d = 1.4$	0.7204	0.3981
	0.8	$d = 2.9$	**0.7229**	**0.3868**
Method 2	0.1	$\delta = 2.5$	**0.7187**	**0.3711**
	0.2	$\delta = 2.5$	0.7187	0.3711
K-means			0.7187	0.3711

Table 6. RI and ARI by K-means, Methods 1 and 2 for hayes-roth

Method	w_1, w	Parameters	RI	ARI
Method 1	0.5	$d = 1.1$	0.5904	0.1343
	0.6	$d = 1.0$	0.5888	0.1207
	0.7	$d = 0.6$	**0.6099**	**0.1734**
	0.8	$d = 1.0$	0.5964	0.1318
Method 2	0.1	$\delta = 0.2$	**0.5868**	**0.1187**
	0.2	$\exists \delta$	0.5849	0.1104
K-means			0.5849	0.1104

5 Conclusion

Two methods of variations of K-means clustering using a lower approximation are proposed. One uses neighborhood which is related to rough sets, of which the details were omitted due to page limitation. Another uses membership matrices, whereby a fuzzy version can be developed. The use of upper approximation is not judged to be useful in our theoretical view. To justify an upper approximation in terms of optimization of an objective function is still difficult.

The examples shown here demonstrated that the proposed methods are superior to the basic K-means but further study should be needed to find good values of the parameters. Moreover the effectiveness and efficiency of the proposed methods should be tested using larger-scale examples.

Acknowledgment. This paper is based upon work supported in part by the Air Force Office of Scientific Research/Asian Office of Aerospace Research and Development (AFOSR/AOARD) under award number FA2386-17-1-4046.

References

1. Bezdek, J.C.: Pattern Recognition with Fuzzy Objective Function Algorithms. Kluwer, Alphen aan den Rijn (1981)
2. Dave, R.N.: Characterization and detection of noise in clustering. Pattern Recog. Lett. **12**(11), 657–664 (1991)
3. Hubert, L., Arabie, P.: Comparing partitions. J. Classif. **2**(1), 193–218 (1985)
4. Kinoshita, N., Endo, Y., Miyamoto, S.: On some models of objective-based rough clustering. In: Proceedings of 2014 IEEE/WIC/ACM International Joint Conference on Web Intelligence and Intelligent Agent Technologies, 11–14 August 2014, Warsaw, Poland (2014)
5. Kohonen, T.: Self-organizing Maps. Springer, Heidelberg (1995). https://doi.org/10.1007/978-3-642-97610-0
6. Lingras, P., West, C.: Interval set clustering of web users with rough k-means. J. Intell. Inf. Syst. **23**, 5–16 (2004)
7. MacQueen, J.B.: Some methods for classification and analysis of multivariate observations. In: Proceedings of 5th Berkeley Symposium on Mathematical Statistics and Probability, vol. 1, pp. 281–297. University of California Press (1967)
8. Miyamoto, S., Ichihashi, H., Honda, K.: Algorithms for Fuzzy Clustering. Springer, Heidelberg (2008). https://doi.org/10.1007/978-3-540-78737-2
9. Pawlak, Z.: Rough sets. Int. J. Comput. Inf. Sci. **11**, 341–356 (1982)
10. Rand, W.M.: Objective criteria for the evaluation of clustering methods. J. Am. Stat. Assoc. **66**(336), 846–850 (1971)
11. Ubukata, S., Notsu, A., Honda, K.: The rough set k-means clustering. In: Proceedings of SCIS-ISIS, pp. 189–193 (2016)

Privacy Preserving Collaborative Fuzzy Co-clustering of Three-Mode Cooccurrence Data

Katsuhiro Honda$^{(\boxtimes)}$, Shotaro Matsuzaki, Seiki Ubukata, and Akira Notsu

Osaka Prefecture University, 1-1 Gakuen-cho, Nakaku, Sakai, Osaka 599-8531, Japan
{honda,subukata,notsu}@cs.osakafu-u.ac.jp

Abstract. Co-cluster structure analysis with three-mode cooccurrence information is a potential approach in summarizing multi-source relational data in such tasks as user-product purchase history analysis. This paper proposes a privacy preserving framework for jointly performing three-mode fuzzy co-clustering under collaboration among two organizations, which independently store object-item cooccurrence information and item-ingredient cooccurrence information, respectively. Even when they cannot mutually share elements of the cooccurrence matrices, the intrinsic co-cluster structures are revealed without publishing each elements of relational data but sharing only the structural information.

Keywords: Fuzzy-clustering · Co-clustering
Three-mode cooccurrence information
Privacy preserving data analysis

1 Introduction

Cluster analysis is a method for finding cluster structures from data, such that mutually similar objects are assigned to a same cluster while dissimilar objects are assigned to different clusters. In recent years, co-cluster analysis based on cooccurrence information among objects and items has become increasingly important in various web data analyses. Co-clustering is a method of extracting a co-cluster structure by simultaneously estimating the degree of membership of both objects and items to each cluster. Fuzzy clustering for categorical multivariate data (FCCM) [1] is an co-clustering extension of fuzzy c-means (FCM) [2], where the fuzzy partition concept [3] was introduced into both object and item partitions. FCCM replaced the FCM criterion with the aggregation degree of objects and items in co-clusters by adopting entropy-based fuzzification [4,5].

In co-clustering of two-mode cooccurrence information among objects and items, it is possible that their co-cluster structure may be distorted by the influence of other implicit factors. For example, in food preference analysis, we may fail to reveal users' preferences considering only user-food (object-item) cooccurrences but can find intrinsic preferences considering implicit relation among

© Springer Nature Switzerland AG 2018
V. Torra et al. (Eds.): MDAI 2018, LNAI 11144, pp. 232–242, 2018.
https://doi.org/10.1007/978-3-030-00202-2_19

users and cooking ingredients (object-ingredient), which compose the foods. By adding the supplemental information on potential elements constituting items, it is expected to improve the reliability of the co-cluster knowledge. Three-mode fuzzy clustering for categorical multivariate data (3-mode FCCM) [6,7] is an extension of FCCM for analyzing three-mode cooccurrence information, where not only object-item cooccurrence information but also item-ingredient cooccurrence information are available and the goal is to extract co-clusters such that mutually familiar pairs of objects and items are assigned to a co-cluster in conjunction with their typical ingredients. Three types of fuzzy memberships for objects, items and ingredients are simultaneously estimated such that the aggregation among them is maximized in three-mode co-clusters.

Besides the development of various clustering algorithms, the awareness of information protection has been increasing in recent years. Privacy preserving data mining (PPDM) [8] is a fundamental approach for utilizing multiple databases including personal or sensitive information without fear of information leaks. Privacy preserving frameworks are necessary in performing clustering of real-world large-scale data, which include personal information. Several models introducing cryptographic mechanisms [9,10] have been proposed for applying the k-Means algorithm [11], in which personal privacy is strictly preserved among multiple organizations. Secure cluster information sharing is realized by conducting analysis such that only the cluster structure and encryption information are shared without disclosing the observation value of each object to other organizations. A similar mechanisms were also applied to the FCCM framework for two-mode cooccurrence information [12,13], where cooccurrence information among common objects and individual items are stored in multiple organizations and cannot be disclosed among them.

In this paper, we newly propose a collaborative fuzzy co-clustering framework of three-mode cooccurrence information data as an extension of three-mode FCCM. It is assumed that two types of cooccurrence information data of *objects × items* and *items × ingredients* are independently collected and accumulated in different organizations and the elements of cooccurrence information data cannot be disclosed each other from the viewpoint of information protection. The goal is to estimate co-cluster structures under collaboration of organizations such that we can extract the same co-cluster structures with the conventional 3-mode FCCM keeping privacy preservation. For example, in food preference analysis, a sales outlet can store the relation among users and food menus while a caterer may know secret information on cooking ingredients of each food. Even if the two organizations cannot mutually disclose their own data elements, we can expect that the intrinsic co-cluster structure among users and ingredients are useful for both the outlet and caterer in improving the quality of food recommendation and menu creation.

The remainder of this paper is organized as follows: Sect. 2 gives a brief review on the conventional FCM-type fuzzy co-clustering models and Sect. 3 proposes a method for applying 3-mode FCCM keeping privacy preservation. The experimental result is shown in Sect. 4 and a summary conclusion is presented in Sect. 5.

2 FCM Clustering and FCM-Type Fuzzy Co-clustering

2.1 Fuzzy c-Means (FCM)

When we have multi-dimensional observation of n objects, FCM divides their multi-dimensional vectors $\boldsymbol{x}_i$ into C fuzzy clusters, whose prototypes are proto-typical centroids $\boldsymbol{b}_c$ (mean vector in cluster c) [2]. Its objective function to be minimized is as follows:

$$L_{fcm} = \sum_{c=1}^{C}\sum_{i=1}^{n} u_{ci}^{\theta}||\boldsymbol{x}_i - \boldsymbol{b}_c||^2, \tag{1}$$

where u_{ci} is the membership of object i to cluster c, and is normalized under the probabilistic constraint of $\sum_{c=1}^{C} u_{ci} = 1$. θ is a parameter for tuning the degree of fuzziness. The larger the weight θ is, the fuzzier the cluster partition is. u_{ci} and $\boldsymbol{b}_c$ are iteratively updated under the iterative optimization principle.

2.2 Fuzzy Clustering for Categorical Multivariate Data (FCCM)

Assume that we have $n \times m$ cooccurrence information $R = \{r_{ij}\}$ among n objects and m items. The goal of co-clustering is to simultaneously estimate fuzzy memberships of objects u_{ci} and items w_{cj} such that mutually familiar objects and items tend to have large memberships in the same cluster considering the aggregation degree of each co-cluster. The objective function for Fuzzy Clustering for Categorical Multivariate data (FCCM) [1] was proposed as:

$$L_{fccm} = \sum_{c=1}^{C}\sum_{i=1}^{n}\sum_{j=1}^{m} u_{ci}w_{cj}r_{ij} - \lambda_u \sum_{c=1}^{C}\sum_{i=1}^{n} u_{ci}\log u_{ci}$$

$$- \lambda_w \sum_{c=1}^{C}\sum_{j=1}^{m} w_{cj}\log w_{cj}. \tag{2}$$

The first term is the aggregation degree to be maximized, which is a modified FCM-type criterion for extracting dense co-cluster. The second and third terms are entropy-like penalty for realizing fuzzy partition under the entropy regularization concept [4]. λ_u and λ_w are fuzzification weights for object and item memberships, respectively. Larger λ_u and λ_w bring fuzzier partitions of objects and items.

Here, object memberships u_{ci} have a similar role to those of FCM under the same condition, such that $\sum_{c=1}^{C} u_{ci} = 1$. On the other hand, if item memberships w_{cj} also obey a similar condition of $\sum_{c=1}^{C} w_{cj} = 1$, the aggregation criterion has a trivial maximum of $u_{ci} = w_{cj} = 1$, $\forall i, j$ in a particular cluster c. Then, in order to avoid trivial solutions, w_{cj} are forced to be exclusive in each cluster, such that $\sum_{j=1}^{m} w_{cj} = 1$, and so, w_{cj} represent the relative typicalities of items in each cluster.

2.3   Three-Mode Fuzzy Clustering for Categorical Multivariate Data (3-Mode FCCM)

Assume that we have $n \times m$ cooccurrence information $R = \{r_{ij}\}$ among n objects and m items, and the items are characterized with other ingredients, where cooccurrence information among m items and p other ingredients are summarized in $m \times p$ matrix $S = \{s_{jk}\}$ with s_{jk} representing the cooccurrence degree of item j and ingredient k. For example, in food preference analysis, R can be an evaluation matrix by n users on m foods and S may be appearance/absence of p cooking ingredients in m foods. The goal of three-mode co-cluster analysis is to extract mutually familiar groups of objects, items and ingredients such that the group object i, item j and ingredient k have large cooccurrence r_{ij} and s_{jk}.

In order to extend the FCCM algorithm to three-mode co-cluster analysis, additional memberships z_{ck} are introduced for representing the membership degree of ingredients k to co-cluster c. Then, the objective function for three-mode FCCM (3FCCM) [6,7] was constructed by modifying the FCCM objective function of (2) as:

$$L_{3fccm} = \sum_{c=1}^{C}\sum_{i=1}^{n}\sum_{j=1}^{m}\sum_{k=1}^{p} u_{ci}w_{cj}z_{ck}r_{ij}s_{jk} - \lambda_u \sum_{c=1}^{C}\sum_{i=1}^{n} u_{ci}\log u_{ci}$$

$$-\lambda_w \sum_{c=1}^{C}\sum_{j=1}^{m} w_{cj}\log w_{cj} - \lambda_z \sum_{c=1}^{C}\sum_{k=1}^{p} z_{ck}\log z_{ck}. \qquad (3)$$

The degree of aggregation in Eq. (2) has been extended to three mode version and a penalty term on z_{ck} was also added. λ_z is the additional penalty weight for fuzzification of ingredient memberships z_{ck}. The larger the value of λ_z is, the fuzzier the ingredient memberships are. As in the same manner to item memberships w_{cj}, ingredient memberships z_{ck} are estimated under the within-cluster constraint of $\sum_{p=1}^{k} z_{ck} = 1, \forall c$ from the view point of typical ingredient selection for characterizing co-cluster features.

The clustering algorithm is an iterative process of updating u_{ci}, w_{cj}, and z_{ck} under the alternative optimization principle. Considering the necessary conditions for the optimality $\partial L_{3fccm}/\partial u_{ci} = 0, \partial L_{3fccm}/\partial w_{cj} = 0$ and $\partial L_{3fccm}/\partial z_{ck} = 0$ under the sum-to-one constraints, the updating rules for three memberships are given as

$$u_{ci} = \frac{\exp\left(\lambda_u^{-1}\sum_{j=1}^{m}\sum_{k=1}^{p} w_{cj}z_{ck}r_{ij}s_{jk}\right)}{\sum_{l=1}^{C}\exp\left(\lambda_u^{-1}\sum_{j=1}^{m}\sum_{k=1}^{p} w_{lj}z_{lk}r_{ij}s_{jk}\right)}, \qquad (4)$$

$$w_{cj} = \frac{\exp\left(\lambda_w^{-1}\sum_{i=1}^{n}\sum_{k=1}^{p} u_{ci}z_{ck}r_{ij}s_{jk}\right)}{\sum_{l=1}^{m}\exp\left(\lambda_w^{-1}\sum_{i=1}^{n}\sum_{k=1}^{p} u_{ci}z_{ck}r_{il}s_{lk}\right)}, \qquad (5)$$

$$z_{ck} = \frac{\exp\left(\lambda_z^{-1} \sum_{i=1}^{n} \sum_{j=1}^{m} u_{ci} w_{cj} r_{ij} s_{jk}\right)}{\sum_{l=1}^{p} \exp\left(\lambda_z^{-1} \sum_{i=1}^{n} \sum_{j=1}^{m} u_{ci} w_{cj} r_{ij} s_{jl}\right)}. \tag{6}$$

Following the above derivation, a sample algorithm is represented as follows:

Algorithm: 3-mode Fuzzy Clustering for Categorical Multivariate data (3-mode FCCM)

Step 1. Given $n \times m$ cooccurrence matrix R and $m \times p$ cooccurrence matrix S, let C be the number of clusters. Choose the fuzzification weights λ_u, λ_w and λ_z.

Step 2. Randomly initialize u_{ci}, w_{cj} and z_{ck} such that $\sum_{c=1}^{C} u_{ci} = 1$, $\sum_{j=1}^{m} w_{cj} = 1$ and $\sum_{k=1}^{p} z_{ck} = 1$.

Step 3. Update u_{ci} with Eq. (4).

Step 4. Update w_{cj} with Eq. (5).

Step 5. Update z_{ck} with Eq. (6).

Step 6. If the convergence determination

$$\max_{c,i} |u_{ci}^{NEW} - u_{ci}^{OLD}| < \varepsilon$$

is satisfied, the process is terminated, and otherwise, the process returns to Step 3.

3   Extension of 3-Mode FCCM for Collaborative 3-Mode FCCM

In this paper, a novel framework for privacy preservation in 3-mode FCCM is proposed, where co-cluster estimation is jointly performed by two organizations. Assume that organization A have $n \times m$ cooccurrence information $R = \{r_{ij}\}$ among n objects and m items, and organization B have $m \times p$ cooccurrence information $S = \{s_{jk}\}$ among m items and p ingredients, respectively. It is expected that we can estimate more informative co-cluster structures by adopting 3-mode FCCM rather than the independent 2-mode FCCM analysis in each organization. However, it may not be possible to disclose each element of cooccurrence matrices from the viewpoint of information protection.

For example, in food preference analysis, a sales outlet would store the preference relation among users and food menus while a caterer may have secret information on cooking ingredients of each food. Now, we can expect that preference tendencies among users and ingredients are useful for both the outlet and caterer in improving the quality of food recommendation and menu creation. However, it may be difficult for the two organizations to share their cooccurrence information matrices due to privacy or business issues. Collaborative framework for achieving three-mode co-clustering without data sharing is expected to bring a new bussiness chance for both organizations under privacy preservation.

In the following, Collaborative 3-mode FCCM is considered, where the two organization have the common goal of extracting co-cluster structures without disclosing each elements of cooccurrence matrices.

Here, the shared and concealed informations in Eqs. (4)–(6) are defined as follows: User memberships u_{ci} and ingredient memberships z_{ck} should be concealed only in organizations A and B, respectively, while item memberships w_{cj} can be shared by them. Then, in u_{ci} calculation, s_{jk} and z_{ck} are not directly available for organization A, and Eq. (4) is rewritten as:

$$
u_{ci} = \frac{\exp\left(\lambda_u^{-1}\sum_{j=1}^{m}w_{cj}r_{ij}\left(\sum_{k=1}^{p}z_{ck}s_{jk}\right)\right)}{\sum_{l=1}^{C}\exp\left(\lambda_u^{-1}\sum_{j=1}^{m}w_{lj}r_{ij}\left(\sum_{k=1}^{p}z_{lk}s_{jk}\right)\right)}
$$

$$
= \frac{\exp\left(\lambda_u^{-1}\sum_{j=1}^{m}w_{cj}r_{ij}\beta_{cj}\right)}{\sum_{l=1}^{C}\exp\left(\lambda_u^{-1}\sum_{j=1}^{m}w_{lj}r_{ij}\beta_{lj}\right)}. \tag{7}
$$

Next, in z_{ck} calculation, r_{ij} and u_{ci} are not directly available for organization B, and Eq. (6) is rewritten as:

$$
z_{ck} = \frac{\exp\left(\lambda_z^{-1}\sum_{j=1}^{m}w_{cj}s_{jk}\left(\sum_{i=1}^{n}u_{ci}r_{ij}\right)\right)}{\sum_{l=1}^{p}\exp\left(\lambda_z^{-1}\sum_{j=1}^{m}w_{cj}s_{jl}\left(\sum_{i=1}^{n}u_{ci}r_{ij}\right)\right)}
$$

$$
= \frac{\exp\left(\lambda_z^{-1}\sum_{j=1}^{m}w_{cj}s_{jk}\alpha_{cj}\right)}{\sum_{l=1}^{p}\exp\left(\lambda_z^{-1}\sum_{j=1}^{m}w_{cj}s_{jl}\alpha_{cj}\right)}, \tag{8}
$$

where α_{cj} and β_{cj} are the following values calculated in organization A and organization B, respectively, which are referred to as shared information matrices A and B.

$$
A = \begin{pmatrix} \alpha_{11} & \cdots & \alpha_{1m} \\ \vdots & \ddots & \vdots \\ \alpha_{C1} & \cdots & \alpha_{Cm} \end{pmatrix} = \begin{pmatrix} \sum_{i=1}^{n}u_{1i}r_{i1} & \cdots & \sum_{i=1}^{n}u_{1i}r_{im} \\ \vdots & \ddots & \vdots \\ \sum_{i=1}^{n}u_{Ci}r_{i1} & \cdots & \sum_{i=1}^{n}u_{Ci}r_{im} \end{pmatrix} \tag{9}
$$

$$
B = \begin{pmatrix} \beta_{11} & \cdots & \beta_{1m} \\ \vdots & \ddots & \vdots \\ \beta_{C1} & \cdots & \beta_{Cm} \end{pmatrix} = \begin{pmatrix} \sum_{k=1}^{p}z_{1k}s_{1k} & \cdots & \sum_{k=1}^{p}z_{1k}s_{mk} \\ \vdots & \ddots & \vdots \\ \sum_{k=1}^{p}z_{Ck}s_{1k} & \cdots & \sum_{k=1}^{p}z_{Ck}s_{mk} \end{pmatrix} \tag{10}
$$

Finally, in w_{cj} calculation, Eq. (5) is rewritten as:

$$
\begin{aligned}
w_{cj} &= \frac{\exp\left(\lambda_w^{-1}\left(\sum_{i=1}^n u_{ci} r_{ij}\right)\left(\sum_{k=1}^p z_{ck} s_{jk}\right)\right)}{\sum_{l=1}^m \exp\left(\lambda_w^{-1}\left(\sum_{i=1}^n u_{ci} r_{il}\right)\left(\sum_{k=1}^p z_{ck} s_{lk}\right)\right)} \\
&= \frac{\exp\left(\lambda_w^{-1} \alpha_{cj}\beta_{cj}\right)}{\sum_{l=1}^m \exp\left(\lambda_w^{-1}\alpha_{cl}\beta_{cl}\right)},
\end{aligned}
\tag{11}
$$

The above modification implies that it is possible to calculate fuzzy memberships in each organization by merely sharing the shared information matrices A and B without disclosing components of cooccurrence information matrices. The components of the shared information matrices A and B are regarded as the cluster structure information of items in each of organizations A and B. α_{cj} represents the typicality of item j in cluster c taking user similarity into consideration while β_{cj} represents the typicality of item j in cluster c taking ingredient similarity into consideration. In these shared information, each of user and ingredient characteristics is kept secret because user similarity and ingredient similarity are shared after summing up in each organization. Actually, even if information matrix A is disclosed, organization B cannot know such knowledge as the number of objects n in organization A, and conversely, even if information matrix B is disclosed, organization A cannot know such knowledge as the number of elements p in organization B.

Following the above consideration, the sample algorithm is represented as follows:

Algorithm: Collaborative Fuzzy Clustering for 3-mode Categorical Multivariate data (Collaborative 3-mode FCCM)

Step 1. [Initialization] Randomly initialize w_{cj} and components of shared information matrix B.

Step 2. In organization A, update u_{ci} with Eq. (7) and calculate information matrix A.

Step 3. In organization A, update w_{cj} with Eq. (11).

Step 4. From organization A to organization B, send w_{cj} and information matrix A.

Step 5. In organization B, update z_{ck} with Eq. (8) and calculate information matrix B.

Step 6. From organization B to organization A, send z_{ck} and information matrix B.

Step 7. In organization A, convergence judgment is performed. If u_{ci} is converged, the clustering process ends. If it has not converged, return to Step 2.

4   Experimental Result

In order to confirm the characteristics of the proposed algorithm, a numerical experiment was performed with the artificial data, which was used in [6,7]. It is an artificially generated three-mode data set, in which 40 objects ($n = 40$) have relational connection with 50 items ($m = 50$) and the items are related to 30 ingredients ($p = 30$). For example, in food preference analysis, 40 users can select 50 foods, which are made of 30 ingredients. The artificial three-mode cooccurrence matrices R and S were generated under the assumption that objects and ingredients have intrinsic (unknown) connections of 40×30 matrix $X = \{x_{ik}\}$ as shown in Fig. 1, where black and white cells represent $x_{ik} = 1$ and $x_{ik} = 0$, respectively.

In this experiment, we assume that organization A has cooccurrence information matrix R on 40 objects $\times$ 50 items as shown in Fig. 2, and organization B has cooccurrence information matrix S on 50 items $\times$ 30 elements as shown in Fig. 3. The goal is to estimate the intrinsic (unknown) cooccurrence information

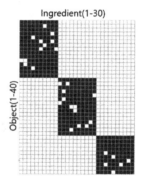

Fig. 1. Intrinsic cooccurrence information matrix X

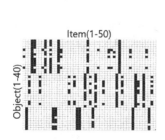

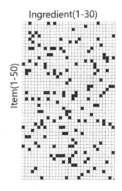

Fig. 2. Cooccurrence information matrix R

Fig. 3. Cooccurrence information matrix S

X without disclosing the components of cooccurrence information matrices R and S among organizations.

Collaborative 3FCCM algorithm was applied to R and S with $C = 3$, $\lambda_u = 0.1$, $\lambda_w = 0.2$ and $\lambda_z = 0.3$. Figure 4 shows object memberships obtained in organization A, Fig. 5 shows ingredient memberships obtained in organization B, and Fig. 6 shows item memberships shared by both organizations. In each figure, membership values are depicted in grayscale such that (white, black) → (0, maximum value). These results successfully imply the intrinsic co-cluster structures of X and are completely equivalent to those given in [6,7].

Here, the shared information of matrices A and B is investigated. Figures 7 and 8 show the final elements of matrices A and B, which reflect item cluster structures induced in each organization. In comparison with the common item memberships w_{cj} shown in Fig. 6, both organization-wise clusters are slightly different from the final item clusters, which are seen to be constructed through

Fig. 4. Object membership u_{ci}

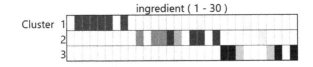

Fig. 5. Ingredient membership z_{ck}

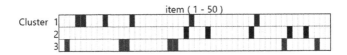

Fig. 6. Item membership w_{cj}

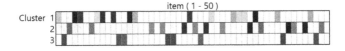

Fig. 7. Shared matrix $A = \{\alpha_{cj}\}$

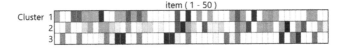

Fig. 8. Shared matrix $B = \{\beta_{cj}\}$

max operation among matrices A and B. So, it was confirmed that sharing of secure co-cluster structures could be realized by the fusion of organization-wise item cluster structures.

From the viewpoint of privacy preservation, matrices A and B are lower dimensional linear transforms of matrices R and S, and then, their informations are degraded from R and S such that all elements of R and S cannot be reconstructed from A and B.

5   Conclusion

In this paper, we proposed a novel method for extracting co-cluster structure from three-mode cooccurrence information stored in two different organization, which is available even when they cannot disclose their cooccurrence information each other. The proposed extension of 3-mode FCCM shares the organization-wise item cluster structures instead of each element of cooccurrence matrices. Numerical experiments confirmed that the proposed method obtains the same results with the conventional method, which does not consider privacy preservation.

A potential future work is to validate the confidentiality of the proposed framework such that it successfully avoids revealing information on objects and ingredients even if other organization intentionally adjusts the shared information matrices A or B. Another direction of future work is to investigate the effects of the quality of the shared information when cooccurrence information includes noise or inconsistent cluster structures.

Acknowledgment. This work was supported in part by Tateisi Science and Technology Foundation through 2017 research grant (A).

References

1. Oh, C.-H., Honda K., Ichihashi, H.: Fuzzy clustering for categorical multivariate data. In: Proceedings of Joint 9th IFSA World Congress and 20th NAFIPS International Conference, pp. 2154–2159 (2001)
2. Bezdek, J.C.: Pattern Recognition with Fuzzy Objective Function Algorithms. Plenum Press, New York (1981)
3. Ruspini, E.H.: A new approach to clustering. Inf. Control **15**(1), 22–32 (1969)
4. Miyamoto, S., Mukaidono, M.: Fuzzy c-means as a regularization and maximum entropy approach. In: Proceedings of the 7th International Fuzzy Systems Association World Congress, vol. 2, pp 86–92 (1997)
5. Miyamoto, S., Ichihashi, H., Honda, K.: Algorithms for Fuzzy Clustering. Springer, Heidelberg (2008). https://doi.org/10.1007/978-3-540-78737-2
6. Honda, K., Suzuki, Y., Nishioka, M., Ubukata, S., Notsu, A.: A fuzzy co-clustering model for three-modes relational cooccurrence data. In: Proceedings of 2017 IEEE International Conference on Fuzzy Systems, #F-0264 (2017)
7. Honda, K., Suzuki, Y., Ubukata, S., Notsu, A.: FCM-type fuzzy coclustering for three-mode cooccurrence data: 3FCCM and 3Fuzzy CoDoK. Adv. Fuzzy Syst. **2017**(9842127), 1–8 (2017)

8. Aggarwal, C.C., Yu, P.S.: Privacy-Preserving Data Mining: Models and Algorithms. Springer, New York (2008). https://doi.org/10.1007/978-0-387-70992-5
9. Vaidya, J., Clifton, C.: Privacy-preserving k-Means clustering over vertically partitioned data. In: Proceedings of 9th ACM SIGKDD International Conference on Knowledge Discovery and Data Mining, pp. 206–215 (2003)
10. Yu, T.-K., Lee, D.T. Chang, S.-M., Zhan, J.: Multi-party k-Means clustering with privacy consideration. In: Proceedings of 2010 International Symposium on Parallel and Distributed Processing with Applications, pp. 200–207 (2010)
11. MacQueen, J.B.: Some methods of classification and analysis of multivariate observations. In: Proceedings of 5th Berkeley Symposium on Mathematical Statistics and Probability, pp. 281–297 (1967)
12. Honda, K., Oda, T., Tanaka, D., Notsu, A.: A collaborative framework for privacy preserving fuzzy co-clustering of vertically distributed cooccurrence matrices. Adv. Fuzzy Syst. **2015**(729072), 1–8 (2015)
13. Oda, T., Honda, K., Ubukata, S., Notsu, A.: Consideration of site-wise confidence in fuzzy co-clustering of vertically distributed cooccurrence data. Int. J. Comput. Sci. Netw. Secur. **16**(2), 15–21 (2016)

Generalized Fuzzy c-Means Clustering and Its Theoretical Properties

Yuchi Kanzawa[1($\boxtimes$)] and Sadaaki Miyamoto[2]

[1] Shibaura Institute of Technology, Tokyo, Japan
kanzawa@sic.shibaura-it.ac.jp
[2] University of Tsukuba, Tsukuba, Japan
miyamoto.sadaaki.fu@u.tsukuba.ac.jp

Abstract. This study shows that a generalized fuzzy c-means (gFCM) clustering algorithm, which covers standard fuzzy c-means clustering, can be constructed if a given fuzzified function, its derivative, and its inverse derivative can be calculated. Furthermore, our results show that the fuzzy classification function for gFCM exhibits similar behavior to that of standard fuzzy c-means clustering.

Keywords: Fuzzy c-means clustering · Fuzzy classification function

1 Introduction

The hard c-means (HCM) clustering algorithm [1] splits objects into well-separated sets of objects, known as clusters, by minimizing the mean squared distance from each object to its nearest cluster center. Fuzzy clustering extends this concept so that object membership is shared among all of the clusters, rather than being constrained to a single cluster.

In fuzzy clustering, the membership degree in the HCM objective function is a nonlinear expression. Specifically, Bezdek's algorithm replaces the linear membership weights with the power of membership and creates cluster centers based on weighted means [2], thereby producing what is commonly known as the fuzzy c-means (FCM) algorithm. To distinguish this from the many variants that have since been proposed, this algorithm is referred to as standard FCM (sFCM) in this paper.

It is important to clarify the features of fuzzy clustering methods. Miyamoto [3] clarified that, theoretically at least, the fuzzy classification function (FCF) of sFCM produces an allocation rule that classifies a brand new object into a Voronoi cell, with the Voronoi seeds being the cluster centers, and that the FCF of sFCM at the infinity point approaches the reciprocal of the given cluster number.

Noting that sFCM fuzzifies clustering result by replacing the membership in the HCM objective function with nonlinear expressions, it may be possible

V. Torra et al. (Eds.): MDAI 2018, LNAI 11144, pp. 243–254, 2018.
https://doi.org/10.1007/978-3-030-00202-2_20

to obtain novel fuzzy clustering algorithms by adopting various nonlinear functions. However, it would be burdensome to investigate the features of each such algorithm.

The present study addresses this issue. First, we consider an optimization problem for generalized fuzzy c-means clustering (gFCM) in which the membership degree in the HCM objective function is replaced by a general nonlinear function. Next, we construct a gFCM algorithm and its associated FCF by solving this optimization problem. We then theoretically show that the FCF of gFCM at the infinity point approaches the reciprocal of the given cluster number. Several numerical examples substantiate our theoretical results.

The remainder of this paper is organized as follows. Section 2 introduces the notation used in this paper and describes the conventional methods. In Sect. 3, we derive the concept of gFCM and discuss its theoretical behavior. Section 4 presents several illustrative examples. Finally, Sect. 5 contains some concluding remarks.

2   Preliminaries

Let $X = \{x_k \in \mathbb{R}^p \mid k \in \{1, \cdots, N\}\}$ be a dataset of p-dimensional points. Consider the problem of classifying the objects in X into C disjoint subsets $\{G_i\}_{i=1}^{C}$ which are termed clusters. The membership degree of x_k with respect to the i-th cluster is denoted by $u_{i,k}$ ($i \in \{1, \cdots, C\}, k \in \{1, \cdots, N\}$) and the set of $u_{i,k}$ is denoted by u, which is known as the partition matrix. The set of cluster centers is denoted by $v = \{v_i \mid v_i \in \mathbb{R}^p, i \in \{1, \cdots, C\}\}$. The squared Euclidean distance between the k-th datum and the i-th cluster center is given by

$$d_{i,k} = \|x_k - v_i\|_2^2. \tag{1}$$

The HCM algorithm iterates the following two steps: (i) calculate the memberships $u_{i,k}$ and (ii) calculated the cluster centers v_i [1]. These update equations of the memberships and cluster centers are obtained by solving the following optimization problem:

$$\underset{u,v}{\text{minimize}} \sum_{i=1}^{C} \sum_{k=1}^{N} u_{i,k} d_{i,k}, \tag{2}$$

$$\text{subject to } \sum_{i=1}^{C} u_{i,k} = 1. \tag{3}$$

The sFCM [2] representation is obtained by solving the following optimization problem:

$$\underset{u,v}{\text{minimize}} \sum_{i=1}^{C} \sum_{k=1}^{N} (u_{i,k})^m d_{i,k}, \tag{4}$$

subject to Eq. (3), where $m > 1$ is an additional weighting exponent. We refer to the nonlinear function $(u_{i,k})^m$ as *fuzzifier*, because this play the role of fuzzificating the clustering results.

A fuzzy classification function [3] (FCF) describes the degree to which any point in the object space is quintessentially attached to a cluster by broadening the membership $u_{i,k}$ to the entire space. The FCF $\{u_i(x)\}_{i=1}^{C}$ with respect to a new object $x \in \mathbb{R}^p$ is defined as the solution to the following optimization problem for sFCM:

$$\underset{u}{\text{minimize}} \sum_{i=1}^{C} (u_i(x))^m d_i(x), \tag{5}$$

subject to

$$\sum_{i=1}^{C} u_i(x) = 1, \tag{6}$$

where

$$d_i(x) = \|x - v_i\|_2^2, \tag{7}$$

and $\{v_i\}_{i=1}^{C}$ are the cluster centers obtained by the corresponding fuzzy clustering algorithms. We define a crisp allocation rule [4] for classifying $\mathbb{R}^p$ using the FCF as

$$x \in G_i \overset{\text{def}}{\equiv} u_i(x) > u_j(x) \text{ for } j \neq i. \tag{8}$$

Theoretically, it has been shown [4] that the FCF of sFCM has the feature that the subsets $\{G_i\}_{i=1}^{C}$ produced from sFCM result in Voronoi sets, because

$$u_i(x) > u_j(x) \text{ for } j \neq i \Leftrightarrow \|x - v_i\|_2 < \|x - v_j\|_2 \text{ for } j \neq i, \tag{9}$$

and that $u_i(x)$ approaches $1/C$ as $\|x\|_2 \to +\infty$.

Noting that sFCM fuzzifies the clustering results by adopting a fuzzifier, we believe that novel fuzzy clustering algorithms could be obtained using other fuzzifiers. However, it would be burdensome to investigate the features of each such algorithm.

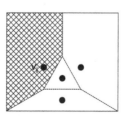

G_i ⬚⬚⬚⬚
Voronoi border ----------
Centers ●

Fig. 1. Example of a Voronoi diagram generated from four cluster centers: the case with $p = 2$, $C = 4$

3   Generalized FCM

3.1   Optimization Problem

The present study considers a gFCM algorithm. For a strictly convex, increasing, non-negative, and smooth function g defined in $[0, 1)$ with $g'(0) = 0$, the gFCM optimization problem is described as follows:

$$\underset{u,v}{\text{minimize}} \sum_{i=1}^{C} \sum_{k=1}^{N} g(u_{i,k}) d_{i,k} \tag{10}$$

subject to Eq. (3). As g is strictly convex, non-negative, and smooth, g' is defined in $[0, 1)$ and is monotonically increasing; hence, g'^{-1} is also monotonically increasing. If $\lim_{\mu \nearrow 1} g'(\mu) < +\infty$, let us define $g'(1) \stackrel{\text{def}}{=} \lim_{\mu \nearrow 1} g'(\mu)$. Then, we consider g' to be defined in $[0, 1]$. Note that gFCM is not an individual method, but a framework of methods. For example, gFCM with $g(u_{i,k}) = u_{i,k}$ reduces to HCM, and gFCM with $g(u_{i,k}) = (u_{i,k})^m$ reduces to sFCM. Thus, gFCM is a generalization of HCM and sFCM. The following subsections discuss the algorithm, FCF, and associated theoretical property.

3.2   Algorithm, FCF, and Its Property

The gFCM algorithm is obtained by solving the optimization problem in Eqs. (10) and (3). We derive the optimal cluster center in the same manner as for HCM and sFCM, i.e.,

$$v_i = \sum_{k=1}^{N} \alpha_k x_k, \tag{11}$$

$$\alpha_k = \frac{g(u_{i,k})}{\sum_{\ell=1}^{N} g(u_{i,\ell})} \tag{12}$$

if g is bounded. In the case where $g(u_{i,k}) \to +\infty$ as $u_{i,k} \nearrow 1$, v_i is obtained as follows. Equation (12) can be written in an equivalent form as

$$\alpha_k = \left[\sum_{\ell=1}^{N} \frac{g(u_{i,\ell})}{g(u_{i,k})} \right]^{-1} = \left[1 + \sum_{\substack{\ell=1 \\ \ell \neq k}}^{N} \frac{g(u_{i,\ell})}{g(u_{i,k})} \right]^{-1}. \tag{13}$$

Thus, we have

$$\alpha_k = \begin{cases} |\{\ell \mid \lim_{u_{i,\ell} \nearrow 1} g(u_{i,\ell}) \to +\infty\}|^{-1} & (\lim_{u_{i,k} \nearrow 1} g(u_{i,k}) \to +\infty) \\ 0 & (\lim_{u_{i,\ell} \nearrow 1} g(u_{i,\ell}) \to +\infty \text{ for } \ell \neq k). \end{cases} \tag{14}$$

To determine the optimal membership, we describe the Lagrangian $L(u)$ as

$$L(u) = \sum_{i=1}^{C} \sum_{k=1}^{N} g(u_{i,k}) d_{i,k} + \sum_{k=1}^{N} \gamma_k \left(1 - \sum_{i=1}^{C} u_{i,k} \right) \tag{15}$$

with the Lagrange multipliers $(\gamma_1, \ldots, \gamma_N)$. The necessary condition for optimal membership is described as

$$\frac{\partial L(u)}{\partial u_{i,k}} = 0, \tag{16}$$

$$\frac{\partial L(u)}{\partial \gamma_k} = 0. \tag{17}$$

We consider two cases: (i) $d_{i,k} > 0$ for all $i \in \{1, \ldots, C\}$ and (ii) there exists one $i \in \{1, \ldots, C\}$ such that $d_{i,k} = 0$.

(i) If $d_{i,k} > 0$ for all $i \in \{1, \ldots, C\}$, the optimal membership is described by Eq. (16) as

$$u_{i,k} = g'^{-1} \left(\frac{\gamma_k}{d_{i,k}} \right). \tag{18}$$

As g'^{-1} is monotonically increasing for γ_k and satisfies

$$\lim_{\gamma_k \to +\infty} \sum_{i=1}^{C} u_{i,k}(\gamma_k) > 1, \quad \sum_{i=1}^{C} u_{i,k}(0) = 0, \tag{19}$$

we have the unique γ_k satisfying condition given in Eq. (17) $\Leftrightarrow$ Eq. (3), which suggests that there is a unique optimal solution $u_{i,k}$ for the optimization problem (10), where $u_{i,k}(\gamma_k)$ represents the value $u_{i,k}$ depending on γ_k.

However, it is difficult to obtain the optimal value of γ_k directly. Thus, the bisection method is utilized. First, we establish $\gamma_k = g'(1/C) \min_{1 \leq i \leq C}\{d_{i,k}\}$ as a lower bound of the γ_k satisfying condition given in Eq. (3), because $\gamma_k = g'(1/C) \min_{1 \leq i \leq C}\{d_{i,k}\} \Leftrightarrow \gamma_k / \min_{1 \leq i \leq C}\{d_{i,k}\} = g'(1/C) \Leftrightarrow g'(\gamma_k / \min_{1 \leq i \leq C}\{d_{i,k}\}) = 1/C \Rightarrow u_{i,k}(\gamma_k) \leq 1/C$ for all $i \in \{1, \cdots, C\}$, which implies $\sum_{i=1}^{C} u_{i,k}(\gamma_k) \leq 1$. Next, we establish $\gamma_k = g'(1/C) \max_{1 \leq i \leq C}\{d_{i,k}\}$ as an upper bound of the γ_k satisfying condition given in Eq. (3), because $\gamma_k = g'(1/C) \max_{1 \leq i \leq C}\{d_{i,k}\} \Leftrightarrow \gamma_k / \max_{1 \leq i \leq C}\{d_{i,k}\} = g'(1/C) \Leftrightarrow g'(\gamma_k / \max_{1 \leq i \leq C}\{d_{i,k}\}) = 1/C \Rightarrow u_{i,k}(\gamma_k) \geq 1/C$ for all $i \in \{1, \cdots, C\}$, which involves $\sum_{i=1}^{C} u_{i,k}(\gamma_k) \geq 1$. Using these bounds, the value of γ_k satisfying condition given in Eq. (3) can be obtained using the following algorithm:

Algorithm 1

STEP 1. Let γ_k^- and γ_k^+ be $g'(1/C) \min_{1 \leq i \leq C}\{d_{i,k}\}$ and $g'(1/C) \max_{1 \leq i \leq C}\{d_{i,k}\}$, respectively.

STEP 2. Let $\tilde{\gamma}_k$ be $(\gamma_k^- + \gamma_k^+)/2$. If $|\gamma_k^- - \gamma_k^+|$ is sufficiently small, terminate the algorithm and let the optimal γ_k be $\tilde{\gamma}_k$.

STEP 3. Calculate $u_{i,k}(\tilde{\gamma}_k)$ using Eq. (18). If $\sum_{i=1}^C u_{i,k}(\tilde{\gamma}_k) > 1$, let $\gamma_k^+ = \tilde{\gamma}_k$. Otherwise, let $\gamma_k^- = \tilde{\gamma}_k$. Go to Step 2.   □

With the resulting value of γ_k, optimal membership is described by Eq. (18).

(ii) If there exists one $i \in \{1, \ldots, C\}$ such that $d_{i,k} = 0$, we heuristically find

$$u_{i,k} = \begin{cases} 1 & (d_{i,k} = 0) \\ 0 & (\text{otherwise}). \end{cases} \tag{20}$$

This is confirmed as follows. Using order statistic notation, $\{d_{i,k}\}_{i=1}^C$ is rewritten as $\{d_{[i],k}\}_{i=1}^C$ where

$$0 = d_{[1],k} < d_{[2],k} \le \cdots \le d_{[C],k}. \tag{21}$$

Denote the membership value corresponding to $d_{[i],k}$ as $u_{[i],k}$. The objective function value with respect to x_k is then described as

$$\sum_{i=2}^C g(u_{[i],k}) d_{[i],k}. \tag{22}$$

Following Eq. (20), this value is

$$\sum_{i=2}^C g(0) d_{[i],k}. \tag{23}$$

If at least one $i \in \{2, \ldots, C\}$ such that $u_{[i],k} > 0$, the value of Eq. (22) is greater than that of Eq. (23) because g is strictly increasing. Therefore, Eq. (20) is the optimal membership when there is one $i \in \{1, \ldots, C\}$ such that $d_{i,k} = 0$.

According to the above discussion, we propose the following gFCM algorithm:

Algorithm 2 (gFCM)

STEP 1. Specify the number of clusters C and the set of initial cluster centers v.

STEP 2. If there exists $i \in \{1, \ldots, C\}$ such that $d_{i,k} = 0$, set u according to Eq. (20). Otherwise, calculate γ_k using Algorithm 1, and set u according to Eq. (18).

STEP 3. Calculate v using Eqs. (11), (12), and (14).

STEP 4. Check the stopping criterion for (γ, u, v). If the criterion is not satisfied, go to Step 2.   □

The FCF of this gFCM is obtained by solving the optimization problem

$$\underset{u}{\text{minimize}} \sum_{i=1}^C g(u_i(x)) d_i(x), \tag{24}$$

with $\sum_{i=1}^C u_i(x) = 1$, where $d_i(x) = \|x - v_i\|_2^2$. This can be accomplished using the following algorithm:

Algorithm 3 (FCF of gFCM)

STEP 1. Inherit v from Algorithm 2, and set $x \in \mathbb{R}^p$.

STEP 2. If there exists $i \in \{1, \ldots, C\}$ such that $d_i(x) = 0$, set u according to Eq. (20), where $u_{i,k}$ is replaced by $u_i(x)$. Otherwise, calculate $\gamma(x)$ using Algorithm 1, where γ_k, γ_k^+, γ_k^-, and $d_{i,k}$ are replaced by $\gamma(x)$, $\gamma(x)^+$, $\gamma(x)^-$, and $d_i(x)$, and set $\{u_i(x)\}_{i=1}^{C}$ using

$$u_i(x) = g'^{-1}\left(\frac{\gamma(x)}{d_i(x)}\right). \qquad (25)$$

$\square$

Next, we determine a property of gFCM from its FCF. First, the crisp allocation rule for the FCF of our gFCM, which classifies $\mathbb{R}^p$ according to Eq. (8), produces Voronoi sets, as in Eq. (9). This is the same as in both HCM and FCM. This is obvious if there exists one $i \in \{1, \ldots, C\}$ such that $d_i(x) = 0$, because we have

$$1 = u_i(x) > u_j(x) = 0 \Leftrightarrow 0 = d_i(x) < d_j(x) \Leftrightarrow \|x - v_i\|_2 < \|x - v_j\|_2, \qquad (26)$$

for $j \neq i$. If $d_i(x) > 0$ for all $i \in \{1, \ldots, C\}$, then we have

$$u_i(x) > u_j(x) \Leftrightarrow g'^{-1}\left(\frac{\gamma(x)}{d_i(x)}\right) > g'^{-1}\left(\frac{\gamma(x)}{d_j(x)}\right) \Leftrightarrow d_i(x) < d_j(x)$$
$$\Leftrightarrow \|x - v_i\|_2 < \|x - v_j\|_2, \qquad (27)$$

for $j \neq i$.

Next, we determine the FCF value as $\|x\|_2$ approaches infinity. We know that

$$\frac{d_j(x)}{d_i(x)} = \frac{\|x - v_j\|_2^2}{\|x - v_i\|_2^2} \to 1 \qquad (28)$$

as $\|x\|_2 \to +\infty$. Denoting

$$\varepsilon_{i,j}(x) \overset{\text{def}}{=} \frac{\|x - v_j\|_2^2}{\|x - v_i\|_2^2} - 1 \qquad (29)$$

for $i, j \in \{1, \ldots, C\}$, this value is bounded as

$$
\begin{aligned}
\varepsilon_{i,j}(x) &= \frac{\|x - v_i + v_i - v_j\|_2^2}{\|x - v_i\|_2^2} - 1 \\
&= \frac{\|x - v_i\|_2^2 + 2(x - v_i)^{\mathsf{T}}(v_i - v_j) + \|v_i - v_j\|_2^2}{\|x - v_i\|_2^2} - 1 \\
&= \frac{2(x - v_i)^{\mathsf{T}}(v_i - v_j) + \|v_i - v_j\|_2^2}{\|x - v_i\|_2^2} \\
&= \frac{1}{\|x - v_i\|_2}\left(\frac{2(x - v_i)^{\mathsf{T}}(v_i - v_j) + \|v_i - v_j\|_2^2}{\|x - v_i\|_2}\right) \\
&\leq \frac{1}{\|x - v_i\|_2}\left(\frac{2\|x - v_i\|_2\|v_i - v_j\|_2 + \|v_i - v_j\|_2^2}{\|x - v_i\|_2}\right) \\
&= \frac{1}{\|x - v_i\|_2}\left(2\|v_i - v_j\|_2 + \frac{\|v_i - v_j\|_2^2}{\|x - v_i\|_2}\right). \qquad (30)
\end{aligned}
$$

Since we have

$$\|x - v_i\|_2 > \|v_i - v_j\|_2 \tag{31}$$

for sufficiently large $\|x\|_2$, the value $\varepsilon_{i,j}(x)$ is further bounded as

$$
\begin{aligned}
\varepsilon_{i,j}(x) &\leq \frac{1}{\|x - v_i\|_2} \left(2\|v_i - v_j\|_2 + \frac{\|v_i - v_j\|_2^2}{\|x - v_i\|_2} \right) \\
&< \frac{1}{\|x - v_i\|_2} \left(2\|v_i - v_j\|_2 + \frac{\|x - v_i\|_2\|v_i - v_j\|_2}{\|x - v_i\|_2} \right) \\
&= \frac{1}{\|x - v_i\|_2} \left(2\|v_i - v_j\|_2 + \|v_i - v_j\|_2 \right) = \frac{3\|v_i - v_j\|_2}{\|x - v_i\|_2}
\end{aligned} \tag{32}
$$

for sufficiently large $\|x\|_2$. Since there exists arbitrary large M such as $\|x - v_i\|_2 > 3M\|v_i - v_j\|_2$, we have

$$\varepsilon_{i,j}(x) < \frac{3\|v_i - v_j\|_2}{\|x - v_i\|_2} < \frac{3\|v_i - v_j\|_2}{3M\|v_i - v_j\|_2} = \frac{1}{M}, \tag{33}$$

which implies that $\varepsilon_{i,j}(x)$ is uniformly bounded for sufficiently small $1/M$ without depending on $i, j \in \{1, \dots, C\}$.

Since we have

$$g'^{-1} \left(\frac{\gamma(x)}{d_j(x)} \right) \leq 1 \tag{34}$$

from Eq. (3), there exists a finite number M' such that

$$\left| \frac{\gamma(x)}{d_j(x)} \right| < M'. \tag{35}$$

Then, Eqs. (29), (33) and (35) imply that

$$
\begin{aligned}
\left| \frac{\gamma(x)}{d_j(x)} - \frac{\gamma(x)}{d_i(x)} \right| &= \left| \frac{\gamma(x)}{d_i(x)}(1 + \varepsilon_{j,i}(x)) - \frac{\gamma(x)}{d_i(x)} \right| = \left| \frac{\gamma(x)}{d_i(x)} \varepsilon_{j,i}(x) \right| \\
&\leq \left| \frac{\gamma(x)}{d_i(x)} \right| |\varepsilon_{j,i}(x)| < \frac{M'}{M}
\end{aligned} \tag{36}
$$

for sufficiently large $\|x\|_2$, thus, there exists sufficiently small $\varepsilon'_{j,i}$ such that

$$g'^{-1} \left(\frac{\gamma(x)}{d_j(x)} \right) = g'^{-1} \left(\frac{\gamma(x)}{d_i(x)} (1 + \varepsilon_{j,i}(x)) \right) \leq g'^{-1} \left(\frac{\gamma(x)}{d_i(x)} \right) + \varepsilon'_{j,i} \tag{37}$$

because g'^{-1} is continuous. Here, we define $\varepsilon' \overset{\text{def}}{=} \max_{1 \leq i,j \leq C} \{\varepsilon'_{j,i}\}$, i.e., we have

$$g'^{-1} \left(\frac{\gamma(x)}{d_j(x)} \right) \leq g'^{-1} \left(\frac{\gamma(x)}{d_i(x)} \right) + \varepsilon'. \tag{38}$$

Then, Eqs. (3) and (38) imply that

$$1 = \sum_{j=1}^{C} u_j(x) = \sum_{j=1}^{C} g'^{-1}\left(\frac{\gamma(x)}{d_j(x)}\right) \leq \sum_{j=1}^{C}\left(g'^{-1}\left(\frac{\gamma(x)}{d_i(x)}\right) + \varepsilon'\right)$$
$$= Cg'^{-1}\left(\frac{\gamma(x)}{d_i(x)}\right) + C\varepsilon', \tag{39}$$

thus, we have

$$g'^{-1}\left(\frac{\gamma(x)}{d_i(x)}\right) \geq \frac{1}{C} - \varepsilon' \tag{40}$$

for all $i \in \{1, \ldots, C\}$, where we note that $|\varepsilon'|$ is sufficiently small for sufficiently large $\|x\|_2$, from which we have

$$u_i(x) = g'^{-1}\left(\frac{\gamma(x)}{d_i(x)}\right) \to \frac{1}{C} \quad (\|x\|_2 \to +\infty) \tag{41}$$

for all $i \in \{1, \ldots, C\}$. This result substantiates the property of FCM.

4 Numerical Examples

This section presents some numerical examples to substantiate the properties of the gFCM (Algorithm 2 discussed in the previous section. We consider three actual fuzzifiers g and one artificial dataset (Fig. 5).

The three fuzzifiers are defined as

$$g_1(u_{i,k}) = (u_{i,k})^2, \tag{42}$$
$$g_2(u_{i,k}) = 2^{u_{i,k}} - 1 - \ln(2)u_{i,k}, \tag{43}$$
$$g_3(u_{i,k}) = \frac{1}{(1 - u_{i,k})^{10^{-15}}} - \frac{1}{10^{15}}u_{i,k} \tag{44}$$

for all $i \in \{1, \ldots, C\}$, $k \in \{1, \ldots, N\}$, where g_1 leads to sFCM with $m = 2$. From the above, we have

$$g_1'(u_{i,k}) = 2u_{i,k}, \tag{45}$$
$$g_2'(u_{i,k}) = \ln(2)(2^{u_{i,k}} - 1), \tag{46}$$
$$g_3'(u_{i,k}) = \frac{1}{10^{15}}\left(\frac{1}{(1 - u_{i,k})^{1+10^{-15}}} - 1\right), \tag{47}$$

and

$$g_1'^{-1}(y) = 0.5y, \tag{48}$$
$$g_2'^{-1}(y) = \frac{1}{\ln(2)}\ln\left(\frac{y + \ln(2)}{\ln(2)}\right), \tag{49}$$
$$g_3'^{-1}(y) = 1 - \frac{1}{(1 + 10^{15}y)^{1/(1+10^{-15})}}. \tag{50}$$

These functions, as well as their derivatives and inverse derivative functions, are shown in Figs. 2, 3 and 4. Obviously, all the functions are defined in $u_{i,k} \in [0,1)$ and are strictly convex, increasing, non-negative, and smooth; all derivatives and their inverse functions are monotonically increasing.

The substantiated property is that Eq. (41) is satisfied using the four-cluster dataset, where each cluster comprises 66 points in a two-dimensional space, as shown in Fig. 5. Partitioning this dataset into four clusters via gFCM with all five fuzzifiers $\{g_q\}_{q=1}^{3}$ elicits the features of the classification rules, where the top cluster and its cluster center are denoted by G_1 and v_1, the bottom left cluster and its cluster center are denoted by G_2 and v_2, the bottom right cluster and its cluster center are denoted by G_3 and v_3, and the mid cluster and its cluster center are denoted by G_4 and v_4. The derived FCFs for G_1 are shown in

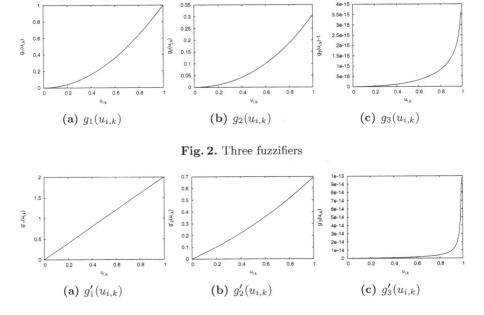

(a) $g_1(u_{i,k})$ (b) $g_2(u_{i,k})$ (c) $g_3(u_{i,k})$

Fig. 2. Three fuzzifiers

(a) $g_1'(u_{i,k})$ (b) $g_2'(u_{i,k})$ (c) $g_3'(u_{i,k})$

Fig. 3. Derivative functions of the three fuzzifiers

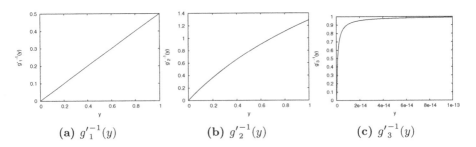

(a) $g_1'^{-1}(y)$ (b) $g_2'^{-1}(y)$ (c) $g_3'^{-1}(y)$

Fig. 4. Inverse derivative functions for the three fuzzifiers

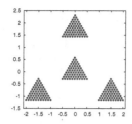

Fig. 5. Artificial dataset.

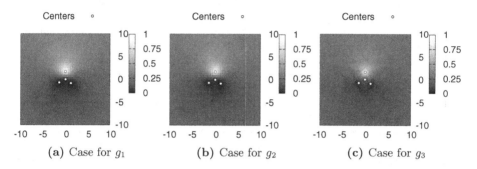

(**a**) Case for g_1 (**b**) Case for g_2 (**c**) Case for g_3

Fig. 6. FCFs of gFCM for an edge cluster

Fig. 6, with the gray-scale indicates the degree of membership to G_1. The circles represent cluster centers. These figures confirm that Eq. (41) holds.

5 Conclusion

The study described in this paper has shown that generalized fuzzy clustering exhibits the same features as sFCM, namely that the FCF at the infinity point approaches the reciprocal of the cluster number. Several numerical examples have been presented to substantiate the above theoretical results.

In future research, we aim to investigate

- a case with $g'(0) > 0$,
- a case with other types of object-cluster dissimilarity than in Eq. (1)
- a case with variable controlling cluster size [5,6], and
- cases of clustering for spherical data [7–12] and for categorical multivariate data [13–18].

References

1. MacQueen, J.B.: Some methods of classification and analysis of multivariate observations. In: Proceedings the 5th Berkeley Symposium on Mathematical Statistics and Probability, pp. 281–297 (1967)
2. Bezdek, J.: Pattern Recognition with Fuzzy Objective Function Algorithms. Plenum Press, New York (1981)
3. Miyamoto, S., Umayahara, K.: Methods in hard and fuzzy clustering. In: Liu, Z.-Q., Miyamoto, S. (eds.) Soft Computing and Human-Centered Machines. Springer, Tokyo (2000). https://doi.org/10.1007/978-4-431-67907-3_5
4. Miyamoto, S., Ichihashi, H., Honda, K.: Algorithms for Fuzzy Clustering. Springer, Heidelberg (2008). https://doi.org/10.1007/978-3-540-78737-2
5. Miyamoto, S., Kurosawa, N.: Controlling cluster volume sizes in fuzzy c-means Clustering. In: Proceedings of SCIS & ISIS 2004, pp. 1–4 (2004)
6. Ichihashi, H., Honda, K., Tani, N.: Gaussian mixture PDF approximation and fuzzy c-means clustering with entropy regularization. In: Proceedings of the 4th Asian Fuzzy System Symposium, pp. 217–221 (2000)
7. Dhillon, I.S., Modha, D.S.: Concept decompositions for large sparse text data using clustering. Mach. Learn. **42**, 143–175 (2001)
8. Miyamoto, S., Mizutani, K.: Fuzzy multiset model and methods of nonlinear document clustering for information retrieval. In: Torra, V., Narukawa, Y. (eds.) MDAI 2004. LNCS (LNAI), vol. 3131, pp. 273–283. Springer, Heidelberg (2004). https://doi.org/10.1007/978-3-540-27774-3_26
9. Mizutani, K., Inokuchi, R., Miyamoto, S.: Algorithms of nonlinear document clustering based on fuzzy set model. Int. J. Intell. Syst. **23**(2), 176–198 (2008)
10. Kanzawa, Y.: On kernelization for a maximizing model of Bezdek-like spherical fuzzy c-means clustering. In: Torra, V., Narukawa, Y., Endo, Y. (eds.) MDAI 2014. LNCS (LNAI), vol. 8825, pp. 108–121. Springer, Cham (2014). https://doi.org/10.1007/978-3-319-12054-6_10
11. Kanzawa, Y.: A maximizing model of bezdek-like spherical fuzzy c-means. J. Adv. Comput. Intell. Intell. Inform. **19**(5), 662–669 (2015)
12. Kanzawa, Y.: A maximizing model of spherical Bezdek-type fuzzy multi-medoids clustering. J. Adv. Comput. Intell. Intell. Inform. **19**(6), 738–746 (2015)
13. Oh, C., Honda, K., Ichihashi, H.: Fuzzy clustering for categorical multivariate data. In: Proceedings IFSA World Congress and 20th NAFIPS International Conference, pp. 2154–2159 (2001)
14. Honda, K., Oshio, S., Notsu, A.: FCM-type fuzzy co-clustering by K-L information regularization. In: Proceedings of 2014 IEEE International Conference on Fuzzy Systems, pp. 2505–2510 (2014)
15. Honda, K., Oshio, S., Notsu, A.: Item membership fuzzification in fuzzy co-clustering based on multinomial mixture concept. In: Proceedings of 2014 IEEE International Conference on Granular Computing, pp. 94–99 (2014)
16. Kanzawa, Y.: Fuzzy co-clustering algorithms based on fuzzy relational clustering and TIBA imputation. J. Adv. Comput. Intell. Intell. Inform. **18**(2), 182–189 (2014)
17. Kanzawa, Y.: On possibilistic clustering methods based on Shannon/Tsallis-entropy for spherical data and categorical multivariate data. In: Torra, V., Narukawa, Y. (eds.) MDAI 2015. LNCS (LNAI), vol. 9321, pp. 115–128. Springer, Cham (2015). https://doi.org/10.1007/978-3-319-23240-9_10
18. Kanzawa, Y.: Bezdek-type fuzzified co-clustering algorithm. J. Adv. Comput. Intell. Intell. Inform. **19**(6), 852–860 (2015)

A Self-tuning Possibilistic c-Means Clustering Algorithm

László Szilágyi[1,2(✉)], Szidónia Lefkovits[3], and Zsolt Levente Kucsván[1]

[1] Computational Intelligence Research Group,
Sapientia - Hungarian Science University of Transylvania, Tîrgu Mureş, Romania
`lalo@ms.sapientia.ro`
[2] Department of Control Engineering and Information Technology,
Budapest University of Technology and Economics, Budapest, Hungary
[3] Department of Informatics, Petru Maior University, Tîrgu Mureş, Romania

Abstract. Most c-means clustering models have serious difficulties when facing clusters of different sizes and severely outlier data. The possibilistic c-means (PCM) algorithm can handle both problems to some extent. However, its recommended initialization using a terminal partition produced by the probabilistic fuzzy c-means does not work when severe outliers are present. This paper proposes a possibilistic c-means clustering model that uses only three parameters independently of the number of clusters, which is able to more robustly handle the above mentioned obstacles. Numerical evaluation involving synthetic and standard test data sets prove the advantages of the proposed clustering model.

Keywords: Fuzzy c-means clustering
Possibilistic c-means clustering · Cluster size sensitivity · Outlier data

1 Introduction

The family of c-means clustering models includes several algorithms that involve fuzzy partitions. The probabilistic fuzzy c-means (FCM) algorithm introduced by Bezdek [3] is very popular in various researches, as it can reliably produce fine – or at least acceptable – fuzzy partitioning of most sets of object vectors with numeric values. However, there are two well-known cases when FCM can crash: (1) at a sufficiently high value of the fuzzy exponent m, cluster prototypes can merge together at the grand mean of the input data; (2) outlier data strongly attract cluster prototypes, and in extreme cases, each outlier may steal a cluster prototype, causing invalid partitioning. Clusters created by FCM are of equal

This research was partially supported by the Institute for Research Programs of the Sapientia University. The work of L. Szilágyi was additionally supported by the Hungarian Academy of Sciences through the János Bolyai Fellowship Program. The work of Sz. Lefkovits was additionally supported by UEFISCDI through grant no. PN-III-P2-2.1-BG-2016-0343, contract no. 114BG.

V. Torra et al. (Eds.): MDAI 2018, LNAI 11144, pp. 255–266, 2018.
https://doi.org/10.1007/978-3-030-00202-2_21

diameter, which represents a serious limitation [5]. Modified probabilistic fuzzy c-means algorithms were introduced to handle clusters of different sizes: Miyamoto and Kurosawa [9] compensated the diameter of the clusters, while the solution of Yang [13] deals with clusters of different weight or cardinality. Another solution based on the conditional FCM algorithm of Pedrycz [10] was recently introduced by Leski [8].

Various versions of c-means clustering algorithms were also proposed to deal with the problem of outliers. These algorithms relaxed the probabilistic constraint of the fuzzy partition. Dave [4] proposed the usage of an extra cluster that attracts noisy data, but this algorithm still creates clusters of equal diameter. The possibilistic c-means (PCM) clustering algorithm [6] addresses the uneven sized clusters as well, via defining a dedicated penalty term for each cluster that would compensate for the variance of the data within the cluster, but frequently creates coincident clusters [2], due to the strong independence of the clusters. Further on, PCM is initialized with a final FCM partition [7], which is only viable if the FCM performed successfully. For example, if an outlier damages the probabilistic partition created by FCM and consequently we set the wrong penalty terms to clusters at the initialization of PCM, we can hardly achieve an accurate partitioning.

In a previous paper [11] we proposed a clustering model that combined the $(c + 1)$-means approach of Dave [4] containing a noise cluster, with the diameter compensation mechanism described by Miyamoto and Kurosawa [9]. That algorithm showed moderately capable to create valid clusters of different sizes. In this paper we propose a modified possibilistic fuzzy c-means clustering approach that incorporates the above mentioned diameter compensation mechanism with the aim of self-tuning the penalty terms of each cluster. While the original PCM algorithm handled the penalty terms as previously set constants, the proposed self-tuning possibilistic c-means algorithm updates the penalty term of each cluster during the iterations of the alternative optimization of the objective function.

2   Background

The Fuzzy c-Means Algorithm. The conventional fuzzy c-means (FCM) algorithm partitions a set of object data $\mathbf{X} = \{\mathbf{x}_1, \mathbf{x}_2, \ldots, \mathbf{x}_n\}$ into a number of c clusters $\Omega_1, \Omega_2, \ldots, \Omega_c$ based on the minimization of a quadratic objective function, defined as:

$$J_{\text{FCM}} = \sum_{i=1}^{c} \sum_{k=1}^{n} u_{ik}^m ||\mathbf{x}_k - \mathbf{v}_i||_{\mathbf{A}}^2 = \sum_{i=1}^{c} \sum_{k=1}^{n} u_{ik}^m d_{ik}^2, \tag{1}$$

where $\mathbf{v}_i$ represents the prototype or centroid of cluster i ($i = 1 \ldots c$), $u_{ik} \in [0, 1]$ is the fuzzy membership function showing the degree to which vector $\mathbf{x}_k$ belongs to cluster i, $m > 1$ is the fuzzyfication parameter, and d_{ik} represents the distance (any inner product norm defined by a symmetrical positive definite matrix $\mathbf{A}$)

between $\mathbf{x}_k$ and $\mathbf{v}_i$. FCM uses a probabilistic partition, meaning that the fuzzy memberships assigned to any input vector $\mathbf{x}_k$ with respect to clusters satisfy the probability constraint $\sum_{i=1}^{c} u_{ik} = 1$. The minimization of the objective function J_{FCM} is achieved by alternately applying the optimization of J_{FCM} over $\{u_{ik}\}$ with $\mathbf{v}_i$ fixed, $i = 1 \ldots c$, and the optimization of J_{FCM} over $\{\mathbf{v}_i\}$ with u_{ik} fixed, $i = 1 \ldots c$, $k = 1 \ldots n$ [3]. Obtaining the optimization formulas involves zero gradient conditions of J_{FCM} and Langrange multipliers. Iterative optimization is applied until cluster prototypes $\mathbf{v}_i$ ($i = 1 \ldots c$) converge.

Relaxing the Probabilistic Constraint. The relaxation of the probabilistic constraint was a necessity provoked by the outlier sensitivity of the FCM algorithm. The most popular of the existing solutions is the possibilistic c-means (PCM) algorithm introduced by Krishnapuram and Keller [6], which optimizes

$$J_{\text{PCM}} = \sum_{i=1}^{c} \sum_{k=1}^{n} \left[t_{ik}^p d_{ik}^2 + (1 - t_{ik})^p \eta_i \right], \tag{2}$$

constrained by $0 \leq t_{ik} \leq 1 \ \forall i = 1 \ldots c$, $\forall k = 1 \ldots n$, and $0 < \sum_{i=1}^{c} t_{ik} < c$ $\forall k = 1 \ldots n$, where $p > 1$ represents the possibilistic exponent, and parameters η_i are the penalty terms that control the diameter of the clusters. The iterative optimization algorithm of PCM objective function is derived from the zero gradient conditions of J_{PCM}. In the probabilistic fuzzy partition, the degrees of membership assigned to an input vector $\mathbf{x}_k$ with respect to cluster i depends on the distances of the given vector to all cluster prototypes: $d_{1k}, d_{2k}, \ldots, d_{ck}$. On the other hand, in the possibilistic partition, the typicality value u_{ik} assigned to input vector $\mathbf{x}_k$ with respect to any cluster i depends on only one distance: d_{ik}. PCM efficiently suppresses the effects of outlier data, at the price of frequently producing coincident cluster prototypes. The latter is the result of the highly independent cluster prototypes [2].

Fuzzy c-Means with Various Cluster Diameters. Komazaki et al. [5] presented a collection of solutions how the FCM algorithm can adapt to different cluster sizes and diameters. From the point of view of this paper, it is relevant to mention the FCMA algorithm by Miyamoto and Kurosawa [9], which minimizes

$$J_{\text{FCMA}} = \sum_{i=1}^{c} \sum_{k=1}^{n} \alpha_i^{1-m} u_{ik}^m d_{ik}^2, \tag{3}$$

subject to the probabilistic constraint of the fuzzy memberships u_{ik} ($i = 1 \ldots c$, $k = 1 \ldots n$), and of the extra terms α_i ($i = 1 \ldots c$): $\sum_{i=1}^{c} \alpha_i = 1$. The optimization algorithm of J_{FCMA} can be derived from zero gradient conditions using Lagrange multipliers. Each iteration updates the probabilistic memberships u_{ik} ($i = 1 \ldots c$, $k = 1 \ldots n$), the cluster prototypes $\mathbf{v}_i$ ($i = 1 \ldots c$), and the cluster diameter terms α_i ($i = 1 \ldots c$) as well. The algorithm stops when cluster prototypes stabilize.

3 Methods

In the following, we propose a modified possibilistic c-means clustering algorithm, which incorporates the cluster diameter compensation mechanism introduced by Miyamoto and Kurosawa [9] for the probabilistic FCM algorithm. Further on, instead of using a dedicated penalty terms to handle the variance of each cluster, here we propose using a single penalty term η. The proposed objective function is:

$$J_{\text{st}-\text{PCM}} = \sum_{i=1}^{c}\sum_{k=1}^{n}[\alpha_i^{1-q}t_{ik}^p d_{ik}^2 + (1-t_{ik})^p\eta], \tag{4}$$

where $d_{ik} = \|\mathbf{x}_k - \mathbf{v}_i\|$ ($\forall i = 1\ldots c$, $\forall k = 1\ldots n$), and $p > 1$ is the possibilistic exponent first introduced by Krishnapuram and Keller [6]. Variables α_i ($i = 1\ldots c$) are intended to tune the algorithm according to clusters diameters, and they satisfy the probabilistic constraint $\sum_{i=1}^{c}\alpha_i = 1$, while the exponent $q > 1$ should be treated independently from the possibilistic exponent p. The minimization formulas of the objective function given in Eq. (4) are obtained using zero gradient conditions and Lagrange multipliers. Let us consider the functional

$$\mathcal{L}_{\text{st}-\text{PCM}} = J_{\text{st}-\text{PCM}} + \lambda_\alpha\left(1 - \sum_{i=1}^{c}\alpha_i\right), \tag{5}$$

where λ_α represents the Lagrange multiplier. The zero gradient conditions with respect to variables α_i ($\forall i = 1\ldots c$) imply

$$\frac{\partial\mathcal{L}_{\text{st}-\text{PCM}}}{\partial\alpha_i} = 0 \quad \Rightarrow \quad \sum_{k=1}^{n}(1-q)\alpha_i^{-q}t_{ik}^p d_{ik}^2 = \lambda_\alpha, \tag{6}$$

and so

$$\alpha_i = \left(\frac{1-q}{\lambda_\alpha}\right)^{1/q}\left(\sum_{k=1}^{n}t_{ik}^p d_{ik}^2\right)^{1/q}. \tag{7}$$

According to the probabilistic constraint $\sum_{i=1}^{c}\alpha_i = 1$, we have:

$$\sum_{j=1}^{c}\alpha_j = 1 \quad \Rightarrow \quad 1 = \left(\frac{1-q}{\lambda_\alpha}\right)^{1/q}\sum_{j=1}^{c}\left[\left(\sum_{k=1}^{n}t_{jk}^p d_{jk}^2\right)^{1/q}\right]. \tag{8}$$

Equations (7) and (8) allow us to eliminate the Lagrange multiplier λ_α from the formula of α_i:

$$\alpha_i = \frac{\alpha_i}{1} = \frac{\alpha_i}{\sum_{j=1}^{c}\alpha_j} = \frac{\left(\sum_{k=1}^{n}t_{ik}^p d_{ik}^2\right)^{1/q}}{\sum_{j=1}^{c}\left[\left(\sum_{k=1}^{n}t_{jk}^p d_{jk}^2\right)^{1/q}\right]}. \tag{9}$$

Algorithm 1. The proposed algorithm

Data: Input data $\mathbf{X} = \{\mathbf{x}_1, \mathbf{x}_2, \ldots, \mathbf{x}_n\}$
Data: Number of clusters c, possibilistic exponent p, cluster diameter
      regulation exponent q, penalty term η, threshold ε
Result: Possibilistic partition t_{ik} ($\forall i = 1 \ldots c$, $\forall k = 1 \ldots n$)
Initialize $\mathbf{v}_i^{(\text{new})}$ ($i = 1 \ldots c$) as random input vectors that have several input
vectors in their close neighborhood, to avoid outliers
$\alpha_i \leftarrow 1/c$, $\forall i = 1 \ldots c$
repeat
   | $\mathbf{v}_i^{(\text{old})} \leftarrow \mathbf{v}_i^{(\text{new})}$, $\forall i = 1 \ldots c$
   | Update partition t_{ik}, ($i = 1 \ldots c$, $k = 1 \ldots n$), according to Eq. (10)
   | Update α_i values for any $i = 1 \ldots c$, according to Eq. (9)
   | Obtain new cluster prototypes $\mathbf{v}^{(\text{new})}$, ($i = 1 \ldots c$), according to Eq. (11)
until $\sum_{i=1}^{c} ||\mathbf{v}_i^{(\text{new})} - \mathbf{v}_i^{(\text{old})}|| < \varepsilon$;

The optimization formula of the possibilistic partition is obtained from the zero gradient condition:

$$\frac{\partial \mathcal{L}_{\text{st-PCM}}}{\partial t_{ik}} = 0 \Rightarrow \alpha_i^{1-q} p t_{ik}^{p-1} d_{ik}^2 - p(1 - t_{ik})^{p-1} \eta = 0$$

$$\Rightarrow \left(\frac{1-t_{ik}}{t_{ik}}\right)^{p-1} = \frac{\alpha_i^{1-q} d_{ik}^2}{\eta}$$

$$\Rightarrow \frac{1}{t_{ik}} - 1 = \left(\frac{\alpha_i^{1-q} d_{ik}^2}{\eta}\right)^{1/(p-1)} = \sqrt[p-1]{\frac{\alpha_i^{1-q} d_{ik}^2}{\eta}} \tag{10}$$

$$\Rightarrow t_{ik} = \left[1 + \left(\frac{d_{ik}^2}{\eta \alpha_i^{q-1}}\right)^{1/(p-1)}\right]^{-1} = \left(1 + \sqrt[p-1]{\frac{d_{ik}^2}{\eta \alpha_i^{q-1}}}\right)^{-1}.$$

Compared to the original PCM algorithm, where the penalty term for cluster Ω_i was η_i ($i = 1 \ldots c$), here we have the penalty term $\eta \alpha_i^{q-1}$ ($i = 1 \ldots c$).

The update formula of cluster prototypes $\mathbf{v}_i$ ($i = 1 \ldots c$) is obtained as:

$$\frac{\partial \mathcal{L}_{\text{st-PCM}}}{\partial \mathbf{v}_i} = 0 \Rightarrow \quad -2\alpha_i^{1-q} \sum_{k=1}^{n} t_{ik}^p (\mathbf{x_k} - \mathbf{v_i}) = 0$$

$$\Rightarrow \quad \sum_{k=1}^{n} t_{ik}^p \mathbf{x_k} = \mathbf{v_i} \sum_{k=1}^{n} t_{ik}^p \tag{11}$$

$$\Rightarrow \quad \mathbf{v_i} = \frac{\sum_{k=1}^{n} t_{ik}^p \mathbf{x_k}}{\sum_{k=1}^{n} t_{ik}^p}.$$

So the cluster prototypes are updated exactly the same way, as in case of the PCM algorithms. If a defuzzyfied partition is desired, any input vector $\mathbf{x}_k$ can be assigned to cluster Ω_j where $j = \arg\max_i \{t_{ik}, i = 1 \ldots c\}$. Any vector $\mathbf{x}_k$ having $\sum_{i=1}^{c} t_{ik} < \theta$ can be declared outliers. The value of θ should be chosen around 0.1, but the appropriate value might be application dependent. The proposed algorithm is summarized in Algorithm 1.

4 Results and Discussion

The proposed method was evaluated on two different data sets, and its behavior was compared to FCM [3] and PCM [6]. In all test scenarios, the PCM cluster prototypes were initialized according to the final outcome by FCM on the same input data, and the penalty terms fixed according to the FCM partition, as indicated by Krishnapuram and Keller [7]:

$$\eta_i = \kappa \left(\sum_{k=1}^{n} u_{ik}^m d_{ik}^2 \right) \left(\sum_{k=1}^{n} u_{ik}^m \right)^{-1} \qquad \forall i = 1 \ldots c, \qquad (12)$$

where the value of κ varied between 1 and 2. In case of FCM and st-PCM, the cluster prototypes were initialized with randomly selected input vectors that had several other input vectors in their proximity, to avoid initialization with outliers. Test data and results will be presented in the following subsections.

4.1 Evaluation Criteria

The evaluation of the algorithms was performed using three cluster validity indexes (CVIs) and the accuracy of the final partitions. Since dedicated CVIs for possibilistic partitions are scarce in the literature, we derived our CVIs from probabilistic ones. The CVIs involved in this study are the following:

$$\text{CVI1} = \frac{1}{n} \sum_{k=1}^{n} \sqrt{\tau_k} \sum_{i=1}^{c} \left(\frac{t_{ik}}{\tau_k} \right)^2 \qquad \text{CVI2} = \frac{-1}{n} \sum_{i=1}^{c} \sum_{k=1}^{n} \left(\frac{t_{ik}}{\tau_k} \right) \log \left(\frac{t_{ik}}{\tau_k} \right)$$

$$\text{CVI3} = \frac{1}{n \min_{i \neq j} \{ \|\mathbf{v}_i - \mathbf{v}_j\| \}} \sum_{i=1}^{c} \sum_{k=1}^{n} \left(\frac{t_{ik}}{\tau_k} \right)^2 \|\mathbf{v}_i - \mathbf{x}_k\|^2, \qquad (13)$$

where $\tau_k = \sum_{i=1}^{c} t_{ik}$. These CVIs are formulated for the possibilistic partition. In case of the FCM algorithm we have $\tau_k = 1$ for any $k = 1 \ldots n$, and we use u_{ik} instead of $\frac{t_{ik}}{\tau_k}$. CVI1 is the indicator of separation, with values between 0 and 1, high accuracy indicated by high values. CVI2 is a modified entropy criterion, while CVI3 is derived from the Xie-Beni index [12]. Low values of CVI2 and CVI3 indicate high accuracy.

4.2 Tests with Two Clusters in Various Scenarios

The first data set consisted of two groups of randomly generated two-dimensional input vectors, situated inside the circle with center at $(-1.75, 0)$ and radius 2, and the circle with center at $(1.75, 0)$ and radius 1, respectively. The first group contained 400, while the second 100 vectors. One such set of input vectors is exhibited in Fig. 1(left). One thousand instances of such data sets were generated and used for all test, so that each algorithm would be repeated within the very same 1000 different settings. These random data sets were involved in numerical

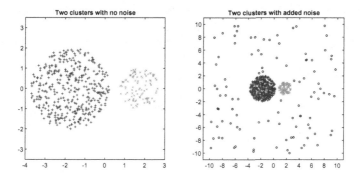

Fig. 1. The input data set with two clusters: (left) two groups of different diameter with no noise or outliers; (right) the same two groups with added noise.

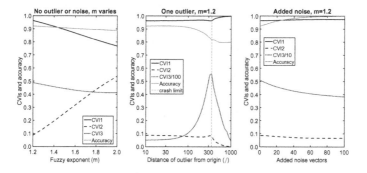

Fig. 2. Results of the evaluation of the FCM algorithm.

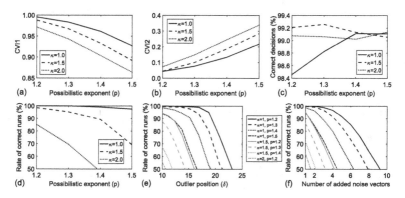

Fig. 3. Results of the evaluation of the PCM algorithm.

evaluation in three different scenarios: (1) no added outlier or noise; (2) one added outlier vector situated at a randomly generated position at distance $\delta \in [10, 1000]$ from the origin; (3) several additional noise vectors situated at random positions within a bounding box $[-10, 10] \times [-10, 10]$, as shown in Fig. 1(right).

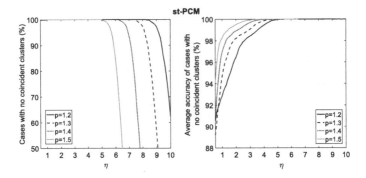

Fig. 4. Robustness and accuracy of the proposed st-PCM algorithm

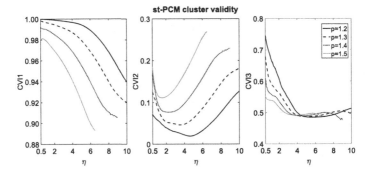

Fig. 5. The three CVIs in case of two clusters with no outlier or noise, for various values of the possibilistic exponent, plotted against penalty term η

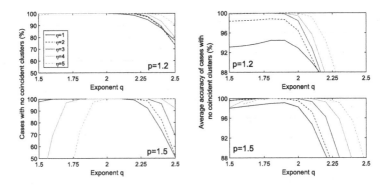

Fig. 6. The effect of the cluster diameter regulating exponent q upon the st-PCM algorithm. Values of q slightly below 2 seem the most promising.

First we tested the FCM algorithm with various values of the fuzzy exponent m. As it was expected, it performed best at low values of m. Details can be found in Fig. 2. In case of a single outlier, the probabilistic partition starts to drastically change around $\delta = 100$, and at a certain limit value around $\delta = 350$, the outlier captures one of the cluster prototypes and all 500 vectors of the two groups will belong to a single cluster. PCM has no chance to perform well, when it is initialized with such final FCM cluster prototypes. On the contrary, our noisy data does not seem an obstacle for FCM, as it creates more valid clusters as the number of added noisy vectors grows. When PCM is initialized with final FCM cluster prototypes, and penalty terms fixed according to Eq. (12), PCM performs well only to very limited extent, see details in Fig. 3. PCM performs best at $\kappa = 1$ and $p = 1.2$, but a single outlier at distance $\delta = 10$ or the presence of a few outlier vectors can significantly damage the possibilistic partition.

The proposed st-PCM algorithm works fine in a wide range of η values. However, the best range of η varies with the value of exponent q. Figures 4, 5 and 6 exhibit the test results with no outlier or additional noisy data. As Fig. 4 indicates, $p = 1.2$ gives best performance at $\eta \in [5, 8]$, while $p = 1.5$ requires $\eta \in [3, 5]$. Figure 5 shows us which parameter settings lead to the most valid clusters. Figure 6 shows us how exponent q influences the robustness and accuracy of st-PCM. There is a relation between best exponents p and q, for larger values of p we need to use larger values of q, but setting them equal to eliminate a parameter would mean a strong limitation. Figures 7 and 8 show the behavior of st-PCM, when a single outlier is present in the input data set. The position of the outlier (regulated by δ) virtually has no effect on the accuracy of the algorithm or the CVI values. Depending on the value of exponent q, η has a wide range where the algorithm is stable and accurate. Figure 9 relates to the case of additional noise vectors in the input data, where the proposed st-PCM algorithm is also stable and reliable within a wide interval of its parameters. While PCM tends to crash in the presence of a dozen noisy data, st-PCM has no problem with its accuracy and stability, even if a hundred noisy data vectors are present.

4.3   Tests with the IRIS Data Set

The second data set employed by the numerical evaluation of the algorithms was the IRIS data set [1], which consist of 150 labeled feature vectors of four dimensions, organized in three groups that contain fifty vectors each. IRIS data vectors were linearly normalized in each dimension into the interval $[-0.5, 0.5]$. The algorithms were evaluated using the IRIS data vectors only, and with additional noise as well. Just as in the previous subsection, 1000 different random noise data ·sets were generated and applied to all algorithms, all noise vectors having coordinates situated in the interval $[-\gamma/2, \gamma/2]$ in all dimensions, where γ is treated as a parameter that regulates the amplitude of added noise. Another parameter is the number of noisy vectors added to the IRIS data. The PCM and st-PCM algorithms were evaluated with values of the possibilistic exponent p ranging between 1.2 and 1.5, and γ ranging from 1 to 10, while the initial cluster prototypes were set according to the final prototypes given by the fuzzy c-means algorithm running at $m = 2$. For all tests of st-PCM, exponent q was set to 1.5. The stability of each algorithm was characterized by the rate of cases when

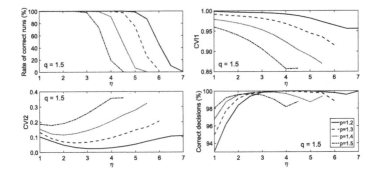

Fig. 7. Results of st-PCM at $q = 1.5$, in the presence of one outlier situated at any distance $\delta \in [10, 1000]$ from the origin. There is no visible difference within this interval.

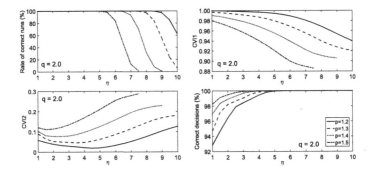

Fig. 8. Results of st-PCM at $q = 2$, in the presence of one outlier situated at any distance $\delta \in [10, 1000]$ from the origin. There is no visible difference within this interval.

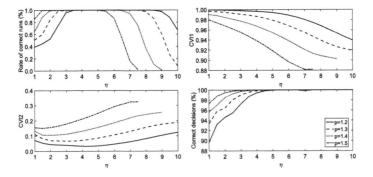

Fig. 9. Results of st-PCM at $q = 2$, in the presence of 100 additional noisy data.

valid clusters were obtained (the algorithm did not crash and no coincident clusters were obtained). Accuracy was defined as the percentage of correct decisions regarding the 150 IRIS data vectors only. The accuracy and CVIs were averaged over those runs out of 1000, where the algorithm proved stable.

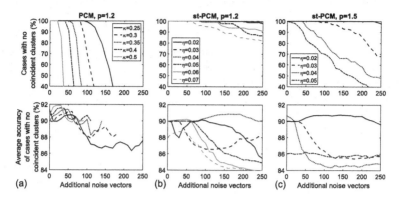

Fig. 10. Reliability (top row) and accuracy (bottom row) of the PCM (a) and st-PCM algorithms (b)–(c), when applied to noisy IRIS data, at $\gamma = 1$.

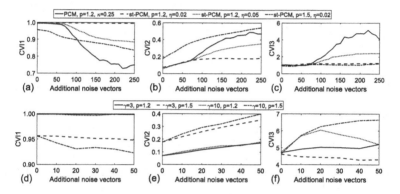

Fig. 11. CVI values of PCM and st-PCM, when applied to noisy IRIS data: (a)–(c) noise amplitude parameter $\gamma = 1$; (d)–(f) st-PCM outcomes when $\gamma \in \{3, 10\}$.

Figure 10 shows the reliability and accuracy the PCM and st-PCM algorithms can be, when applied to noisy IRIS data. In case of $\gamma = 1$, PCM can be stable in the presence of up to 100 noisy data vectors, while st-PCM can work fine at $p = 1.2$ in a wide range of the joint penalty term η. We also need to remark that the fine setting for the κ parameter that influences the initialization of PCM was found far outside the recommended range. Figure 11 relates on the validity indexes of clusters produced by st-PCM, in the presence of low and high amplitude noise. As the noise amplitude grows, the number of tolerated noise vectors reduces. However, fifty additional noisy vectors at $\gamma = 3$ or $\gamma = 10$ amplitude, or 200 noisy vectors within the normalized bounding box of the IRIS data ($\gamma = 1$), are well tolerated by the proposed st-PCM algorithm.

5 Conclusions

This paper proposed a self-tuning possibilistic c-means clustering model with a possibly reduced number of parameters (3 instead of $c + 1$) that can robustly handle distant outlier or noisy data. The behavior of the proposed algorithm was numerically validated using synthetic and standard test data sets. The proposed st-PCM algorithm proved more stable and more accurate, and in most cases provides more valid clusters than the conventional PCM. Based on the tests executed so far, we can suggest setting the cluster diameter regulation parameter $q = 2$. Further detailed numerical test could enable us to propose parameter selection strategies that would grant even finer accuracy and robustness.

References

1. Anderson, E.: The Irises of the Gaspe Peninsula. Bull. Am. Iris Soc. **59**, 2–5 (1935)
2. Barni, M., Capellini, V., Mecocci, A.: Comments on a possibilistic approach to clustering. IEEE Trans. Fuzzy Syst. **4**, 393–396 (1996)
3. Bezdek, J.C.: Pattern Recognition with Fuzzy Objective Function Algorithms. Plenum, New York (1981)
4. Dave, R.N.: Characterization and detection of noise in clustering. Patt. Recogn. Lett. **12**, 657–664 (1991)
5. Komazaki, Y., Miyamoto, S.: Variables for controlling cluster sizes on fuzzy c-means. In: Torra, V., Narukawa, Y., Navarro-Arribas, G., Megías, D. (eds.) MDAI 2013. LNCS (LNAI), vol. 8234, pp. 192–203. Springer, Heidelberg (2013). https://doi.org/10.1007/978-3-642-41550-0_17
6. Krishnapuram, R., Keller, J.M.: A possibilistic approach to clustering. IEEE Trans. Fuzzy Syst. **1**, 98–110 (1993)
7. Krishnapuram, R., Keller, J.M.: The possibilistic c-means clustering algorithm: insights and recommendation. IEEE Trans. Fuzzy Syst. **4**, 385–393 (1996)
8. Leski, J.M.: Fuzzy c-ordered-means clustering. Fuzzy Sets Syst. **286**, 114–133 (2016)
9. Miyamoto, S., Kurosawa, N.: Controlling cluster volume sizes in fuzzy c-means clustering. In: SCIS and ISIS, Yokohama, Japan, pp. 1–4 (2004)
10. Pedrycz, W.: Conditional fuzzy c-means. Patt. Recogn. Lett. **17**, 625–631 (1996)
11. Szilágyi, L., Szilágyi, S.M.: A possibilistic c-means clustering model with cluster size estimation. In: Mendoza, M., Velastín, S. (eds.) CIARP 2017. LNCS, vol. 10657, pp. 661–668. Springer, Cham (2018). https://doi.org/10.1007/978-3-319-75193-1_79
12. Xie, X.L., Beni, G.A.: Validity measure for fuzzy clustering. IEEE Trans. Pattern Anal. Mach. Intell. **3**, 841–846 (1991)
13. Yang, M.S.: On a class of fuzzy classification maximum likelihood procedures. Fuzzy Sets Syst. **57**, 365–375 (1993)

k-CCM: A Center-Based Algorithm for Clustering Categorical Data with Missing Values

Duy-Tai Dinh$^{(\boxtimes)}$ (ID) and Van-Nam Huynh

School of Knowledge Science, Japan Advanced Institute of Science and Technology, Nomi, Ishikawa, Japan
{taidinh,huynh}@jaist.ac.jp

Abstract. This paper focuses on solving the problem of clustering for categorical data with missing values. Specifically, we design a new framework that can impute missing values and assign objects into appropriate clusters. For the imputation step, we use a decision tree-based method to fill in missing values. For the clustering step, we use a kernel density estimation approach to define cluster centers and an information theoretic-based dissimilarity measure to quantify the differences between objects. Then, we propose a center-based algorithm for clustering categorical data with missing values, namely k-CCM. An experimental evaluation was performed on real-life datasets with missing values to compare the performance of the proposed algorithm with other popular clustering algorithms in terms of clustering quality. Generally, the experimental result shows that the proposed algorithm has a comparative performance when compared to other algorithms for all datasets.

Keywords: Data mining · Partitional clustering · Categorical data
Missing values · Incomplete dataset · Decision tree-based imputation

1 Introduction

Data clustering or clustering is one of the most important topics in data mining. The goal of clustering is to assign data points with similar properties to the same groups and dissimilar data points to different groups [6]. From a machine learning perspective, clusters correspond to hidden patterns, the search for clusters is unsupervised learning, and the resulting system represents a data concept. Therefore, clustering is unsupervised learning of a hidden data concept [2].

In general, clustering algorithms can be classified into two categories: hierarchical clustering and partitional clustering. A hierarchical clustering is a set of nested clusters where lower-level clusters are sub-clusters of higher-level clusters [15]. Unlike hierarchical algorithms, partitional algorithms create a one-level non-overlapping partitioning of the data points. For large datasets, hierarchical methods become impractical because the complexity of hierarchical algorithms

© Springer Nature Switzerland AG 2018
V. Torra et al. (Eds.): MDAI 2018, LNAI 11144, pp. 267–279, 2018.
https://doi.org/10.1007/978-3-030-00202-2_22

are $\mathcal{O}(N^3)$ for CPU time and $\mathcal{O}(N^2)$ for memory space, while non-hierarchical methods generally have a time and space complexity of order N, where N is the number of data points in the dataset [16]. Moreover, partitional clustering algorithms have shown their efficiency because their computational complexities are linearly proportional to the size of the datasets, they often terminate at a local optimum. Clustering algorithms are very highly associated with data types. Categorical data, which is also referred to as nominal data, appears popularly in many real-life applications. Categorical attributes are simply used as name, gender, age group, and educational level, etc. Designing partitional clustering algorithms for categorical data has attracted the attention of many researchers over the last two decades.

It can be easily seen that many categorical data from the UCI Machine Learning Repository[1] and the CMU datasets archive[2] contain missing values. Moreover, some existing frameworks for clustering categorical data such as the k-modes implementation[3] strongly suggest that users should consider filling in the missing data themselves in a way that makes sense for the problem at hand. This is especially important in case of many missing values. This observation motivates us to design an algorithm for clustering categorical data with missing values. Generally, there are two ways to facilitate a clustering algorithm run over categorical datasets with missing values. The first way is to preprocess datasets so that they only consist of complete values and then run the clustering algorithm. The second way is to develop a clustering algorithm that can deal with incomplete datasets. In this research, we focused on the latter way. More specifically, we developed a center-based algorithm that can run over incomplete categorical datasets without a preprocessing procedure. The key contributions of this paper are as follows:

- Based on the imputation method proposed in [4], we design a new measure to quantify the similarity between an object with missing values and an object with no missing values, namely MCS. By using our proposed measure and the IS measure [14], we can find the most similar object with no missing values for an object with missing values. From that, the appropriate values can be chosen for imputation.
- We design an integrated framework that combines the imputation step and clustering step into a common process.
- We propose a new categorical clustering algorithm named k-CCM that takes into account the advantages of missing values imputation to improve the performance of the clustering algorithm.
- We carry out an extensive experimental evaluation on benchmark datasets from the UCI Machine Learning Repository and the CMU datasets archive to evaluate the performance of the proposed algorithm in terms of the clustering quality.

---

[1] https://archive.ics.uci.edu/ml/datasets.html.
[2] http://lib.stat.cmu.edu/datasets.
[3] https://github.com/nicodv/kmodes.

The rest of this paper is organized as follows. In the second section, related work is reviewed. In the third section, preliminaries and problem statement are introduced. In the fourth section, a new clustering algorithm for categorical data with missing values is proposed. Next, the fifth section describes an experimental evaluation. Finally, the last section draws a conclusion.

2   Related Work

2.1   Partitional Clustering for Categorical Data

The k-means algorithm (MacQueen, 1976) is one of the most used clustering algorithms. It was designed to cluster numerical data in which each cluster has a center called the mean. Working only on numerical data restricts some applications of the k-means algorithm. More specifically, it cannot be applied directly to categorical data, which is very popular in many real-life applications nowadays.

To address this limitation, several studies have been made in order to remove the numerical-only limitation of the k-means algorithm and make it applicable to clustering for categorical data. In 1997, Huang proposed k-modes and k-prototypes algorithms [7]. The k-modes algorithm is very popular for clustering categorical data. It has some important properties [6]: it is efficient for clustering large datasets, it also produces locally optimal solutions that are dependent on initial modes and the order of objects in the data set, it works only on categorical data. The k-prototypes [7] integrates k-means and k-modes algorithms to allow for clustering objects with mixed numeric and categorical attributes. In the k-prototypes algorithm, the prototype is the center of a cluster, just as the mean and mode are the centers of a cluster in the k-means and k-modes algorithms, respectively. The k-prototypes algorithm is practically more useful because frequently encountered objects in real world databases are mixed type objects.

In 2004, San et al. proposed a k-means-like algorithm named k-representatives [13]. The k-representatives uses the Cartesian product and union operations for the formation of cluster centers based on the notion of means in the numerical setting. It uses the dissimilarity measure based on the relative frequencies of categorical values within the cluster and the simple matching measure between categorical values. The algorithmic structure of k-representatives is formed in the same way as the k-modes [7]. In 2005, Kim et al. proposed k-populations [8] that uses the notion of the population to represent the centroid of each cluster. The population is a set of pairs that contain category values and their confidence degrees for each attribute.

Recently, Chen et al. proposed a kernel-density-based clustering algorithm named k-centers [3]. The k-centers uses the kernel density estimation method to define the center of a categorical cluster, called the probabilistic center. It incorporates a built-in feature weighting in which each attribute is automatically assigned with a weight to measure its individual contribution for the clusters.

More recently, Nguyen et al. proposed three extensions of k-representatives [11]. The first extension, namely Modified 1, uses the information theoretic-based dissimilarity measure instead of the simple matching dissimilarity measure to quantify the distance between objects. The Modified 2 combines the new dissimilarity measure with the concept of cluster centers proposed by Chen et al. [3] to form clusters. The Modified 3 uses the new information theoretic-based dissimilarity measure and a modified representation of cluster centers using the kernel density based estimate. In this research, the proposed clustering method is based on the scheme of Modified 1.

2.2   Imputation Methods for Categorical Data with Missing Values

Imputation of missing values is an important task for improving the quality of the data mining result. Some of these methods are expectation maximization imputation (EMI), decision tree based methods, similarity based imputation, k-decision tree based imputation, k-nearest neighbor based imputation, genetic algorithm and correlation based imputation [4]. For imputation categorical data with missing values, Fujikawa et al. proposed two algorithms named Natural Cluster Based Mean-and-Mode (NCBMM) and attribute Rank Cluster Based Mean-and-Mode (RCBMM) [5]. The NCBMM can be applied to supervised data where missing value attributes can be either categorical or numeric. The RCBMM can be applied to both supervised and unsupervised data by filling up missing values for categorical attributes independently with the class attribute.

In 2013, Rahman proposed DMI and SiMI algorithms [12]. The DMI uses the decision tree and majority class voting method in the decision tree leaves to impute for categorical missing values. The SiMI uses the decision forest algorithm and the most frequent values method to impute for categorical missing values.

In 2016, Deb et al. proposed an imputation method named DSMI that exploits the within-record and between-record correlations to impute missing data of numerical or categorical values. The DSMI algorithm first utilizes the decision tree to find the set of correlated records. Then, it uses the IS measure and the weighted similarity measure to exploit the correlation between missing and non-missing attributes within a record. The missing values are imputed by random sampling from a list of potential imputed values based on their degree of affinity. By modifying upon this imputation method, we integrate the imputation step into clustering step to make it applicable to clustering categorical data with missing values. The next section introduces preliminary definitions and problem statement.

3   Preliminaries and Problem Statement

The problem of clustering for categorical data has been the subject of several prior studies [3,7,11,13,17]. Let $D = \{D_1, D_2, \ldots, D_m\}$ be a set of m distinct categorical attributes where the dth attribute ($1 \leq d \leq m$) takes a unique finite set O_d that contains $|O_d|$ (> 1) discrete values as its domain. A categorical object (record) is a tuple of the form $\langle id, X \rangle$ where id is its unique identifier and X is represented by a tuple $t \in O_1 \times O_2 \times \cdots \times O_m$. For the simplicity, a categorical object X having $id = k$ is denoted as X_k. A categorical dataset $S = \{X_1, X_2, \ldots, X_n\}$ is a set of n categorical objects where $X_k = \{x_{1k}, x_{2k}, \ldots, x_{mk}\}$ is a set of m categorical values at the kth element of S. On the other hand, S is a $m \times n$ matrix $(n \gg m)$, where m and n are the number of attributes and objects in dataset S, respectively. The element at position (i, j) $(0 \leq i < n, 0 \leq j < m)$ of the matrix stores the value of the object i^{th} at the attribute j^{th}, such that $x_{ij} \in O = \bigcup_{d=1}^{m} O_d$. Note that if a categorical value in S is a missing value, then it is represented as "?" or " " (empty). For the sake of brevity, we denote a categorical object/dataset without missing values as complete object/dataset, while a categorical object/dataset with missing values is denoted as incomplete object/dataset.

Definition 1 (Clusters). *Let $C = \{C_1, C_2, \ldots, C_k\}$ be the set of k disjoint subset. C_α $(1 \leq \alpha \leq k)$ is called a cluster of S iff for every $C_i \in C$ $(1 \leq i \leq k \wedge i \neq \alpha)$, $C_\alpha \cap C_i = \emptyset$ and $S = \bigcup_{\alpha=1}^{k} C_\alpha$. The number of data objects in the cluster C_α is denoted by n_α.*

Definition 2 (Relative frequency in a cluster). *Given a cluster C_α and a categorical value appearing in C_α at d^{th} attribute o_l^d $(1 \leq l \leq n_\alpha)$, the relative frequency of o_l^d in C_α is denoted and defined as:*

$$f_\alpha(o_l^d) = \frac{\#_\alpha(o_l^d)}{n_\alpha} \tag{1}$$

where $\#_\alpha(o_l^d)$ is the number of o_l^d appearing in the cluster C_α.

In order to define the centers of clusters, Chen et al. [3] and Nguyen et al. [11] used the kernel density estimation (KDE) method. Specifically, they used a variation on Aitchison & Aitken's kernel function [1] to estimate the probability density function of each attribute in the center. In this research, we also use the KDE to define centers.

Definition 3 (Kernel density estimation for categorical data [1,3,11]). *Given a cluster C_α. Let X^d be a random variable associated with observations x_i^d $(1 \leq i \leq n_\alpha)$ appearing in cluster C_α at d^{th} attribute, and it's probability of density is denoted as $p(X^d)$. Let O_α^d be the set of categorical values in C_α such that $O_\alpha^d = \bigcup_{i=1}^{|C_\alpha|} x_i^d$ and $\lambda_\alpha \in [0, 1]$ be the unique smoothing bandwidth for cluster C_α. For each value o_l^d in O_α^d $(1 \leq l \leq n_\alpha)$, the variation on Aitchison & Aitken's kernel function is denoted and defined as:*

$$K(X^d, o_l^d, \lambda_\alpha) = \begin{cases} 1 - \frac{|O_\alpha^d| - 1}{|O_\alpha^d|} \lambda_\alpha & if \ X^d = o_l^d \\ \frac{1}{|O_\alpha^d|} \lambda_\alpha & otherwise \end{cases} \tag{2}$$

Note that the kernel function of a categorical value at d^{th} attribute is defined in terms of the cardinality of the domain O_α^d of the cluster C_α as [11] instead of the cardinality of the whole domain O_d as [3]. The kernel density estimation of the $p(X_d)$ is denoted and defined as:

$$\hat{p}(X^d, \lambda_\alpha, C_\alpha) = \sum_{o_l^d \in O_\alpha^d} f_\alpha(o_l^d) K(X^d, o_l^d, \lambda_\alpha) \tag{3}$$

Definition 4 (Smoothing bandwidth parameter [3,11]). *The parameter λ is a unique smoothing bandwidth that uses the least square cross validation to minimize the total error of the resulting estimation over all the data objects. The optimal smoothing parameter for cluster C_α is defined as:*

$$\lambda_\alpha = \frac{1}{n_\alpha - 1} \frac{\sum_{d=1}^{|D|}(1 - \sum_{o_l^d \in O_\alpha^d}[f_\alpha(o_l^d)]^2)}{\sum_{d=1}^{|D|}(\sum_{o_l^d \in O_\alpha^d}[f_\alpha(o_l^d)]^2 - \frac{1}{|O_\alpha^d|})} \tag{4}$$

Definition 5 (Center of cluster). *Let there be a cluster $C_\alpha = \{X_1, X_2, \ldots, X_p\}$ where $X_i = (x_{1i}, x_{2i}, \ldots, x_{mi})$, $m = |D|$. Then, the center of C_α is defined as:*

$$V_\alpha = \{v_\alpha^1, v_\alpha^2, \ldots, v_\alpha^m\} \tag{5}$$

where the dth element v_α^d $(1 \le d \le m)$ is a probability distribution on O_α^d estimated by a kernel density estimation method using Eq. (3), which is defined as:

$$v_\alpha^d = [P_\alpha^d(o_1^d), P_\alpha^d(o_2^d), \ldots, P_\alpha^d(o_{|O_\alpha^d|}^d)] \tag{6}$$

where the probabilistic value of a categorical value o_l^d $(1 \le l \le |O_\alpha^d|)$ can be estimated based on Eqs. (1), (2) and (3) as:

$$P_\alpha^d(o_l^d) = \begin{cases} \lambda_\alpha \frac{1}{|O_\alpha^d|} + (1 - \lambda_\alpha) f_\alpha(o_l^d) & \text{if } o_l^d \in O_\alpha^d \\ 0 & \text{if } o_l^d \notin O_\alpha^d \end{cases} \tag{7}$$

There are many methods to measure the dissimilarity between a categorical data object and its center such as the simple matching dissimilarity measure [7,13], the Euclidean norm [3] and the information theoretic-based dissimilarity measure [11]. In 1998, Lin proposed an information theoretic definition of similarity [9] that is applicable as long as the domain has a probabilistic model. In this work, we also use the information theoretic based dissimilarity measure to compute the distance between categorical objects and cluster centers.

Definition 6 (Information theoretic based dissimilarity measure [11]). *Given two categorical values o_l^d and $o_{l'}^d$ at the dth attribute. The similarity between them is defined as:*

$$sim^d(o_l^d, o_{l'}^d) = \frac{2 \log f_\alpha(o_l^d, o_{l'}^d)}{\log f_\alpha(o_l^d) + \log f_\alpha(o_{l'}^d)} \tag{8}$$

where $f_\alpha(o_l^d, o_{l'}^d)$ is the relative frequency of categorical objects in dataset S that receives the value belonging to $\{o_l^d, o_{l'}^d\}$ at the dth attribute.

Then, the dissimilarity measure between two categorical values o_l^d and $o_{l'}^d$ at the dth attribute can be defined as:

$$dsim^d(o_l^d, o_{l'}^d) = 1 - sim^d(o_l^d, o_{l'}^d) = 1 - \frac{2\log f_\alpha(o_l^d, o_{l'}^d)}{\log f_\alpha(o_l^d) + \log f_\alpha(o_{l'}^d)} \quad (9)$$

Definition 7 (Dissimilarity between a data object and a cluster). *Let there be a cluster C_α with its center is $V_\alpha = \{v_\alpha^1, v_\alpha^2, \ldots, v_\alpha^{|D|}\}$ and a categorical data object $X_i = (x_{1i}, x_{2i}, \ldots, x_{|D|i})$. The dissimilarity between X_i and V_α at d^{th} attribute can be defined as:*

$$dis^d(X_i, V_\alpha) = \sum_{o_l^d \in O_\alpha^d} P_\alpha^d(o_l^d) dsim^d(x_{di}, o_l^d) \quad (10)$$

That is, the dissimilarity between X_i and V_α is measured by accumulating the probability distribution on O_α^d and the dissimilarity between d^{th} component x_{di} of the object X_i and the dth component v_α^d of the center V_α.

Then the dissimilarity between data object X_i and cluster center V_α can be defined as follows:

$$dis(X_i, V_\alpha) = \sum_{d=1}^{|D|} dis^d(X_i, V_\alpha) \quad (11)$$

Definition 8 (IS measure [4]). *Given a categorical dataset $S = \{X_1, X_2, \ldots, X_n\}$. Let $S_C = \{C_1, C_2, \ldots, C_p\}$ be a set of attributes with non-missing values (complete attributes) and $S_M = \{M_1, M_2, \ldots, M_q\}$ be a set of attributes with missing values (incomplete attributes). The IS measure measures the degree of associations between two sets S_C and S_M as follows:*

$$IS(S_C, S_M) = \frac{Support(S_C, S_M)}{\sqrt{Support(S_C) * Support(S_M)}} \quad (12)$$

where $Support(S_C, S_M) = |S_C, S_M|/|S|$, $|S_C, S_M|$ is the number of categorical objects that contain both attributes in S_C and S_M, $|S|$ is the size of the dataset.

Definition 9 (MCS measure). *Given a set T that contains both complete and incomplete objects. Let there be two categorical values o_l^d and $o_{l'}^d$ appearing in T at the dth attribute. The similarity between them is defined as:*

$$sim_{mis}^d(o_l^d, o_{l'}^d) = \begin{cases} \frac{2\log f_T(o_l^d, o_{l'}^d)}{\log f_T(o_l^d) + \log f_T(o_{l'}^d)} & if\, o_l^d \neq ?\, and\, o_{l'}^d \neq ? \\ 0 & otherwise \end{cases} \quad (13)$$

where $f_T(o_l^d) = \frac{\#_T(o_l^d)}{n_T}$, $\#_T(o_l^d)$ and n_T are respectively the number of o_l^d appearing in T and the number of objects in T.

Let $X_c = (x_{1c}, x_{2c}, \ldots, x_{|D|c})$ and $X_m = (x_{1m}, x_{2m}, \ldots, x_{|D|m})$ be the complete object and incomplete object, respectively. Then, the MCS between X_c and X_m

is defined as follows :

$$MCS(X_c, X_m) = \sum_{d=1}^{|D|} sim_{mis}^d(x_{dc}, x_{dm}) \qquad (14)$$

Based on these definitions, the clustering algorithm for categorical datasets with missing values now aims to minimize the following objective function:

$$J(U, V) = \sum_{\alpha=1}^{k} \sum_{i=1}^{n} u_{i,\alpha} \times dis(X_i, V_\alpha) \qquad (15)$$

subject to

$$\begin{cases} \sum_{\alpha=1}^{k} u_{i,\alpha} = 1 & 1 \le i \le n \\ u_{i,\alpha} \in \{0,1\} & 1 \le \alpha \le k,\ 1 \le i \le n \end{cases} \qquad (16)$$

where $U = [u_{i,\alpha}]_{n \times k}$ is the partition matrix ($u_{i,\alpha}$ take value 1 if object X_i is in cluster C_α and 0 otherwise).

4 The Proposed k-CCM Algorithm

The proposed k-CCM algorithm is based on the general framework depicted in Fig. 1. According to this model, the k-CCM initially scans the categorical database S once to divide it into two sub-datasets, namely S_1 and S_2, which are the complete and incomplete datasets, respectively. First, the k-CCM randomly initiates k cluster centers from S_1. Each cluster center is formed by the Eq. (5). Two sets, namely Set_{DT} and U, are respectively used to store decision trees and clusters, while t is used as a counter for the number of the iterations of the clustering process. In the next step, the k-CCM scans all objects in S_2 to impute missing values and assign objects to clusters. For each object X_i in S_2, the algorithm finds attributes containing missing values (called as missing attributes) and puts them into the set Set_{attr}. For each missing attribute A in Set_{attr}, the k-CCM checks if there exists any decision tree (DT) that uses A as a class attribute. If there is no such DT, then a DT that uses the missing attribute A as the class attribute is constructed from the complete dataset S_1. This process is repeated until all missing attributes in A have their corresponding

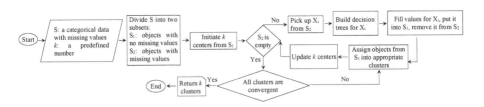

Fig. 1. The flowchart of k-CCM algorithm

DTs. After the tree construction step, object X_i is assigned to the leaf of the tree with the same class attribute as the missing attribute. Once X_i is assigned to the appropriate leaves, each leaf consists of objects from S_1 and X_i that are correlated. Each leaf node in DT is represented as a list of objects. In the k-CCM, we use the same manner as the $DSMI$ algorithm [4]. That is, if an object has more than one missing values fallen into multiple leaves, the algorithm will merge these leaves and group objects into the same collection. To impute the missing values in X_i, the algorithm searches for objects in the node which have the maximum number of non-missing attributes in common to the complete object. Then, the attribute values in these objects corresponding to the missing attributes in the missing object are taken to be the possible imputed values. For each such object, the k-CCM finds possible imputed values and calculate the IS and MCS measures for these values. The affinity degree of possible imputed values is given by the average of the IS and MCS measures computed for each possible imputed values. Once affinity degrees of possible imputed values are obtained, the k-CCM assigned actual imputed values by random sampling from the list of possible imputed values based on their affinity degrees. Based on the results from [4], random sampling according to affinity degree ensures that uncertainty and randomness in attribute values are accounted for and helps to reduce systematic bias in the imputed dataset. After missing values are imputed, the k-CCM puts X_i into S_1 and removes X_i from S_2. In the next step, the k-CCM assigns objects in S_1 into appropriate clusters and updates center of clusters. A similar process is performed for all incomplete objects in the S_2. Then, the k-CCM performs clustering step until all clusters are convergent. Finally, it returns k clusters as the desired output.

5   Comparative Experiment

Experiments were performed to evaluate the performance of k-CCM on an HPC cluster[4]. Each node is equipped with an Intel Xeon E5-2680v2@2.80 GHz×20, 64 GB of RAM, running Red Hat Enterprise Linux 6.4. The proposed algorithm was implemented in Python using PyCharm. The performance of the proposed k-CCM algorithm is compared with five partitional clustering algorithms: k-modes [7], k-representatives [13], Modified 1, Modified 2 and Modified 3 [11]. Standard benchmark datasets were used for the experiment. The characteristics of these datasets are shown in Table 1. They have varied characteristics. By using these datasets, the performance of the proposed algorithm and the compared algorithms are evaluated for the main types of data encountered in real-life world. For each algorithm, we ran 300 times per dataset. In this research, we use three metrics: Purity, Normalized Mutual Information (NMI) and Adjusted Rand Index (ARI) to evaluate the performance of the proposed algorithm. These metrics use class information in original datasets and clustering results generated by the algorithm to evaluate how well the clustering matches the original classes.

---

[4] https://www.jaist.ac.jp/iscenter/en/mpc/.

Table 1. Characteristics of the experimental datasets

Dataset	#instances	#attributes	#miss values	#classes	source
Breast cancer	286	9	9	2	UCI
Gsssex survey	159	9	6	5	CMU
Mushroom	8,124	22	2,480	2	UCI
Negotiation	92	6	26	6	CMU
Runshoes	60	10	14	7	CMU
Soybean	307	35	712	19	UCI
Sponge	76	45	22	12	UCI
Voting records	435	16	392	2	UCI

Given a categorical dataset S with N objects. Let there be a set of clusters $C = \{C_1, ..., C_K\}$ generated by a clustering algorithm from S, and $P = \{P_1, ..., P_J\}$ is the set of partitions which are inferred by the original class information in S. Purity is a simple and transparent evaluation measure. To compute purity value of a clustering result, each cluster is assigned to the class which is most frequent in the cluster. Then, the accuracy of this assignment is measured by counting the number of correctly assigned objects and dividing by the number of objects in the dataset. Bad clusterings have purity values close to 0, a perfect clustering has a purity of 1.

$$Purity(C, P) = \frac{1}{N} \sum_k \max_j |C_k \cap P_j| \qquad (17)$$

In Tables 2, 3 and 4, bolded numbers indicate the best performers in each categorical dataset. The purity results of the k-CCM and compared algorithms are shown in Table 2.

Table 2. Purity results of clustering algorithms

Dataset	k-modes	k-representatives	Modified 1	Modified 2	Modified 3	k-CCM
Breast Cancer	0.7028	0.7028	0.7030	0.7028	0.7028	**0.7098**
Gsssex survey	0.7799	0.7814	0.7802	0.7803	0.7803	**0.7816**
Mushroom	0.8818	**0.8876**	0.8858	0.7235	0.7312	0.8861
Negotiation	0.4692	0.4829	0.4469	0.4517	0.4645	**0.5002**
Runshoes	0.4613	0.4591	0.4725	0.4725	0.4798	**0.4851**
Soybean	0.6107	0.7139	0.6955	0.6924	**0.7181**	0.6957
Sponge	0.9211	0.9211	0.9211	0.7159	0.9211	**0.9214**
Voting records	0.8581	0.8764	0.8713	0.8760	0.8775	**0.8805**

Table 3. NMI results of clustering algorithms

Dataset	k-modes	k-representatives	Modified 1	Modified 2	Modified 3	k-CCM
Breast Cancer	0.0040	0.0018	0.0047	0.0042	0.0040	**0.0057**
Gsssex survey	0.0536	0.0606	0.0430	0.0428	0.0630	**0.0634**
Mushroom	0.5446	0.5383	0.5310	0.1937	0.2077	**0.5492**
Negotiation	0.0939	0.1342	0.1193	0.1039	0.1193	**0.1353**
Runshoes	0.2158	0.2192	0.2224	0.2246	0.2267	**0.2289**
Soybean	0.6085	0.7552	0.7314	0.7243	0.7509	**0.7555**
Sponge	0.0668	0.0638	0.0765	0.0748	**0.0887**	0.0770
Voting records	0.4359	0.4990	0.4961	0.4950	0.4947	**0.5002**

High purity is easy to achieve when the number of clusters is large. Thus, we cannot use purity to trade off the quality of the clustering against the number of clusters [10]. A measure that allows us to make this trade-off is Normalized Mutual Information (NMI). NMI is computed as the average mutual information between any pairs of clusters and classes. Because NMI is normalized, we can use it to compare clustering with different numbers of clusters. NMI is always a number between 0 and 1. This measure takes its maximum value when the clustering partition matches completely the original partition.

$$NMI(C,P) = \frac{\sum_{k=1}^{K}\sum_{j=1}^{J}|C_k \cap P_j|\log\frac{N|C_k \cap P_j|}{|C_k||P_j|}}{\sqrt{\sum_{k=1}^{K}|C_k|\log\frac{|C_k|}{N}\sum_{j=1}^{J}|P_j|\log\frac{|P_j|}{N}}} \tag{18}$$

The NMI results of the proposed k-CCM and other five algorithms are shown in Table 3.

In 1985, Hubert and Arabie proposed the adjusted Rand index (ARI) that ranges between -1 and 1 and is 0 if there is only chance agreement between clusters and classes. Let n be the number of object pairs belonging to the same cluster in C and to the same class in P. This metric captures the deviation of n from its expected value corresponding to the hypothetical value of n obtained when C and P are two random, independent partitions. The expected value of n is defined and denoted as:

$$E[n] = \frac{\pi(C)\pi(P)}{N(N-1)/2} \tag{19}$$

where $\pi(C)$ and $\pi(P)$ denote respectively the number of object pairs from the same clusters in C and from the same classes in P. The maximum value for n is defined as:

$$max(n) = \frac{1}{2}(\pi(C) + \pi(P)) \tag{20}$$

The agreement between C and P can be estimated by the ARI as follows:

$$ARI(C,P) = \frac{n - E[n]}{max(n) - E[n]} \tag{21}$$

when $ARI(C, P) = 1$, we have identical partitions. The ARI results are shown in Table 4.

Table 4. Adjusted Rand Index results of clustering algorithms

Dataset	k-modes	k-representatives	Modified 1	Modified 2	Modified 3	k-CCM
Breast Cancer	0.0019	−0.0030	**0.1351**	0.0021	0.0055	**0.1351**
Gsssex survey	0.0066	0.0135	0.0168	0.0140	**0.0171**	**0.0171**
Mushroom	0.5924	0.5952	0.5963	0.2357	0.2527	**0.6009**
Negotiation	0.0143	**0.0267**	0.0105	0.0111	0.0193	0.0155
Runshoes	0.0381	0.0320	0.0385	0.0335	0.0385	**0.0392**
Soybean	0.3759	**0.4767**	0.4163	0.4167	0.4686	0.4167
Sponge	−0.0173	−0.0176	−0.0024	−0.0030	**0.0190**	−0.0007
Voting records	0.5119	0.5658	0.5504	0.5540	0.5644	**0.5779**

6 Summary and Future Work

In this paper, we have proposed an algorithm named k-CCM for clustering categorical datasets with missing values. The proposed algorithm integrates the imputation step and clustering step into a common process. By this way, all incomplete objects are first imputed and then assigned into appropriate clusters. In particular, we have extended a decision tree-based imputation method [4] to fill in missing values. For clustering, we use a kernel density estimation approach to define cluster centers and an information theoretic-based dissimilarity measure to quantify the differences between objects. An extensive experimental evaluation is conducted on benchmark categorical datasets to evaluate the performance of the proposed algorithm. According to the experimental results, the designed algorithm has a comparative result in terms of clustering quality when compared to the other five algorithms. Thus, the imputation step has improved the quality of the clustering.

In future work, we will consider clustering mixed data with missing values. In addition, we will study clustering for large-scale and high-dimensional datasets with missing values.

Acknowledgment. This paper is based upon work supported in part by the Air Force Office of Scientific Research/Asian Office of Aerospace Research and Development (AFOSR/AOARD) under award number FA2386-17-1-4046.

References

1. Aitchison, J., Aitken, C.G.: Multivariate binary discrimination by the kernel method. Biometrika **63**(3), 413–420 (1976)
2. Berkhin, P.: A survey of clustering data mining techniques. In: Kogan, J., Nicholas, C., Teboulle, M. (eds.) Grouping Multidimensional Data, pp. 25–71. Springer, Heidelberg (2006). https://doi.org/10.1007/3-540-28349-8_2
3. Chen, L., Wang, S.: Central clustering of categorical data with automated feature weighting. In: IJCAI, pp. 1260–1266 (2013)
4. Deb, R., Liew, A.W.C.: Missing value imputation for the analysis of incomplete traffic accident data. Inf. Sci.s **339**, 274–289 (2016)
5. Fujikawa, Y., Ho, T.B.: Cluster-based algorithms for dealing with missing values. In: Chen, M.-S., Yu, P.S., Liu, B. (eds.) PAKDD 2002. LNCS (LNAI), vol. 2336, pp. 549–554. Springer, Heidelberg (2002). https://doi.org/10.1007/3-540-47887-6_54
6. Gan, G., Ma, C., Wu, J.: Data Clustering: Theory, Algorithms, and Applications. SIAM (2007)
7. Huang, Z.: Extensions to the k-means algorithm for clustering large data sets with categorical values. Data Mining Knowl. Discov. **2**(3), 283–304 (1998)
8. Kim, D.W., Lee, K., Lee, D., Lee, K.H.: A k-populations algorithm for clustering categorical data. Pattern Recogn. **38**(7), 1131–1134 (2005)
9. Lin, D., et al.: An information-theoretic definition of similarity. In: ICML, vol. 98, pp. 296–304. Citeseer (1998)
10. Manning, C.D., Raghavan, P., Schütze, H.: Introduction to Information Retrieval. Cambridge University Press, New York (2008)
11. Nguyen, T.-H.T., Huynh, V.-N.: A k-means-like algorithm for clustering categorical data using an information theoretic-based dissimilarity measure. In: Gyssens, M., Simari, G. (eds.) FoIKS 2016. LNCS, vol. 9616, pp. 115–130. Springer, Cham (2016). https://doi.org/10.1007/978-3-319-30024-5_7
12. Rahman, M.G., Islam, M.Z.: Missing value imputation using decision trees and decision forests by splitting and merging records: Two novel techniques. Knowl. Based Syst. **53**, 51–65 (2013)
13. San, O.M., Huynh, V.N., Nakamori, Y.: An alternative extension of the k-means algorithm for clustering categorical data. Int. J. Appl. Math. Comput. Sci. **14**, 241–247 (2004)
14. Tan, P.N., Kumar, V.: Interestingness measures for association patterns: a perspective. In: Proceedings of Workshop on Postprocessing in Machine Learning and Data Mining (2000)
15. Tan, P.N., Steinbach, M., Kumar, V.: Introduction to Data Mining, 1st edn. Addison-Wesley Longman Publishing Co., Inc., Boston (2005)
16. Zaït, M., Messatfa, H.: A comparative study of clustering methods. Fut. Gener. Comput. Syst. **13**(2–3), 149–159 (1997)
17. Thanh-Phu, N., Duy-Tai, D., Van-Nam, H: A new context-based clustering framework for categorical data. Pacific Rim International Conference on Artificial Intelligence, pp. 697–709. Springer (2018)

Data Privacy and Security

WEDL-NIDS: Improving Network Intrusion Detection Using Word Embedding-Based Deep Learning Method

Jianjing Cui, Jun Long[✉], Erxue Min, and Yugang Mao

Department of Computer Science,
National University of Defense Technology, Changsha 410005, China
junlong@nudt.edu.cn

Abstract. A Network Intrusion Detection System (NIDS) helps system administrators to detect security breaches in their organization. Current research focus on machine learning based network intrusion detection method. However, as numerous complicated attack types have growingly appeared and evolved in recent years, obtaining high detection rates is increasingly difficult. Also, the performance of a NIDS is highly dependent on feature design, while a feature set that can accurately characterize network traffic is still manually designed and usually costs lots of time. In this paper, we propose an improved NIDS using word embedding-based deep learning (WEDL-NIDS), which has the ability of dimension reduction and learning features from data with sophisticated structure. The experimental results show that the proposed method outperforms previous methods in terms of accuracy and false alarm rate, which successfully demonstrates its effectiveness in both dimension reduction and practical detection ability.

Keywords: Network intrusion detection · Deep neural networks
Word embedding · Long short-term memory networks

1 Introduction

Network Intrusion Detection System (NIDS) is an important part of protecting computers and networks against inner or outer intruders. In recent years, machine learning methods have been widely explored to solve the intrusion detection tasks.

However, there are two issues that make an efficient and flexible NIDS a big challenge when detecting unknown attacks. Firstly, an important part of machine learning methods is "feature engineering", which costs lots of time and needs professional knowledge. As attack scenarios are continuously changing and evolving, the features selected for one type of attack may not work well for another type. Secondly, it is very difficult to train a model with low overheads when high dimensional features are fed into the training procedure.

© Springer Nature Switzerland AG 2018
V. Torra et al. (Eds.): MDAI 2018, LNAI 11144, pp. 283–295, 2018.
https://doi.org/10.1007/978-3-030-00202-2_23

To address the first challenge, deep learning has been demonstrated to be good at replacing handcrafted features with efficient algorithms [1,2]. There are many researchers studying this problem by deep learning methods, which can avoid feature engineering. Moreover, a deep learning based NIDS can be designed as an end-to-end system (input: raw traffic, output: detection result). To address the second challenge, it is worth noticing that word embedding [3] can effectively reduce the dimension of features while keeping the similarity relationships in semantics and syntax.

Therefore, we propose an improved NIDS using word embedding-based deep learning (WEDL-NIDS). Specifically, WEDL-NIDS first reduces the dimension of a packet's payload via word embedding and learns the local contentful features of network traffic using deep convolutional neural networks (CNNs) [4]. Then it adds the head features and learns global temporal features using long short-term memory (LSTM) networks [5]. Comparative experiments demonstrate that the proposed method can achieve significant performance improvement compared with previous methods in terms of accuracy and detection rate. Moreover, the proposed method can obtain quite impressive performance when being applied to detect malicious traffics.

The remainder of this paper is organized as follows. Section 2 describes related work and introduces motivation of this work. Section 3 describes the design and implementation of the proposed method. Section 4 shows and analyzes experimental results. Finally, Sect. 5 presents conclusions and future work.

2   Related Work

In recent years, deep learning has become increasingly popular and studies have shown that deep learning completely surpasses traditional methods in the fields of computer vision and natural language processing. Therefore, deep learning has been widely applied for intrusion detection. The current deep learning based intrusion detection methods can be divided into 3 types:

- **Deep learning as classifiers**. In this type, researchers first extracted predefined features from raw data, then used deep learning methods to train their models as classifiers. Tang et al. [6] utilized a deep learning approach based on a deep neural network for flow-based anomaly detection, and the experimental results showed that deep learning can be applied for anomaly detection in software defined networks. Salama et al. [7] proposed a deep learning approach with Deep Belief Network (DBN) as a feature selector and SVM as a classifier for intrusion detection. This approach resulted in an accuracy of 92.84% when applied on training data. Fiore et al. [8] used a similar, however, semi-supervised learning approach. They used real-world trace for training, and evaluated their approach on real-world and KDD Cup 99 traces. These work show the effect of deep learning for intrusion detection. But they ignore that the strongest ability of deep learning is to learn from the raw data.

- **Deep learning as feature extractor**. In this type, researchers first used deep learning methods to extract features automatically from raw data, then used simple classification algorithms based on these features. Most of these work used auto-encoder (AE) as their preprocess method. Wang et al. [9] proposed a sparse auto-encoder (SAE) based deep learning approach for network traffic identification. However, they performed TCP based unknown protocols identification in their work instead of network intrusion detection. Javaid et al. [10] proposed a deep learning based approach using self-taught learning (STL) on the benchmark NSL-KDD dataset in a network intrusion detection system. When comparing its performance with those observed in previous studies, the method was shown to be more effective. Yu et al. [11,12] obtained quite impressive performance through applying stacked denoising autoencoders (SDA) based deep learning architecture to detect botnet traffics.
- **Deep learning as an end-to-end model**. In this type, researchers used deep learning methods directly on the raw data and trained their models for classification tasks. Most of these work used CNNs or RNNs to learn spatial and temporal features. Wang et al. [1] used a CNN to learn the spatial features of network traffic and achieved malware traffic classification using the image classification method. Yin et al. [13] used the RNN model to perform classification directly and researched both the binary and multi-class classification. Wang et al. [2] combined the CNN and LSTM networks to learn hierarchical spatial-temporal features. The result they got was quite well but the limitations of their work are that they always transform the traffic data into pictures before their study, which we think is out of nature of the traffic data. Also, they ignore the highly normalized head features of the traffic data and mix the entire data as the input before transforming into pictures.

The main differences among these methods are: the first type used deep learning methods as classifiers only on the predefined features; the second type used deep learning methods as feature extractor only on the raw traffic data; the third type used deep learning methods as an end-to-end model. The WEDL-NIDS model belongs to the third type but with some improvements.

As a result of the uselessness of transforming traffic data into pictures, we put forward a point of view that the network traffic data could be regarded as text. And we noticed that Mikolov et al. [3,14] proposed the method of word embedding to reduce the dimension of the text feature while keeping the similarity in semantics and syntax at the same time. Following this line of thinking, we reduced the dimension via word embedding and then input the result into the deep learning model to get a high-performance NIDS.

3   Methodology

Real traffic data can be regarded as a collection of binary numbers 0 and 1. As a result, we tried to introduce the knowledge in the field of natural language processing (NLP). Suppose that a flow is a paragraph of text, each packet in flow is the sentence in the text. The original task of classifying a flow as normal or attack can be converted into classifying whether the text is malicious or kind.

The goals of the WEDL-NIDS is to reduce the dimensions of features by word embedding and automatically learn the features of raw network traffic data using deep neural networks, and finally to improve the effectiveness of the NIDS. Generally, this model is composed of 3 modules: data preprocessing module, dimension reduction module and deep learning architecture module. Specifically, data preprocessing module transformed raw network traffic into 2 parts: head features and payload text. Then the payload text were input into dimension reduction module, which transformed payload text into word vectors. Finally, the head features and word vectors were emerged together and input into the deep learning architecture. The whole process is shown in Fig. 1, and the various stages of the WEDL-NIDS are described below.

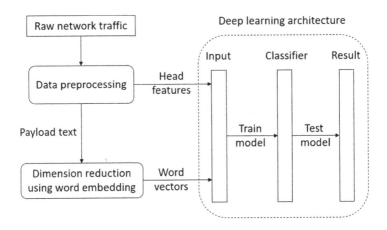

Fig. 1. The whole process of WEDL-NIDS.

3.1   Data Preprocessing

In this stage, the input raw network traffic data are transformed into the text sequences and the head features are extracted. There are 2 steps to transform raw network traffics into trainable samples: packet head features election, packet payload extraction.

Packet Head Features Selection. We choose 11 features from the head of packets: protocol, source port, destination port, icmp_type, icmp_code, length, flags, ack_num, urgptr, window_size, option_num. All the features are described in Table 1. All the integer numbers then are normalized to float numbers from −1 to 1.

Table 1. Descriptions of the head features in our experiments

Feature name	Description	Example
Protocol	TCP, UDP, ICMP	110
Port	Source port and destination port for TCP or UDP packets	51032
icmp_type	Value of type field of ICMP packet	3
icmp_code	Value of code field of ICMP packet	3
Length	Total length of the packet	141
Flags	6 TCP flags of packets; 0 for UDP and ICMP packets	010000
ack_val	Value of ACK for TCP packets; 0 for UDP and ICMP packets	3745556791
urgptr_val	Value of urgptr for TCP packets; 0 for UDP and ICMP packets	0
window_size	Value of window_size for TCP packets; 0 for UDP and ICMP packets	256
option_num	Number of options for TCP packets; 0 for UDP and ICMP packets	5

Packet Payload Extraction. Each packet payload is split in 8 bits part. Then each 8 bits were token as a "word" in text. The length of one text sequence is 100, which means that we choose the first 100 words from each packet's payload. If the length of a payload is less than 100, zeroes are padded. Correspondingly, the extra part is truncated. At the end of this step, we use one-hot encoding to transform each word in text sequences into a 256 dimension vector.

3.2   Dimension Reduction

The dimension reduction of WEDL-NIDS is based on word embedding, which can keep the similarity in semantics and syntax at the same time. The two popular methods of learning word embedding from text include: Word2Vec [15] and GloVe [16]. We choose Word2Vec as our tool because of its convenience, high efficiency and performance. The theory of Word2Vec is CBoW model and Skip-gram model. The model we used in the experiment is based on Skip-gram model. The difference of CBoW and Skip-gram is that, given context, the CBoW predicts input word; while Skip-gram predicts the context when given input word (Fig. 2).

The Skip-gram model is actually divided into two parts. The first part is to establish the model, and the second part is to obtain the embedded word vector through the model. The modeling process of Skip-gram is very similar to auto-encoder, which reconstruct the data based on a neural network when training the model. But we do not use the trained model to deal with the new task, what we really need is the parameters of the model learned from the training data (e.g. the weight matrix of hidden layer). In Skip-gram, these weights are actually called

INPUT   PROJECTION OUTPUT

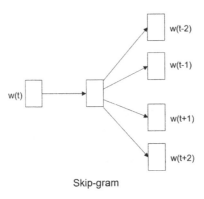

Skip-gram

Fig. 2. The architecture of Skip-gram model.

the "word vectors". The forward calculation process of the Skip-gram model can be written as a mathematical form:

$$p(\omega_0 \mid \omega_i) = \frac{e^{U_0 \cdot V_i}}{\sum_j e^{U_j \cdot V_i}}. \tag{1}$$

in which V_i is the column vector in the embedding layer matrix (also called input vector of ω_i), U_j is the row vector in the softmax layer matrix (also called input vector of ω_j). The embedding layer matrix is actually the hidden layer weight matrix in Algorithm 1. The softmax layer matrix is a $p \times n$ matrix, in which p is the number of labels, n is the dimension of data.

Algorithm 1 shows the process of word embedding and the construction of training samples is described as follows.

Training Samples Construction. Let S be the set of all the words from the training data, $w_i \in S$ denotes the i-th word in text. Suppose a packet's payload is $w_{1:n} = w_1, w_i, ..., w_n$. We first choose a word in text as the input word (e.g. w_i). Then we define a parameter named "skip_window", which denotes the number of words we choose from the side of the current input word (left or right). For skip_window $= k$, the window is $[w_{i-k}, ..., w_i, ..., w_{i+k}]$. Another parameter is called "num_skips", which denotes how many different words we choose from the whole window as our output word. For num_skips $= p$, we will get p groups of training word pairs: $(w_i, w_{i-1}), (w_i, w_{i+1}), ..., (w_i, w_{i-\frac{p}{2}}), (w_i, w_{i+\frac{p}{2}})$.

3.3 Deep Learning Architecture

The architecture of the deep learning model we used is that we first learns the local contentful features of network traffic using deep convolutional neural networks (CNNs) and then adds the head features and learns global temporal features using long short-term memory (LSTM) networks. Finally, the softmax

---

Algorithm 1. Dimension reduction based on word embedding

Input: Network traffic text sequences
Output: Network traffic word vector sequences
Step 1: Create word embedding model
1: Create word pairs from the network traffic text sequences of length l with skip_window k_1 and num_skips p_1.
2: One-hot encoding every word in the training data.
3: Add a hidden layer with w neurons, the weight matrix of which is a $l \times w$ matrix.
4: Add a dense layer, the activation of which is softmax, which predicts words at the nearby position of the input word.
Step 2: Train and validate model
5: **while** early termination condition is not met **do**
6:   **while** training dataset is not empty **do**
7:     Prepare the mini-batch dataset as the model input.
8:     Compute the categorical cross-entropy loss function $H(p,q) = -\sum_x p(x) \log(q(x))$, p=true_dist and q=predict_dist.
9:     Update the weights and biases using the RMSprop gradient descent optimization algorithm.
10:   **end while**
11: **end while**
Step 3: Get the word vector
12: Output the hidden layer weight matrix, the shape of which is $l \times w$. Match each word with its word vector, the dimension of which is w.
13: **return** The network traffic word vector sequences

---

classifiers will classify the traffic as normal or attack. Details of the deep learning architecture are presented below.

Payload Feature Learning. CNNs are used to learn the local features of the word vector sequences which have already been calculated by word embedding. Inspired by the architectures of CNNs in the field of computer vision [17], we used two convolution filters with different sizes and concatenated two outputs together as the final vector. This method can obtain better results from different granularity levels. Then we merge the head features with the output of CNN as packet vectors P. Figure 3 shows the generation of packet vectors.

Global Feature Learning. LSTMs are used in this part to learn the global features of the packets. The input is the packet vector sequences $[P_1, P_2, ..., P_n]$ and the output is a single vector F which combines the local features in a single packet and the global features in a flow. We use bidirectional LSTMs in which each direction includes two sequence-to-sequence Recurrent Neural Layers [18]. Recurrent Neural Layers help us trace the history from previous network packets. In specific, LSTM aims to overcome vanishing gradient problem of RNN and uses a memory cell to present the previous timestamp [5]. The softmax classifier is used to determine whether the input traffic is normal or malware based on the flow vector. Softmax is a commonly used multi-class classification method in the field of machine learning. Figure 4 shows the details of this step.

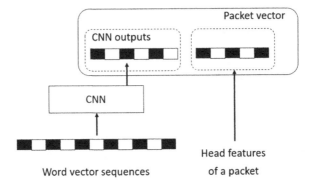

Fig. 3. The payload feature learning and generation of packet vectors.

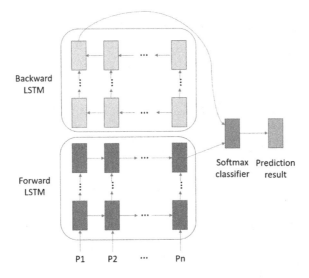

Fig. 4. The global feature learning and classifier.

4   Experiment Results and Discussion

This section evaluates the performance of the proposed WEDL-NIDS by performing various experiments on ISCX2012, a commonly used public standard intrusion detection datasets. Specifically, these experiments can be divided into 2 parts:

- Multi-class classification using WEDL-NIDS
- Comparison WEDL-NIDS with other methods

4.1   Experimental Methodology

1. **Experiment setup**

   The Keras (using tensorflow as backend) is used as experiment software framework which runs on CentOS 7.2 64bit OS, with 2 Xeon e5 CPUs with 10 cores and 64 GB memory. An Nvidia Tesla K80 GPU is used as accelerator. The mini-batch size is 64 and the initial learning rate is 0.0001, training time is about 50 epochs.

2. **Evaluation metrics**

   Four metrics are used to evaluate the performance of the WEDL-NIDS: accuracy (ACC), detection rate (DR), false alarm rate (FAR) and F1 score. They can be obtained by a confusion matrix. By definition, entry i, j in a confusion matrix is the number of observations actually in group i, but predicted to be in group j. The definitions of these metrics are presented below: $Accuracy\,(ACC) = (TP + TN)/(TP + FP + FN + TN)$; $DetectionRate\,(DR) = TP/(TP + FN)$; $FalseAlarmRate\,(FAR) = FP/(FP + TN)$; $F1\_score\,(F_1) = 2 \cdot (PR \cdot DR)/(PR + DR)$

where TP is the number of instances correctly classified as X, TN is the number of instances correctly classified as Not-X, FP is the number of instances incorrectly classified as X, and FN is the number of instances incorrectly classified as Not-X.

4.2   Dataset

Most public intrusion detection datasets, such as NSL-KDD [19] and Kyoto2009 [20], do not contain raw traffic data. What's more, DARPA1998 [21] are not appropriate to simulate actual network systems according to [22]. As a result, we choose ISCX2012 [23] as our experiment datasets.

ISCX2012 recorded 7 days' network traffic (legitimate and malicious). During those 7 days, 4 types of attacks happened on 4 days respectively. Considered about the training time and the computer memory, we collect all attack packets and choose one day's normal ones (16/6/2010) to match the number of attack packets. We divide the whole dataset into training and test datasets using a ratio of 60% to 40%, respectively. Table 2 presents the distribution of traffic records in the final dataset.

4.3   Multi-classification Using WEDL-NIDS

The results of multi-classification using WEDL-NIDS are presented in Table 3. Specifically, the approach achieved 99.97% overall accuracy rate, which strongly proved its performance. The detection rate was 96.39% on the whole dataset, but it dropped to 90.00% when considered about the Infiltrating attack. It's probably because the training samples of HttpDoS is the least in the whole training set. The FAR was pretty good, some class (BFSSH, DDoS) even achieved 0. The overall F_1 score was 0.94, which meant the model had a quite good comprehensive performance.

Table 2. Distribution of traffic records in our dataset

Dataset	Training	Test	Total
Normal	241,951	161,301	403,252
BFSSH	2,974	1,982	4,956
Infiltrating	6,057	4,038	10,095
HttpDoS	2,110	1,406	3,516
DDoS	13,577	9,050	22,627
Total	266,669	177,777	444,446

(Normal: normal traffic; BFSSH: brute force SSH attacks; Infiltrating: infiltrating attacks; HttpDoS: HttpDoS attacks; DDoS: DDoS attacks)

Table 3. Multi-classification results of WEDL-NIDS for ISCX2012

Dataset	ACC	DR	FAR	F_1
Normal	99.97%	99.97%	0.05%	0.99
BFSSH	99.99%	99.79%	0.00%	0.99
Infiltrating	99.97%	92.38%	0.02%	0.83
HttpDoS	99.96%	90.00%	0.02%	0.88
DDoS	99.96%	99.79%	0.00%	0.99
Overall	99.97%	96.39%	0.02%	0.94

4.4 Comparison with Other Methods

As mentioned above, researchers have proposed many intrusion detection methods. Thus, we compare the experimental results of the WEDL-NIDS with those of other published methods.

Table 4 shows a comparison of the experimental results for the ISCX2012 dataset. The first four methods listed in Table 4 all use traditional machine learning methods. The DeepDefense used deep learning method but they do

Table 4. Comparison with other published methods for ISCX2012

Method	ACC	Attack-DR	FAR
MHCVF [24]	99.50%	68.20%	0.03%
ALL-AGL [25]	95.40%	93.20%	0.30%
KMC+NBC [26]	99.00%	99.70%	2.20%
AMGA2-NB [27]	94.50%	92.70%	7.00%
DeepDefense [28]	97.61%	97.83%	2.39%
WEDL-NIDS	99.97%	95.49%	0.02%

experiments only on the detection of DoS attack, and they only used DDoS attacks data and normal data as their training set. Except the DeepDefense, all the methods in Table 4 used the same set of data and same set of attacks. The overall accuracy, the DR of attack traffic and overall FAR are used as the evaluation metrics.

Table 4 shows that the WEDL-NIDS method achieves the best performances regarding the overall accuracy and overall FAR. Noticed that the DR of attack is lower than the state-of-art by 4.21%, but their method gets a higher FAR than us (2.20% compared to 0.02%). Similarly, the DR of attack for DeepDefense is also higher than the WEDL-NIDS. However, we should consider about their results are performed only on the detection of DoS attack, and their FAR is higher than us.

5   Conclusions and Future Work

As a result of the difficulty of hand-designing accurate traffic features in the field of intrusion detection, we propose a word embedding-based network intrusion detection model (WEDL-NIDS). The experimental results show that the WEDL-NIDS effectively improves the accuracy and DR compared to other published methods.

Our contributions are that we effectively reduce the dimensions of features by word embedding and automatically extract the suitable features by deep learning. Two problems require further study in future work. The first involves the interpretability of the word vectors we get after embedding. The second problem involves the ability of finding unknown attacks. As a result of the fact that our method belongs to supervised learning, we can't promise the performance of our model when unknown attacks occur.

Acknowledgement. This research work is supported by National Natural Science Foundation of China under grant number 61105050.

References

1. Wang, W., Zhu, M., Zeng, X., et al.: Malware traffic classification using convolutional neural network for representation learning. In: International Conference on Information Networking, pp. 712–717. IEEE (2017)
2. Wang, W., Sheng, Y., Wang, J., et al.: HAST-IDS: learning hierarchical spatial-temporal features using deep neural networks to improve intrusion detection. IEEE Access **6**, 1792–1806 (2017)
3. Mikolov, T., Yih, W.T., Zweig, G.: Linguistic regularities in continuous space word representations. In: HLT-NAACL (2013)
4. Krizhevsky, A., Sutskever, I., Hinton, G.E.: ImageNet classification with deep convolutional neural networks. In: International Conference on Neural Information Processing Systems, pp. 1097–1105. Curran Associates Inc. (2012)
5. Hochreiter, S., Schmidhuber, J.: Long short-term memory. Neural Comput. **9**(8), 1735–1780 (1997)

6. Tang, T.A., Mhamdi, L., McLernon, D., et al.: Deep learning approach for network intrusion detection in software defined networking. In: 2016 International Conference on Wireless Networks and Mobile Communications (WINCOM), pp. 258–263. IEEE (2016)

7. Salama, M.A., Eid, H.F., Ramadan, R.A., Darwish, A., Hassanien, A.E.: Hybrid intelligent intrusion detection scheme. In: Gaspar-Cunha, A., Takahashi, R., Schaefer, G., Costa, L. (eds.) Soft Computing in Industrial Applications. AINSC, vol. 96, pp. 293–303. Springer, Heidelberg (2011). https://doi.org/10.1007/978-3-642-20505-7_26

8. Fiore, U., Palmieri, F., Castiglione, A., et al.: Network anomaly detection with the restricted Boltzmann machine. Neurocomputing **122**, 13–23 (2013)

9. Wang, Z.: The applications of deep learning on traffic identification. BlackHat USA (2015)

10. Javaid, A., Niyaz, Q., Sun, W., et al.: A deep learning approach for network intrusion detection system. In: Proceedings of the 9th EAI International Conference on Bio-inspired Information and Communications Technologies (formerly BIONETICS). ICST (Institute for Computer Sciences, Social-Informatics and Telecommunications Engineering), pp. 21–26 (2016)

11. Yu, Y., Long, J., Cai, Z.: Session-based network intrusion detection using a deep learning architecture. In: Torra, V., Narukawa, Y., Honda, A., Inoue, S. (eds.) MDAI 2017. LNCS (LNAI), vol. 10571, pp. 144–155. Springer, Cham (2017). https://doi.org/10.1007/978-3-319-67422-3_13

12. Yu, Y., Long, J., Cai, Z.: Network intrusion detection through stacking dilated convolutional autoencoders. Secur. Commun. Netw. **2017**, 1–10 (2017)

13. Yin, C., Zhu, Y., Fei, J., et al.: A deep learning approach for intrusion detection using recurrent neural networks. IEEE Access **5**, 21954–21961 (2017)

14. Mikolov, T., Le, Q.V., Sutskever, I.: Exploiting similarities among languages for machine translation. arXiv preprint arXiv:1309.4168 (2013)

15. Goldberg, Y., Levy, O.: word2vec Explained: deriving Mikolov et al.'s negative-sampling word-embedding method. arXiv preprint arXiv:1402.3722 (2014)

16. Pennington, J., Socher, R., Manning, C.: GloVe: global vectors for word representation. In: Proceedings of the 2014 Conference on Empirical Methods in Natural Language Processing (EMNLP), pp. 1532–1543 (2014)

17. Gu, J., Wang, Z., Kuen, J., et al.: Recent advances in convolutional neural networks. arXiv preprint arXiv:1512.07108 (2015)

18. Sutskever, I., Vinyals, O., Le, Q.V.: Sequence to sequence learning with neural networks. In: Advances in Neural Information Processing Systems, pp. 3104–3112 (2014)

19. Tavallaee, M., Bagheri, E., Lu, W., et al.: A detailed analysis of the KDD CUP 99 data set. In: IEEE Symposium on Computational Intelligence for Security and Defense Applications, 2009, CISDA 2009, pp. 1–6. IEEE (2009)

20. Song, J., Takakura, H., Okabe, Y.: Description of Kyoto University benchmark data. http://www.takakura.com/Kyoto_data/BenchmarkData-Description-v5.pdf

21. Lippmann, R., Cunningham, R.K., Fried, D.J., et al.: Results of the DARPA 1998 offline intrusion detection evaluation. In: Recent Advances in Intrusion Detection, vol. 99, pp. 829–835 (1999)

22. Mchugh, J.: Testing Intrusion detection systems: a critique of the 1998 and 1999 DARPA intrusion detection system evaluations as performed by Lincoln Laboratory. ACM Trans. Inf. Syst. Secur. **3**(4), 262–294 (2000)

23. Shiravi, A., Shiravi, H., Tavallaee, M., et al.: Toward developing a systematic approach to generate benchmark datasets for intrusion detection. Comput. Secur. **31**(3), 357–374 (2012)

24. Akyol, A., Hacibeyoglu, M., Karlik, B.: Design of multilevel hybrid classifier with variant feature sets for intrusion detection system. IEICE Trans. Inf. Syst. **E99.D**(7), 1810–1821 (2016)

25. Sallay, H., Ammar, A., Saad, M.B., et al.: A real time adaptive intrusion detection alert classifier for high speed networks. In: IEEE International Symposium on Network Computing and Applications, pp. 73–80. IEEE (2013)

26. Yassin, W., Udzir, N.I., Muda, Z., et al.: Anomaly-based intrusion detection through K-Means clustering and Naives Bayes classification (2013)

27. Tan, Z., Jamdagni, A., He, X., et al.: Detection of denial-of-service attacks based on computer vision techniques. IEEE Trans. Comput. **64**(9), 2519–2533 (2015)

28. Yuan, X., Li, C., Li, X.: DeepDefense: identifying DDoS attack via deep learning. In: IEEE International Conference on Smart Computing, pp. 1–8. IEEE (2017)

Anonymization of Unstructured Data via Named-Entity Recognition

Fadi Hassan, Josep Domingo-Ferrer, and Jordi Soria-Comas[✉]

Department of Computer Science and Mathematics, CYBERCAT-Center
for Cybersecurity Research of Catalonia, UNESCO Chair in Data Privacy,
Universitat Rovira i Virgili, Av. Països Catalans 26, 43007 Tarragona, Catalonia
{fadi.hassan,josep.domingo,jordi.soria}@urv.cat

Abstract. The anonymization of structured data has been widely studied in recent years. However, anonymizing unstructured data (typically text documents) remains a highly manual task and needs more attention from researchers. The main difficulty when dealing with unstructured data is that no database schema is available that can be used to measure privacy risks. In fact, confidential data and quasi-identifier values may be spread throughout the documents to be anonymized. In this work we propose to use a named-entity recognition tagger based on machine learning. The ultimate aim is to build a system capable of detecting all attributes that have privacy implications (identifiers, quasi-identifiers and sensitive attributes). In particular, we present a proof of concept focused on the detection of confidential attributes. We consider a case study in which confidential values to be detected are disease names in medical diagnoses. Once these confidential attribute values are located, one can use standard statistical disclosure control techniques for structured data to control disclosure risk.

Keywords: Anonymization · Unstructured data
Named-entity recognition · Conditional random fields

1 Introduction

Nowadays, large amounts of data are being collected from very diverse sources, quite often without the affected individuals being aware of it. Such a systematic data collection, coupled with new data analysis techniques, has given rise to big data. Although sometimes qualified as a buzzword, big data entail a significant change in the way data are managed. In this work, we are concerned with the privacy implications of big data, in particular unstructured big data.

In the traditional setting, data were mainly collected through surveys or from other administrative data sources. As a result, they usually had a structured nature (a table). The wide variety of data sources in the current big data context (e.g. emails sent and received, participation in social networks, etc.) forces us to consider other types of data, such as semi-structured or unstructured data

© Springer Nature Switzerland AG 2018
V. Torra et al. (Eds.): MDAI 2018, LNAI 11144, pp. 296–305, 2018.
https://doi.org/10.1007/978-3-030-00202-2_24

(free text). Already in 2005, it was claimed in [9] that as many as 80% of the business and medical data were stored in unstructured form. In the health-care context, a proper use of such data is critical for research and policy-making purposes, and useful for related industries such as health insurance.

The new European General Data Protection Regulation (GDPR, [8]) states that explicit consent from the affected individuals is needed to use personally identifiable information (PII) for secondary purposes (different from the primary purpose that motivated the collection, such as healthcare or service billing). Ideally, the data collector should strive to gather such consent. However, in practice this may not be feasible. It may be difficult to contact individuals to obtain their consent. Additionally, individuals with rare conditions are more likely be concerned about their privacy, which makes them less prone to grant consent for their data to be used. Due to these shortcomings, the resulting data sets will probably be biased.

To avoid the need for consent, data used for secondary purposes should no longer be personally identifiable. Anonymization, also known as statistical disclosure control (SDC), provides a way to turn PII into information that cannot be linked to a specific identified individual any more and hence is not subject to privacy regulations.

There is a substantial amount of literature on SDC for the case of structured data [4,5,10]. Structured data are those that can be described as a set of records each of which corresponds to an individual and contains the values of a fixed set of attributes for that individual. A common approach to anonymize structured data is to remove attributes that are identifiers and then mask quasi-identifier attributes. The latter are attributes that are not identifiers but together might allow linking the record with some external data source containing identifiers, and therefore might allow re-identifying the individual to whom a record corresponds. Alternatively, instead or in addition to masking quasi-identifiers, one can mask the confidential attributes, to introduce uncertainty about the confidential attribute values.

Once a decision has been made on which attributes are quasi-identifiers and which are confidential ones, anonymization of structured data can be fully automated. (Admittedly, in some cases the above decision may be unclear, as it depends on the background information that is assumed to be available to an intruder.) However, automation of unstructured data anonymization is much more difficult, because there is no database schema that can be followed to classify the data into identifiers, quasi-identifiers and confidential attributes. As a result, anonymizing unstructured data remains today a largely manual task.

In fact, it can be argued that unstructured textual data are the ones for which anonymization is hardest. Other types of data that might seem more difficult at first sight can be either reduced to unstructured text by using tools for automated semantics extraction (as it occurs with video and audio) or are not amenable to anonymization because their semantics is not yet sufficiently understood (as is the case for genetic data).

Contribution and Plan of This Paper

The purpose of this work is to automate the extraction of quasi-identifier and/or confidential attributes from unstructured textual data. That is, we want to be able to automatically identify attributes such as passport number, name, location, age, birth date, etc. For the sake of concreteness, in this work, the focus will be on medical diagnosis reports. Once this automatic identification of the relevant attributes is completed, we can apply some of the methods designed for anonymizing structured data. To identify attributes, we will take advantage of a named-entity recognition (NER) tagger [7].

In Sect. 2, we briefly introduce some concepts that are important to understand this work. In Sect. 3, previous work on document anonymization is recalled. In Sect. 4, we describe our proposal. Experiments are presented in Sect. 5 and conclusions and future work ideas are gathered in Sect. 6.

2 Background

2.1 Named-Entity Recognition

Named entity recognition (NER) is the task of locating and categorizing important terms in a text [17]. Named-entity recognition is a source of information for different natural language processing applications. NER has been used to improve the performance of many applications, such as answering questions [12], automatic text translation [1], information retrieval [23], and sentiment analysis of tweets [11].

NER is also useful in the anonymization of unstructured data (e.g. free text documents). In particular, it can detect those terms that might be used to re-identify an individual and those terms that contain sensitive information. Once these terms have been located, they constitute structured information that can be anonymized as usual using SDC methods (e.g. generalization, supression, etc.) to keep the disclosure risk under control.

There are many tagging schemes for NER. In this work we use the IOB2 tagging scheme [21]. In IOB2, each word in the text is labeled using one of three possible tags: I, O, or B, which indicate if the word is inside, outside, or at the beginning of a named entity. Usually, in the IOB2 tagging scheme, the B and I letters come as prefix and are followed by the category name of the named entity to distinguish between the B and I tags of different entities, e.g. in our case B-DIS refers to beginning of named entity Disease and I-DIS means within entity Disease.

2.2 Conditional Random Fields

In natural language processing, there are two common models used to solve NER tasks: hidden Markov models (HMMs), used in works such as [16,27], and conditional random fields (CRFs), used in works such as [3,6,11]. NER using

CRFs is widely used and applied, and usually gives the best results in many domains, so in this work we design our model using CRFs.

CRFs [15] are conditionally trained undirected graph models often applied in pattern recognition. These models are used to calculate the conditional probability of values on designated output nodes given values assigned to other designated input nodes.

3   Related Work

Several techniques to anonymize unstructured textual data have been proposed. Most of them can be classified into one the following two categories: dictionary-based techniques and machine learning techniques [19].

In the past, document anonymization was carried out by manual search and replacement of the named entities. Sweeney [24] proposed the Scrub method that relies on the definition of some templates for the named entities, like location, name and country. Once these entities are found, the related value is masked.

Neamatullah et al. [18] proposed a software for document anonymization that uses lexical look-up tables, regular expressions and simple heuristics that perform context checks to locate named entities. After that, they replace these entities by non-indexed category values (e.g. replace "New York" by "[**Location**]").

Vico and Calegari [26] proposed a software architecture for document anonymization. The key idea is to recognize the named entities with an architecture of multiple natural language processing tools. After that, they replace the sensitive entities by a generic indexed category value (e.g. replace "Fever" by "generic_term_1").

In 2016, the United Kingdom Data Archive (UKDA) released a text anonymization helper tool [22]. This tool identifies numbers and words starting with a capital letter, and replaces them with "XXX".

Kleinberg et al. [13] designed Netanos, a tool to allow researchers to anonymize large texts. They use machine learning to recognize named entities (e.g. persons, locations, times and dates). Then, they replace them by a privacy-preserving indexed category value (e.g. "Location_1", "Person_1").

4   Methodology

The aim of this work is to locate terms in an unstructured text that can have privacy implications, either because they can be used to re-identify an individual or because they contain confidential information.

4.1   General Approach

Formally, given a collection of text documents $D_1, \ldots, D_n$, we want to locate supersets of all the privacy-relevant attributes they contain. Specifically, we want to come up with a superset of identifier attributes $\mathcal{ID} = \{ID_1, \ldots, ID_p\}$, a

superset of quasi-identifier attributes $\mathcal{QID} = \{QID_1, \ldots, QID_q\}$, and a superset of confidential attributes $\mathcal{C} = \{C_1, \ldots, C_r\}$. The set $\mathcal{ID}$ should contain the
identifier attributes that appear in at least one of the documents; for exam-
ple, $\mathcal{ID}$ will contain "Passport no." if at least one of the documents contains
a passport number (even if the other documents contain no passport number).
Similarly, the set $\mathcal{QID}$ should contain the quasi-identifier attributes that appear
in at least one document, and the set $\mathcal{C}$ the confidential attributes that appear
in at least one document.

Once the above supersets have been determined, the collection of documents
can be viewed as a *structured* data set with records $D_1, \ldots D_n$ and attributes
that are the elements of $\mathcal{ID} \cup \mathcal{QID} \cup \mathcal{C}$. Obviously, this structured data set is
likely to be a sparse one, as not all attributes take values in all documents. To
anonymize this data set, we proceed as usual in the case of structured data sets.
The values of attributes in $\mathcal{ID}$ should be suppressed from all records/documents
and masking should be applied to attributes in $\mathcal{QID}$ and/or $\mathcal{C}$. Depending on
the type of masking used, it may be necessary to deal first with the missing
attribute values in some documents; imputing them by partial synthesis is a
possibility [5,10].

Thus, the problem of anonymizing unstructured data reduces to locating the
appearances of the various privacy-relevant attributes in the collection of documents and then anonymizing the resulting structured data set. We can tackle
the task of locating attribute appearances by building several machine learning
models, each of them recognizing a different type of named entity. For example, a first model to recognize identifier attributes (e.g. passport number, social
security number, etc.), a second model to recognize quasi-identifier attributes
(e.g. location, birth date, age, postal code, etc.), and a third model to recognize
confidential attributes (e.g. disease names, etc.).

4.2 Proof of Concept

As a proof of concept, we focus on locating confidential data within medical
diagnoses. We propose a model based on conditional random fields to extract
the disease names from a given medical record. For a given text, this model
predicts a sequence of corresponding IOB2 tags.

Once we have the predicted sequence of IOB2 tags for every token in the
medical record, we can interpret this sequence of labels and extract the Disease
entity entity. For instance, if we have the sentence "Retinopathy was assessed
by ophthalmoscopy" and the corresponding IOB2 tags sequence {B-DIS, O, O,
O}, we move through the IOB2 sequence tags and every word corresponding
to a B-DIS label is considered as the beginning of a disease entity and every
word corresponding to an I-DIS label is considered as being within a disease
entity. Thus, a B-DIS word with all directly following I-DIS words forms one
disease entity. In fact, B-DIS and I-DIS labels do the same job but B-DIS has
the particular job of distinguishing between two consecutive disease entities.

Figure 1 shows the structure of the proposed model for disease name recognition. It consists of three steps:

- The first step is the tokenizer, which splits a sentence into tokens.
- The second step is the feature extractor; in this step, we use a window of three words (the current word, the previous word and the next word), and we extract the features of these words. Table 1 explains all the features we considered.
- The third step uses a CRF model, which takes the features from the second step and produces a sequence of tags for the whole sentence.

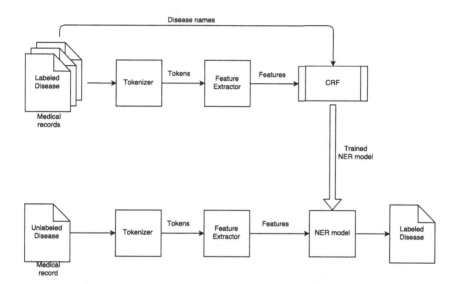

Fig. 1. Architecture of the named-entity recognition tagger

Table 1. Feature extraction

Feature	Explanation
Word stem	E.g. the stem of "illness" is "ill". We extract stems using SnowballStemmer from the nltk library [2].
Word length	The length of the word
Word shape	The shape of the word, which can be 'lowercase', 'uppercase', 'capitalized', 'mixed'
Word POS	Part of speech for the word. We use the Stanford POS tagger to extract this feature [25]

5   Experimental Results

In this section we describe the experimental results of the above-mentioned proof of concept. We programmed the experiments in Python, and we used sklearn-crfsuite for CRF [14] and SnowballStemmer for word stemming [2].

5.1 Data Set

In our experiments, we took advantage of medical texts that were labeled to study the relation between diseases and treatments. These files were obtained from MEDLINE 2001 using the first 100 titles and the first 40 abstracts from the 59 files medline01n*.xml, that are available in [20].

These data contain 3,654 labeled sentences. The labels are: "DISONLY", "TREATONLY", "TREAT PREV", "DIS PREV", "TREAT SIDE EFF", "DIS SIDE EFF", "DIS VAG", "TREAT VAG", "TREAT NO" and "DIS NO". As we were only interested in diseases, we only kept the 629 sentences with the "DISONLY" labels.

5.2 Evaluation Metrics

We used three metrics to evaluate the performance of the proposed model for the recognition of diseases:

– *Precision.* Number of diseases correctly identified by the classifier divided by the total number of identified diseases:

$$\text{Precision} = \frac{|S \cap T|}{|S|},$$

where S is the set of all diseases identified by the classifier and T is the set of correct diseases according to the original dataset.
– *Recall.* Number of diseases correctly identified by the classifier divided by the number of correct diseases in the original dataset:

$$\text{Recall} = \frac{|S \cap T|}{|T|}.$$

– *F1.* Harmonic mean of precision and recall:

$$F_1 = 2 \cdot \frac{\text{Precision} \cdot \text{Recall}}{\text{Precision} + \text{Recall}}.$$

5.3 Results and Discussion

We did the experimental evaluation in two phases: model training and model testing. Out of the 629 samples of labeled sentences, 503 were devoted to model training (80% of the samples), and 126 to model testing (20% of the samples).

The training phase was performed via 10-fold cross-validation, as follows. We partitioned the training data set into 10 equal-size subsamples. Out of the 10 subsamples, a single subsample was retained as validation data for testing the model while in the training phase, and the remaining 9 subsamples were used in training.

While most words in the data set were labeled as O (outside disease), we were interested in words labeled as B-DIS (beginning of disease) and I-DIS (in disease). Thus, we computed the precision, the recall and the F1 score only for

Table 2. Evaluation of the model on the test dataset at word level

	Precision	Recall	F1-score
B-DIS	0.766	0.677	0.719
I-DIS	0.789	0.709	0.747
Avg/total	0.778	0.693	0.733

B-DIS and I-DIS. For example, if we have the sentence "Diagnostic evaluation of the patient with high blood pressure", its word tokens are {"Diagnostic", "evaluation", "of", "the", "patient", "with", "high", "blood", "pressure"} and the corresponding labels are {O, O, O, O, O, O, B-DIS, I-DIS, I-DIS}. The named entity here contains three words "high blood pressure". Table 2 shows the evaluation of the predicted tags against the correct tags at the word level (separately for each word). In contrast, Table 3 reports the same evaluation metrics for whole entities. That is, in the previous example, Table 2 would separately refer to the three words "high", "blood" and "pressure", while Table 3 would refer to the entity "high blood pressure"; in the latter case, unless *all three* words of the entity were correctly labeled, the whole entity would be considered as misclassified.

Table 3. Evaluation the model on the test dataset at entity level

	Precision	Recall	F1-score
Disease entity	0.742	0.660	0.698

According to Table 3, our model performed significantly better regarding the precision than regarding the recall. It is very likely that the recall can be increased by using more training samples. Nonetheless, we consider the above results to be promising, as a recall similar to manual labeling is achieved. Indeed, the authors of [18] asked 14 clinicians to detect and anonymize named entities in approximately 130 patient notes: the result of this manual procedure varied from clinician to clinician, with recall ranging between 0.63 and 0.94 on the data they used.

6 Conclusions and Future Work

In this work, we have dealt with the anonymization of unstructured textual data. As a proof of concept, we have focused on locating disease names (i.e. sensitive attributes) in medical records. Once located, these sensitive attributes can be protected using common SDC techniques for structured data.

The main contribution of this work relates to the architecture of the recognizer for named entities. The proposed model is based on machine learning and outperforms dictionary-based NER approaches. Specifically, it avoids the out-of-dictionary problem that arises when the entities to be located are not in the dictionary being used.

As future work, we plan to extend the presented proof of concept to the detection of identifiers and quasi-identifiers. This will require investing substantial effort to generate annotated datasets for attributes such as name, location, age, etc. These annotated data sets will subsequently be used to train the identifier and the quasi-identifier detection models.

Acknowledgments and Disclaimer. The following funding sources are gratefully acknowledged: European Commission (projects H2020-644024 "CLARUS" and H2020-700540 "CANVAS"), Government of Catalonia (ICREA Acadèmia Prize to J. Domingo-Ferrer) and Spanish Government (projects TIN2014-57364-C2-1-R "Smart-Glacis" and TIN2015-70054-REDC). The views in this paper are the authors' own and do not necessarily reflect the views of UNESCO or any of the funders.

References

1. Babych, B., Hartley, A.: Improving machine translation quality with automatic named entity recognition. In: Proceedings of the 7th International EAMT Workshop on MT and Other Language Technology Tools, Improving MT Through Other Language Technology Tools: Resources and Tools for Building MT (EAMT 2003), pp. 1–8. Association for Computational Linguistics (2003)
2. Bird, S., Klein, E., Loper, E.: Natural Language Processing with Python – Analyzing Text with the Natural Language Toolkit. O'Reilly (2009). The Natural Language Toolkit software (NLTK): https://www.nltk.org
3. Culotta, A., Bekkerman, R., McCallum, A.: Extracting Social Networks and Contact Information from Email and the Web. Computer Science Department Faculty Publication Series, no. 33. University of Massachusetts-Amherst, 2004
4. Domingo-Ferrer, J., Sánchez, D., Soria-Comas, J.: Database Anonymization: Privacy Models, Data Utility, and Microaggregation-Based Inter-model Connections. Morgan & Claypool, San Rafael (2016)
5. Drechsler, J.: Synthetic Datasets for Statistical Disclosure Control. LNS, vol. 201. Springer, New York (2011). https://doi.org/10.1007/978-1-4614-0326-5
6. Ekbal, A., Haque, R., Bandyopadhyay, S.: Bengali part of speech tagging using conditional random field. In: Proceedings of the Seventh International Symposium on Natural Language Processing (SNLP 2007) (2007)
7. Finkel, J.R., Grenager, T., Manning, C.: Incorporating non-local information into information extraction systems by Gibbs sampling. In: Proceedings of the 43rd Annual Meeting of the Association for Computational Linguistics (ACL 2005), pp. 363–370. Association for Computational Linguistics (2005)
8. EU General Data Protection Regulation, 2016/679. https://gdpr-info.eu
9. Grimes, S.: Structure, models and meaning. Intelligent Enterprise, March 2005
10. Hundepool, A., et al.: Statistical Disclosure Control. Wiley, New York (2012)
11. Jabreel, M., Hassan, F., Moreno, A.: Target-dependent sentiment analysis of tweets using bidirectional gated recurrent neural networks. In: Hatzilygeroudis, I., Palade, V. (eds.) Advances in Hybridization of Intelligent Methods, pp. 39–55. Springer, Cham (2018). https://doi.org/10.1007/978-3-319-66790-4_3

12. Khalid, M.A., Jijkoun, V., de Rijke, M.: The impact of named entity normalization on information retrieval for question answering. In: Macdonald, C., Ounis, I., Plachouras, V., Ruthven, I., White, R.W. (eds.) ECIR 2008. LNCS, vol. 4956, pp. 705–710. Springer, Heidelberg (2008). https://doi.org/10.1007/978-3-540-78646-7_83

13. Kleinberg, B., Mozes, M., van der Toolen, Y., Verschuere, B.: NETANOS - Named Entity-based Text Anonymization for Open Science. Open Science Framework, 31 January 2018. https://osf.io/w9nhb

14. Korobov, M.: sklearn-crfsuite (2015). https://sklearn-crfsuite.readthedocs.io/en/latest/

15. Lafferty, J., McCallum, A., Pereira, F.C.N.: Conditional random fields: probabilistic models for segmenting and labeling sequence data. In: Proceedings of the 18th International Conference on Machine Learning 2001 (ICML 2001), pp. 282–289. ACM (2001)

16. Morwal, S., Jahan, N., Chopra, D.: Named entity recognition using hidden Markov model (HMM). Int. J. Nat. Lang. Comput. **1**(4), 15–23 (2012)

17. Nadeau, D., Sekine, S.: A survey of named entity recognition and classification. Linguisticae Investigationes **30**(1), 3–26 (2007)

18. Neamatullah, I., et al.: Automated de-identification of free-text medical records. BMC Med. Inform. Decis. Making **8**(1), 32 (2008)

19. Pérez-Laínez, R., Iglesias, A., de Pablo-Sánchez, C.: Anonimytext: anonymization of unstructured documents. Universidad Carlos III de Madrid (2009). https://e-archivo.uc3m.es/handle/10016/19829

20. Rosario, B., Hearst, M.A.: Classifying semantic relations in bioscience texts. In: Proceedings of the 42nd Annual Meeting of the Association for Computational Linguistics (ACL 2004). Association for Computational Linguistics (2004). Data: http://biotext.berkeley.edu/dis_treat_data.html

21. Sang, E.F., Veenstra, J.: Representing text chunks. In: Proceedings of the 9th Conference of the European Chapter of the Association for Computational Linguistics, pp. 173–179. Association for Computational Linguistics (1999)

22. United Kingdom Data Service: Text Anonymization Helper Tool. https://bitbucket.org/ukda/ukds.tools.textanonhelper/wiki/Home. Accessed 24 Mar 2018

23. Sundheim, B.M.: Overview of results of the MUC-6 evaluation. In: Proceedings of the TIPSTER Text Program: Phase II, pp. 423–442. Association for Computational Linguistics (1996)

24. Sweeney, L.: Replacing personally-identifying information in medical records, the Scrub system. In: Proceedings of the AMIA Annual Fall Symposium, p. 333. American Medical Informatics Association (1996)

25. Toutanova, K., Klein, D., Manning, C., Singer, Y.: Feature-rich part-of-speech tagging with a cyclic dependency network. In: Proceedings of HLT-NAACL 2003, pp. 252–259 (2003)

26. Vico, H., Calegari, D.: Software architecture for document anonymization. Electron. Notes Theor. Comput. Sci. **314**(C), 83–100 (2015)

27. Zhou, G., Su, J.: Named entity recognition using an HMM-based chunk tagger. In: Proceedings of the 40th Annual Meeting of the Association for Computational Linguistics (ACL 2002), pp. 473–480. Association for Computational Linguistics (2002)

On the Application of SDC Stream Methods to Card Payments Analytics

Miguel Nuñez-del-Prado[1] and Jordi Nin[2]($\boxtimes$)

[1] Universidad del Pacífico, Av. Salaverry 2020, Lima, Peru
m.nunezdelpradoc@up.edu.pe
[2] BBVA Data & Analytics, Barcelona, Catalonia, Spain
jordi.nin@bbvadata.com

Abstract. Banks and financial services have to constantly innovate their online payment services to avoid large digital companies take the control of online card transactions, relegating traditional banks to simple payments carriers. Apart from creating new payment methods (*e.g.* contact-less cards, mobile wallets, etc.), banks offers new services based on historical payments data to endow traditional payments methods with new services and functionalities. In this latter case, it is where privacy preserving techniques play a fundamental role ensuring personal data is managed full-filling all the applicable laws and regulations. In this paper, we introduce some ideas about how SDC stream anonymization methods could be used to mask payments data streams. Besides, we also provide some experimental results over a real card payments database.

Keywords: Statistical Disclosure Control
General Data Protection Regulation (GDPR)
Payment Service Directive (PSD2) · Stream mining

1 Introduction

There has been a lot of discussion and debate in various forums regarding how the European Union (EU) General Data Protection Regulation (GDPR), 2016/679/EU law [12] comed into force on May 2018. The GDPR will widen the definition of 'personal data' to include data that relates to an 'identifiable' natural person, as opposed to just an 'identified' person. This means data may be 'personal' even if the organisation holding the data cannot itself identify a natural person. It also brings in provisions for the 'right to data portability', allowing an end user to request all data held about them; and a 'right to be forgotten'.

Besides, banks and financial institutions need to implement the Second Payment Services Directive (PSD2) which makes it clear that customers have a right to use what are termed Payment Initiation Service Providers (PISPs) and Account Information Service Providers (AISPs) where the payment account is accessible online and where they have given their explicit consent.

© Springer Nature Switzerland AG 2018
V. Torra et al. (Eds.): MDAI 2018, LNAI 11144, pp. 306–318, 2018.
https://doi.org/10.1007/978-3-030-00202-2_25

Apart from these two legal and regulatory issues, banks are into a competition with big digital companies to offer new online tools to add extra value to traditional banking business, such as credit card payments. For instance, BBVA launched C360[1] in 2016 or Banc Sabadell created the Kelvin Retail system[2] last year. To have a clearer idea about the wide spectrum of online services offered by banks check [11]. Due to these reasons, privacy preserving techniques are in the eye of the storm of all online banking services. In this paper we describe the application of the streaming statistical disclosure control (SDC) methods presented in [9] to a real card payments database. Specifically, we will compare online versions of noise addition, microaggregation, rank swapping and differential private microaggregation. Our results illustrates the viability of these techniques in different scenarios where banks could be interested in sharing data with third parties in an anonymous way. The rest of this paper is organized as follows. In Sect. 2, we introduce some basic concepts about the SDC streaming scenario. Then, Sect. 3 describes the SDC methods used in the experiments and its metrics. Later, Sect. 4 describes the experiments carried out in this paper. Finally, we conclude the work in Sect. 5, where we also propose some possible future research lines.

2   Preliminaries

The main problem when dealing with streams is the high throughput of data being analyzed, under computational resources constraints. Usually, stream mining technologies require to modify traditional data mining methods to enable their use on data streams. Such modifications include *approximation algorithms*, *sliding window methods* and *algorithm output granularity*. In the present work we rely on *sliding window methods*, where a window of specified length l moves over the data, sample by sample, and each SDC method is computed over the data in the window. We also introduce SDC and differential privacy concepts here.

2.1   Statistical Disclosure Control

The purpose of Statistical Disclosure Control (*SDC*) is to prevent that confidential information can be linked to specific individuals. There are two categories of *SDC* algorithms to achieve a certain privacy level [5], such as *Perturbative* and *Non-perturbative* methods. The former methods add some noise to samples in order to make re-identification more difficult, and the latter do not transform records, they suppress samples partially or reduces its details.

---

[1] BBVA C360 – https://www.bbva.es/autonomos/banca-online/commerce360/index.jsp.

[2] Banc Sabadell Kelvin Retail – https://www.bancsabadell.com/cs/Satellite/SabAtl/Kelvin-Retail/6000019696135/es/.

When assessing Disclosure Risk (DR) of a released data stream, one must consider that a sample is composed by different kind of variables [16], namely:

- **Identifiers.** They are variables which unambiguously identify the individual, for example, the passport number.
- **Quasi-identifiers.** They are variables which can identify the individual when some of those attributes are combined. For example, the combination (age = 16, city = NY, job = photographer in the 'Daily Bugle') unequivocally identifies Peter Parker, Spider-man.
- **Confidential.** They are variables which contain sensitive information about the individual. For example, salary.

Therefore, a data set X is defined as $X = X_{id}||X_{nc}||X_c$, where X_{id} are the identifiers, X_{nc} are the non-confidential quasi-identifier values, and X_c are the confidential values. Normally, before releasing a data set X containing confidential data, a protection method ρ is applied, leading to a protected data set X'. Indeed, we will assume the following typical scenario: (i) identifiers in X are either removed or encrypted, therefore we will write $X = X_{nc}||X_c$; (ii) confidential values X_c are not modified, and so we have $X'_c = X_c$; (iii) the protection method itself is applied to non-confidential quasi-identifier values, in order to preserve the privacy of the individuals whose confidential data is being released. Therefore, we have $X'_{nc} = \rho(X_{nc})$. This scenario allows third parties to have precise information on confidential data without revealing to whom the confidential data belongs to.

Besides, when DR is assessed, data practitioners must consider several privacy breaches such as *identity disclosure* or *attribute disclosure*. A common way to measure individual Disclosure Risk (DR) is using *Record Linkage* (RL) methods [2]. Thus, once a SDC method has been used to anonymize a sample, the RL procedure is applied to the original and masked samples. This *linkage* attempts to identify, for each sample in the masked stream, the corresponding sample in the original stream.

Another critical measurement concerning data protection is Information Loss (IL) or *Data Utility*, which could be defined as the amount of useful statistical information that is lost along the data masking process [1,10]. On one hand, a good anonymization method should minimize IL, in order to provide optimally useful data to the legitimate users of such data. On the other hand, it is also interesting to keep a low disclosure risk.

2.2   Differential Privacy

The main technique in differential privacy [4] is to add noise which is calibrated to the global sensitivity of a query, being the maximal amount by which the query result may change if one adds to the database a single record, this idea is defined as *neighbour datasets*.

The Laplace mechanism is a relatively extended differential privacy method. However, it can only be applied to functions that provide a *numerical* answer.

We define the *global sensitivity* of a numerical function $f : \mathcal{D} \rightarrow \mathbb{R}^w$, with $w \in \mathbb{N}^+$, over the universe of datasets $\mathcal{D}$, as:

$$\Delta(f) = \max_{\substack{D_1, D_2 \in \mathcal{D} \\ |D_1 \Delta D_2| = 1}} \|f(D_1) - f(D_2)\|_1 \tag{1}$$

Where the norm is defined by: $\|x\|_p = \sqrt[p]{x_1^p + x_2^p + \cdots + x_n^p}$

Therefore, given a database $D \in \mathcal{D}$ and a function $f : \mathcal{D} \rightarrow \mathbb{R}^w$, with $w \in \mathbb{N}^+$ and global sensitivity Δ, a ε-differential privacy mechanism $\mathcal{M}$ for releasing f is to publish

$$\mathcal{M}(D) = f(D) + L \tag{2}$$

where L is a vector of random variables each drawn from a Laplace distribution $Lap(0, \frac{\Delta(f)}{\varepsilon})$. This mechanism ensures that ε-differential privacy is achieved for the release function f [7,15].

3   Adapting SDC Methods to the Streaming Setting

Once we have introduced some basics about SDC, now we focus on how to adapt traditional SDC methods to the stream environment using sliding windows.

3.1   Noise Addition

Noise Addition [5] adds uncorrelated noise to the values of the attributes of a sample x. We use a *Gaussian variable* to estimate the properties of a Gaussian distributed set of samples. We denote by x_i the value of the i-th attribute of the sample x and by x_i' its masked counterpart. The masked values are calculated as $x_i' = x_i + \beta \cdot \sigma \cdot \epsilon$ where $\beta \in [0, 1]$ is an input parameters, σ is the standard deviation estimate, obtained from the attribute's *Gaussian Estimator* and, finally, ϵ is drawn from a Gaussian random variable $\varepsilon \sim N(0, 1)$.

3.2   Microaggregation

Microaggregation [3] methods are one of the best performing methods regarding both speed and disclosure risk versus information loss trade-off. Three main issues are involved when microaggregation on a data stream: the need for a *sliding window*, a *partition*, and an *aggregation* steps:

- *Sliding window.* It is evident that no partition can be made by just processing a single instance at a time. We need a historical knowledge of the previous or future records that the algorithm will process in order to cluster them into groups. Therefore, the last b instances of the stream are stored in a window, being $b \in \mathbb{N}^+$ an input parameter.
- *Partition.* Microaggregation, even for fixed-sized methods, has a time complexity equal to $O(n^2)$. To reduce this cost, we use a k-Nearest Neighbours

(KNN) algorithm to *continuously* partition the sliding window to provide anonymized instances much faster, by changing just *one* cluster each time a new instance is requested to be released. The records in this single cluster are then aggregated, and the target instance is returned.

- *Aggregation.* After a cluster has been obtained, its instances are aggregated, and the values of their attributes are replaced by the values of the *centroid* of the cluster. For each attribute, the arithmetic mean (in the case that the attribute is numeric) or the mode (if the attribute is nominal) are calculated over the instances of the cluster.

3.3   Rank Swapping

The streaming version of rank swapping [13,14] uses the same approach than *microaggregation*. As depicted in Fig. 1, it works as follows: The first row is the raw data (a). Next, the not already swapped values of the attribute are filtered from the samples in the buffer W (b) and are ranked, i.e., sorted (c). A maximum swap range is calculated using the p parameter (d) and a value within this range is selected to perform the swap (e). Finally, the vector of values is returned in the original order they were in the buffer (f).

Fig. 1. Rank swap of a single attribute for a target instance τ.

3.4   Differential Privacy

The main drawback of differential privacy is that it was defined for interactive *query-response* environments, which is not the scenario we encounter in stream data mining. However, we can find in the literature many efforts to bring differential privacy to non-interactive settings [7,8]. Here, we are interested in a function that returns the attribute values corresponding to the r-th instance of a stream X, *i.e.* a "identity" function but enforcing differential privacy.

As it was proven in [15], *insensitive microaggregation* enforces differential privacy by using a function M, where the global sensitivity of its composition with I_r is $\Delta(I_r \circ M) \leq \Delta(I_r)/k$, being k the minimum size of the clusters returned by M and I_r the identity function of sample r. The condition that such an *insensitive* algorithm must fulfill is:

Definition 1. *Let X be the set of samples of a window W, M a microaggregation algorithm, and let $\{C_1, ..., C_n\}$ be the set of clusters that result from running M on X. Let X^* be a neighbour dataset of X, differing in a single sample, and $\{C_1^*, ..., C_n^*\}$ the clusters that result from running M on X^*. We say that M is insensitive to the input samples if there is a bijection between the set of clusters $\{C_1, ..., C_n\}$ and the set of clusters $\{C_1^*, ..., C_n^*\}$ such that each corresponding pair of clusters differs at most in a single sample.*

However, microaggregation algorithms are very sensitive to the input data. A minimum change in a single sample can cause the generation of entirely different clusters. To overcome this issue, it is possible to use an *order relation consistent* metric in the partition step.

One way to achieve such a consistent distance function is to define a total order relation among the window samples X. Given a *reference point* $R \in X$, for a pair of elements $x, y \in X$, we say that $x \leq y$ if $d(R, x) \leq d(R, y)$, where d is a function $d : Dom(X) \times Dom(X) \rightarrow \mathbb{R}$ (the Euclidean distance between samples of X, for example). Furthermore, in order to increase the *within-cluster* homogeneity, this reference point R should be located at the boundaries of $Dom(X)$.

The adaptation of the insensitive microaggregation algorithm to a stream processing follows the same scheme presented for the *microaggregation*, the only difference being the use of a *reference* point in order to achieve a total ordering relation between the instances of the stream and accomplishing the insensitivity condition. The reference "point" $\mathcal{R}$ is incrementally built as *new* instances are processed by the filter independently updated of the clustering step, when a new sample is added to the buffer.

The Laplace-distributed noise addition final step of the mechanism is performed by a noise adder that works in a very similar fashion to the *Noise Addition* algorithm, with the addition of the scale parameter estimation. Finally, the generation of a random variable Λ following a Laplace distribution is shown in the following equation: $\Lambda \sim Lap(\mu, b) \iff \Lambda = \mu - b \; \mathrm{sgn}(U) \; \ln(1 - 2|U|)$ where U is another random variable drawn from a uniform distribution constrained to the $(-0.5, 0.5]$ interval.

4   Experiments

In this section, we evaluate the SDC streaming methods, presented in Sect. 3, performance with a real credit card payments dataset. Therefore, we present the dataset in Subsect. 4.1. Then, we specify how to compute information loss and disclosure risk in Subsect. 4.2. Finally, Subsect. 4.3, describes our findings.

4.1   Data Description

The database used contains information concerning credit and debit card transactions done by BBVA continental clients using debit and credit cards from 01/06/2016 to 31/10/2017. Transactions are associated with purchases carried out throughout Peru. Nevertheless, for this study, we analyze only transactions performed in the capital of Peru (Lima). The filtered dataset contains about 17 millions of transactions.

Concerning privacy risks, from credit card owners perspective, releasing such data without a proper anonymization could enable an adversary to re-identify some individuals. Thus, the adversary could estimate the users' spending patterns and associate them a socio-economical category [6]. From the point of selling (POS) perspective, if an adversary is able to re-identify a POS, he could estimate its earnings and share such information with their competence.

In the experiments, we assumed that credit card id (PAN number) and 'client_id' are identifiers, so they were removed. Purchase imports are considered as confidential values, so following the scenario described in Sect. 2.1, no modification is applied to purchase imports. Finally, 'shop_id' and their corresponding geographic locations (ubigeo) are considered as non-confidential, therefore, we ran our anonymization methods over these values.

4.2   Metrics for IL and DR

To compare the performance of the different SDC methods, we have modified the standard distance-based record linkage and the Sum of Square Errors (SSE) estimators.

Buffered Record Linkage. It uses a record linkage approach to estimate the risk of records re-identification. The estimator holds a buffer W of the last b original samples. Each time that a $\langle x, x' \rangle$ pair is passed into the estimator, it adds the original sample x to the buffer, deletes the oldest one, and performs a record linkage trying to re-identify x' with any sample in the buffer. Once a sample at distance $d < \delta$ is found, all samples from G are removed and δ is updated. Finally, the algorithm checks if the target instance is in G. The *linkage probability* for an anonymized instance x' is calculated as

$$P(x') = \begin{cases} 0 & \text{if } x \notin G \\ \frac{1}{|G|} & \text{if } x \in G \end{cases} \tag{3}$$

Finally, being X the set of all the instances already processed and $|X| = n$, the Disclosure Risk is estimated in a $[0, 1]$ range as

$$DR = \frac{\sum_{x \in X} P(x')}{n} \tag{4}$$

Sum of Square Errors Estimator. The aim of this measure is to provide a way to compare the information loss produced by different SDC methods. The estimation is based on SSE value between the original x and anonymized x' samples.

$$SSE = \sum_{x \in X} \sum_{x' \in X'} (\text{dist}(x, x'))^2 \tag{5}$$

The main drawback in using this approach, besides it being harder to make comparisons due to not being a bounded measure, is that categorical attributes are over-weighted, thus distorting the validity of the estimation.

Table 1. SDC stream anonymization methods parameters used in the experiments, where sw_size stands for sliding window size.

Method	Parameters	Method	Parameters
Noise addition	$\beta = \{0.1, 0.2, 0.3, 0.4, 0.5\}$	Rank swapping	$p = \{10, 25\}$ sw_size $= \{10^1, 10^2, 10^3, 10^4, 10^5\}$
Microaggregation	$k = \{3, 10, 100, 1000\}$ sw_size $= \{10^1, 10^2, 10^3, 10^4, 10^5\}$	Differential privacy	$k = \{3, 10, 100, 1000\}$ $\epsilon = \{0.1, 0.01, 0.001\}$ sw_size $= \{10^1, 10^2, 10^3, 10^4, 10^5\}$

Table 2. Noise addition time, IL and DR estimations for all considered β parameterizations.

β	TotalTime [s]	DR	IL
0.1	645.203	0.037	19
0.2	977.368	0.033	20
0.3	1,318.560	0.031	20
0.4	1,686.119	0.030	20
0.5	2,030.960	0.029	21

4.3 Results

In the current section, we describe the experiments performed using the SDC methods introduced in Sect. 3. For each method, we used different parameters setups as shown in Table 1. Where, *Noise Addition* receives as parameter β scaling factor of the noise added to attributes and class variables. *Microaggregation* takes the cluster size k and the sliding windows size *sw_size*. *Rank Swapping* accepts as input the maximum swap range, as a percentage of the sliding windows size p and the sliding windows size *sw_size*. Finally, *Differential* Privacy admits the cluster size k, the sliding windows size *sw_size* and ϵ differential privacy.

Noise Addition. Table 2 illustrates the increment of the processing time, while β becomes larger. Regarding Fig. 2, the DR decreases really fast as the number of processed instances grow up. Later on, when arriving to the 12 millions of processed instances, DR slightly increases from 2% to 4%. Concerning IL, it behaves on the contrary way of DR, it shows an exponential growth when the number of instances increases, this effect is because we do not apply any correction effect on the scaling factor, so it always grows, adding a larger noise any time a new instance is processed.

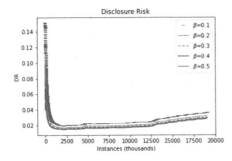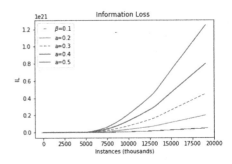

Fig. 2. Disclosure risk and information loss evolution with regard to the number of processed instances for noise addition. a stands for the β parameter.

Table 3. Microaggregation time, IL and DR estimations for all considered parameterizations

k	sw_size	Time[s]	DR	IL	k	sw_size	Time[s]	DR	IL
3	10^5	57,568.05	0.101	16	100	10^5	4,523.93	0.028	17
33	10^3	748.20	0.101	16	100	10^3	378.15	0.028	17
3	10	247.38	0.101	16	100	10	315.33	0.067	16
10	10^5	23,53.768	0.067	16	1000	10^5	2,312.65	0.012	18
10	10^3	387.63	0.067	16	1000	10^3	353.21	0.012	18
10	10	226.93	0.067	16	1000	10	311.34	0.07	16

Microaggregation. Figure 3 shows how the sliding window parameter affects to the execution time of microaggregation. This fact makes this method quite unpractical for real time applications, specially when the considered window is big.

From Table 3, we observe DR becoming really small when k and w_size increase, making microaggregation a really secure method. Regarding the IL metric, as expected, it increases when the number of processed instances grows.

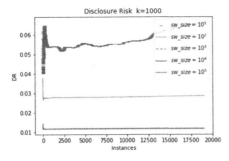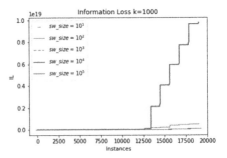

Fig. 3. Disclosure risk and information loss evolution with regard to the number of processed instances for microaggregation.

Table 4. Rank swapping time, IL and DR estimations for all considered parameterizations.

p	sw_size	Time[s]	DR	IL	p	sw_size	Time[s]	DR	IL
10	10^5	698.491	0.793	21	25	10^5	665,313.875	0.757	21
10	10^3	6,889.998	0.406	19	25	10^3	5,628.211	0.236	19
10	10	360.476	0.08	17	25	10	282.637	0.08	17

Rank Swapping. Regarding Table 4, execution time increases when w_size increases. Therefore, sw_size is the main parameter to consider when execution time is crucial.

Besides, comparing the influence of sw_size and p parameters from Fig. 4, sw_size has greater impact over DR. Therefore, it is possible to combine a small sliding window size with a high maximum swapping range to find a good trade off between execution time and DR *vs.* IL ratio.

Differential Privacy. As shown in Fig. 5, sw_size, k and e parameters do not increase the DR significantly. Besides, IL metric is really large compare

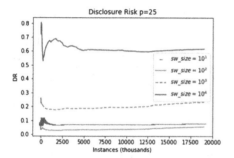
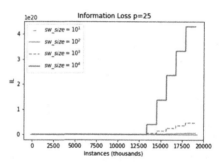

Fig. 4. Disclosure risk and information loss evolution with regard to the number of processed instances for rank swapping.

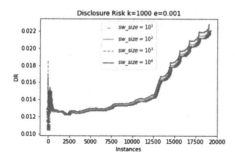
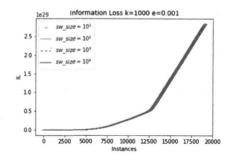

Fig. 5. Disclosure risk and information loss evolution with regard to the number of processed instances for differential privacy.

Table 5. Differential Privacy time, IL and DR estimations for all considered parameterizations

k	e	sw_size	Time[s]	DR	IL	k	e	sw_size	Time[s]	DR	IL
3	0.001	10^5	52,727.56	0.022	29	100	0.001	10^5	3,548.229	0.022	29
3	0.001	10^3	918.74	0.022	29	100	0.001	10^3	267.706	0.022	29
3	0.001	10	319.58	0.022	29	100	0.001	10	237.036	0.022	29
3	0.01	10^5	60,900.72	0.022	27	100	0.01	10^5	3,534.526	0.022	27
3	0.01	10^3	979.30	0.022	27	100	0.01	10^3	273.523	0.022	27
3	0.01	10	341.08	0.022	27	100	0.01	10	238.374	0.022	27
3	0.1	10^5	62,726.36	0.023	25	100	0.1	10^5	3,781.847	0.023	25
3	0.1	10^3	884.88	0.023	25	100	0.1	10^3	290.408	0.023	25
3	0.1	10	328.44	0.023	25	100	0.1	10	252.479	0.023	25
10	0.001	10^5	18,196.16	0.022	29	1000	0.001	10^5	1,705.869	0.023	29
10	0.001	10^3	399.68	0.022	29	1000	0.001	10^3	254.825	0.023	29
10	0.001	10	237.90	0.022	29	1000	0.001	10	238.885	0.022	29
10	0.01	10^5	18,180.21	0.022	27	1000	0.01	10^5	1,737.341	0.023	27
10	0.01	10^3	397.42	0.022	27	1000	0.01	10^3	258.627	0.023	27
10	0.01	10	239.22	0.022	27	1000	0.01	10	239.62	0.022	27
10	0.1	10^5	19,275.56	0.023	25	1000	0.1	10^5	1,833.291	0.023	25
10	0.1	10^3	421.99	0.023	25	1000	0.1	10^3	267.674	0.023	25
10	0.1	10	258.63	0.023	25	1000	0.1	10	257.638	0.023	25

to the other methods. Regarding the processing time, window size determines the required execution time for processing the dataset independently of the other parameters (*c.f.*, Table 5).

4.4   Discussion

In this work, we executed several stream anonymization methods to sanitize credit card payments. From the results presented in Sect. 4.3, we observe the rank swapping method is the less performing method regarding DR value showing values up to 0.9. In terms of IL, it is also the one, together with the differential privacy technique, that generates the largest IL values. Regarding noise addition technique, it improves the DR and IL values with regards to rank swapping. Thus, The highest values of DR and IL are 0.14 and 1.2, respectively. Concerning microaggregation, it decreases noise addition DR values but at the cost of increasing IL in the worst scenario. Therefore, in an optimistic scenario microaggregation reduces both IL and DR values of noise addition making it a good candidate to be applied in real scenarios. Finally, differential privacy outperforms the aforementioned methods in terms of DR, which reaches a really low value (0.022) in the worst scenario. Nonetheless, the cost of IL is the largest one in most of the cases.

5   Conclusion

In this paper, we sum up how to implement four SDC methods (noise addition, rank swapping, microaggregation and differential privacy) in a data streaming fashion. We studied their performance using a real data stream consisting of all card payments done in Lima (Peru) between 01/06/2016 and 31/10/2017. As future work, we plan to further develop these techniques to ensure that multiple streams can be anonymized at the same time. Besides, we would like to study new information loss and disclosure risk metrics for the streaming scenario to improve the evaluation performed in this work.

References

1. Domingo-Ferrer, J., Torra, V.: Disclosure control methods and information loss for microdata. In: Confidentiality, Disclosure, and Data Access: Theory and Practical Applications for Statistical Agencies, pp. 91–110. Elsevier Science (2001)
2. Domingo-Ferrer, J., Torra, V.: Disclosure risk assessment in statistical data protection. J. Comput. Appl. Math. **164**, 285–293 (2003)
3. Domingo-Ferrer, J., Sebé, F., Solanas, A.: A polynomial-time approximation to optimal multivariate microaggregation. Comput. Math. Appl. **55**(4), 714–732 (2008)
4. Dwork, C.: Differential privacy. In: Bugliesi, M., Preneel, B., Sassone, V., Wegener, I. (eds.) ICALP 2006. LNCS, vol. 4052, pp. 1–12. Springer, Heidelberg (2006). https://doi.org/10.1007/11787006_1
5. Hundepool, A., et al.: Statistical Disclosure Control. Wiley, New York (2012)
6. Leo, Y., Karsai, M., Sarraute, C., Fleury, E.: Correlations of consumption patterns in social-economic networks. In: 2016 IEEE/ACM International Conference on Advances in Social Networks Analysis and Mining (ASONAM), pp. 493–500. IEEE (2016)
7. Leoni, D.: Non-interactive differential privacy: a survey. In: Proceedings of the First International Workshop on Open Data, WOD 2012, pp. 40–52. ACM, New York (2012)
8. Li, N., Lyu, M., Su, D., Yang, W.: Differential Privacy: From Theory to Practice. Synthesis Lectures on Information Security, Privacy, vol. 8, pp. 1–138 (2016)
9. Martínez-Rodríguez, D., Nin, J., Nuñez-del-Prado, M.: Towards the adaptation of SDC methods to stream mining. Comput. Secur. **70**, 702–722 (2017)
10. Mateo-Sanz, J.M., Domingo-Ferrer, J., Sebé, F.: Probabilistic information loss measures in confidentiality protection of continuous microdata. Data Min. Knowl. Discov. **11**(2), 181–193 (2005)
11. BBVA API Market. https://www.bbvaapimarket.com/
12. Information Commissioner's Office Guide to the General Data Protection Regulation (GDPR). https://ico.org.uk/for-organisations/guide-to-the-general-data-protection-regulation-gdpr/whats-new/
13. Navarro-Arribas, G., Torra, V.: Rank swapping for stream data. In: Torra, V., Narukawa, Y., Endo, Y. (eds.) MDAI 2014. LNCS (LNAI), vol. 8825, pp. 217–226. Springer, Cham (2014). https://doi.org/10.1007/978-3-319-12054-6_19
14. Nin, J., Herranz, J., Torra, V.: Rethinking rank swapping to decrease disclosure risk. Data Knowl. Eng. **64**(1), 346–364 (2008)

15. Soria-Comas, J., Domingo-Ferrer, J., Sánchez, D., Martínez, S.: Enhancing data utility in differential privacy via microaggregation-based k-anonymity. Very Large Data Base J. **23**(5), 771–794 (2014)
16. Templ, M., Meindl, B., Kowarik, A.: Introduction to statistical disclosure control (SDC). http://cran.r-project.org/web/packages/sdcMicro/vignettes/sdc_guidelines.pdf

Correction to: Modeling Decisions for Artificial Intelligence

Vicenç Torra, Yasuo Narukawa, Isabel Aguiló,
and Manuel González-Hidalgo

Correction to:
V. Torra et al. (Eds.):
Modeling Decisions for Artificial Intelligence, **LNAI 11144,**
https://doi.org/10.1007/978-3-030-00202-2

The original versions of chapters "Graded Logic Aggregation" and "Implicative Weights as Importance Quantifiers in Evaluation Criteria" have been revised; minor errors in the text have been corrected at the request of the author.

The updated online version of these chapters can be found at
https://doi.org/10.1007/978-3-030-00202-2_1
https://doi.org/10.1007/978-3-030-00202-2_16

Author Index

Printed in the United States
By Bookmasters